Invitation to Cybersecurity

Invitation to Cybersecurity

Seth Hamman, PhD

Cedrus Press is the digital and print-on-demand publishing service of CedarCommons, the institutional repository of Cedarville University. Though not an official university press, the work of Cedrus Press is authorized by Cedarville University and thus submissions for publication must be in harmony with the mission and doctrinal statements of the university. Publication by the Press does not represent the endorsement of the University unless specified otherwise. The opinions and sentiments expressed by the authors do not necessarily reflect the views of Cedar-Commons, the Centennial Library, or Cedarville University. The author is solely responsible for the content of this work.

Cedrus Press
251 N. Main Street
Cedarville, OH 45314

Invitation to Cybersecurity

Ed. 1.02

ISBN (print version): 979-8-9862831-8-0
ISBN (electronic version): 979-8-9862831-7-3

Cover Art by Anna Mullinger.
Scripture quotations are from Holy Bible, New International Version®, NIV® Copyright ©1973, 1978, 1984, 2011 by Biblica, Inc.®

Dedication

"If I were giving a young man advice as to how he might succeed in life, I would say to him, pick out a good father and mother, and begin life in Ohio."
- Wilbur Wright

This book is dedicated to my mom and dad. They have loved and supported me all my life and have been my greatest role models and teachers. They have been a fount of wisdom and a constant source of comfort and encouragement. God gave me the blessing of wonderful parents, and from this first blessing innumerable others have followed.

Table of Contents

Foreword

Cyberspace is vital to human progress in the 21st century. Networked computing and the digital devices that provide us access enable the ingenuity of the human mind to expand our understanding of the world in which we live and make it more comfortable and more engaging for more of the world's population. The vitality that flows in and through cyberspace, however, is challenged by an inherent vulnerability that enables, simultaneously, misuse in the form of unauthorized access to systems, software, and computer hardware. The goal of cybersecurity is to ensure that the vitality of cyberspace is not overwhelmed by its vulnerabilities.

The title of this volume reveals how we can achieve this positive outcome. The author has asked the reader not to be a passive absorber of information but rather a partner in a common effort. This is the difference between an introduction to something and an invitation to participate in a way of doing things. Professor Hamman has done the reader a great service in *inviting* us to do our part to actively reduce cyber insecurity in our roles as individuals, individuals who work in organizations, and individuals who come together nationally and even internationally, since cyberspace is both ubiquitous in our lives and global in its reach. His invitation is grounded on fundamentals rather than pure technical details or contemporary examples that, given the nature of cyberspace, will be fleeting.

To contribute as a partner in cybersecurity, we all do not need to be technical experts, but we do need a base understanding of how to operate our digital technology responsibly, ethically, safely, and securely. I do not need to know how to build a car in order to drive it, but I do need to know how to fuel it, how to maintain its basic functions (put air in tires and change my oil) and how to turn it on and get it to propel me to where I want to go (of course in a few years, computers will handle this driving for me and I'll need to know something about my algorithm-chauffeur). But society does also need those who know how to build the car itself and mechanics who can handle the most complex of car system updates and repairs. So too in cybersecurity. We need more cyber aware citizens and more cyber ninjas. This volume will help produce both.

For those seeking cyber awareness, this volume is accessible and, again, inviting. You will be ready to contribute to more security. For those seeking cyber expertise, this volume will deepen your thinking, excite your interests, and drive you to learn more. Accept the *invitation to cybersecurity* and we all will benefit from a more secure digital world.

Professor Richard J. Harknett

About This Text

"Education is what remains after you have forgotten everything you learned in school."
- Anonymous

This is a book about the foundations of cybersecurity. Foundations are stable, last a long time, and are meant to be built upon. In the fast-paced world of cybersecurity, it is difficult to stay up-to-date. This text aims to avoid becoming out-of-date by identifying and staying focused on the unchanging fundamentals of cybersecurity.

Because of its focus on foundations, this book's emphasis is less on training and more on education (see Table 1). Training focuses on competency. It is concerned with how to do certain tasks. At the end of the day, in order to accomplish work of practical value, you must be competent, and competency comes through training. Without question, training is good and appropriate and has a vital role to play in cybersecurity education. However, training has a limited shelf-life due to its specificity. This book, as an introductory text, will avoid training as much as possible, and instead lay a firm foundation for training to follow.

The main focus of this text is on education. Education is abstract and conceptual. It is about learning how to think and how to contextualize the subject matter. While not as immediately practical, education is valuable because it imparts the ability to think critically, to solve problems, and to innovate and adapt. Strong cybersecurity education is important for future-proofing our cyber workforce. When done well, it makes the training that comes afterward more efficient and effective.

Because of its emphasis on fundamentals, the latest and greatest technologies are not highlighted, and to the extent possible, specific examples are limited. References that risk becoming out of date are minimized. Even recent breakthroughs in artificial intelligence (AI) are not addressed in any substantial way. AI is obviously important and is changing the practice of cybersecurity, perhaps tilting the balance of power between cyber defenders and cyber attackers, but even in a future world permeated with AI, the fundamental constructs and challenges of cybersecurity remain the same. AI will enhance both cyber defense and offense. It will change how the topics covered in this book are implemented. But only in a world where AI runs everything and people are free to pursue a full-time life of leisure will the study of cybersecurity as embodied in this text become irrelevant. In the meantime, learning the basic constructs and philosophy of cybersecurity is the best foundation for innovating new technology, learning cutting edge skills, and applying AI to solve problems.

Table 1

A comparison of training and education.

Training focuses on...	Education focuses on...
Content	Context
Concrete details	Abstract ideas
Technologies	Principles
Practical	Theoretical
How	Why
Competency	Understanding
Applying	Evaluating
Answering questions	Asking questions

Furthermore, this book provides a broad overview of cybersecurity and views cybersecurity from the perspective of computer science. Cybersecurity spans several technical and non-technical disciplines (see Figure 1). However, it was birthed out of the discipline of computer science, and computer science remains the foundation of cybersecurity. To really understand cybersecurity, a person has to understand how computers and the Internet work. This book does not delve too deeply into details, but it does get technical in sections to help the reader understand the world of cyberspace. This technical background is necessary to enable the reader to contextualize the tools, principles, and practices of cybersecurity.

This book builds cybersecurity from the ground up. It starts with a brief examination of how cyberspace is fundamentally different from physical space and how this creates a difficult security context (Chapter 1). Then it takes a deeper dive into how cyberspace works because cyberspace is the context of cybersecurity (Chapter 2). Then it moves to the reasons why cybersecurity is necessary by examining cyber adversaries (Chapter 3) and their activities (Chapter 4). Then it explores how to approach cybersecurity at a macro level through an introduction to cyber risk management (Chapter 5). Next, it focuses on the hallmark of cybersecurity, adversarial thinking (Chapter 6). This aspect of cybersecurity makes it unique among other technical disciplines. Then it explores cryptography (Chapter 7). Cryptography is the bedrock of cybersecurity because it is how data is kept secure in cyberspace. Next, because cybersecurity is primarily concerned with preventing unauthorized access, it addresses the means of cybersecurity, access control (Chapter 8). Then the book transitions to basic cybersecurity principles and practices (Chapter 9), and the ethical and legal boundaries of cybersecurity (Chapter 10). It concludes with a brief reflection on the societal impact of cybersecurity (Chapter 11).

Figure 1

The spectrum of the academic discipline of cybersecurity.

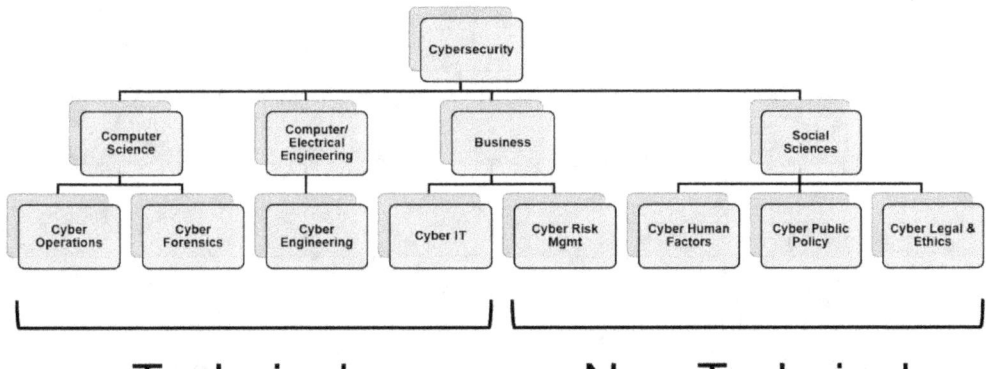

Hands-on lab exercises accompany each chapter. The labs drive home the main points of each chapter and include supplemental material. They provide students with practical skills more closely aligned with training outcomes as outlined above. The labs are designed to be completed in a custom VM pod meant to be hosted on an Internet-accessible cyber range. This allows students to perform cybersecurity tasks on Windows and Linux virtual machines from within web browsers using their own computers. The labs follow a storyline about Alice and Bob, two college freshmen, providing a fun backstory to the technical content. The labs walk students through meaningful cybersecurity exercises involving cyber attacks, cyber defenses, cryptography, and more, and provide step-by-step instructions with illustrations to ensure students can complete the tasks successfully.

This material is meant to be covered in a semester-long undergraduate introductory cybersecurity course. Chapters 1-6 should be covered by midterm. Chapters 7-11 are lengthier and should be covered over the second half of the semester.

For convenience only, male pronouns are used throughout this book to refer to the generic subject. Please see the text's website for supplemental materials and to provide feedback: cedarville.edu/invitationtocyber

Chapter 1

1. Introduction: The Hacker Advantage

"When a general, unable to estimate the enemy's strength, allows an inferior force to engage a larger one, or hurls a weak detachment against a powerful one, and neglects to place picked soldiers in the front rank, the result must be rout."
- The Art of War *by Sun Tzu*

In the beginning, God created the heavens and the earth, and in the 20th century AD, mankind created the Internet. So in a sense, today, we inhabit two worlds: on the one hand, the natural world, which we call physical space, and on the other, an artificial world, which we call cyberspace. Increasingly, our way of life depends on a secure cyberspace.

This is disconcerting, because the news is filled with stories of cyber-insecurity. What is it about cyberspace that lends itself to criminal activity run amok? Obviously, cyberspace is very different from physical space, but perhaps not so obvious is that some of these differences have significant security implications. Specifically, cyberspace is a *distanceless*, *digital*, and *dynamic* world, and each of these fundamental features of cyberspace tips the scales in the attacker's favor.[1]

1.1 Distanceless

Even though real distances are involved and time does elapse as we send and receive signals over the Internet, we experience cyberspace as a distanceless world. This distanceless property is what makes the Internet so powerful. Every computer is within instant reach of every other computer, making everything online immediately accessible. On the Internet, physical distance is immaterial.

Being a distanceless world has major implications for security. Take, for example, a brick-and-mortar bank in a rural Midwestern town. On any given day, this bank has to worry about threat actors like bank robbers, but the number of potential threats is minimal. In

---

[1] This chapter is based on the author's "Guardians of the Cyber Galaxy" TEDx talk: https://www.ted.com/talks/seth_hamman_guardian_of_the_cyber_galaxy

order for a person to rob the bank, the person needs to be in the physical proximity of the bank. Therefore, at any given moment, the number of people who pose an imminent threat to the bank is small. In fact, almost the entire human population is eliminated as a threat.

However, if this bank allows online banking, the circumstances change dramatically. Now, not only can legitimate customers reach the bank at any time from anywhere in the world, but so can attackers. Now, every person on the Internet is a potential threat to the bank. These days, that includes almost the entire human population. Due to being online, this small Midwestern bank now has to worry about a variety of threat actors from all over the world.

To make matters worse, the fact that cyberspace is a distanceless world increases *social distance*. Bad actors are removed from their victims not only physically but also psychologically. They do not witness the havoc they wreak on people's lives. Because they do not see the devastation they cause, they can feel like they are committing victimless crimes. This diminishes the role of the conscience in inhibiting their behavior.

Therefore, the fact that the Internet is a distanceless world increases the number of threat actors capable of attacking at any given moment, and it makes it easier for them to rationalize doing so.

1.2 Digital

Cyberspace is also a digital world. We live in physical space, and our physical senses process the world on a continuous scale. Every physical thing is distinct upon close enough examination. A significant level of expertise is needed to convincingly impersonate another person or to create counterfeit objects. Most people do not possess the skills to do these types of things, and this makes many types of crimes rare in the real world. Additionally, objects in physical space have weight and mass—this, too, creates barriers to crime.

But cyberspace is a digital world. It operates not on a continuous scale but on a discrete one, and the cyberspace scale is binary. This means that the cyberspace periodic table has only two elements: the 1 and the 0. All digital artifacts are composed of 1s and 0s, and it is trivially easy to perfectly replicate them. In fact, that is what we do every time we copy a file, load a webpage, or send an email. Unlike physical space where everything is unique, nothing in cyberspace is unique. This lowers the barriers to committing identity theft and makes it easy to create counterfeit websites and personas.

And unlike physical space, digital artifacts have no weight or mass. Think again about the brick-and-mortar bank from the previous section. There is a significant physical aspect to robbing the bank. Cash has to be moved from the bank to someplace else. In general, the physicality of stealing in the real world creates logistical barriers that complicates committing crimes.

But in cyberspace, all the assets are digital. This means it is trivial to move cyberspace artifacts from any point A to any point B over the Internet. There are few logistical barriers to stealing digital assets from an online bank. This is also true for intellectual property, personal information, and every other kind of data stored in cyberspace. All it takes is just a few keystrokes to move enormous bank account balances or volumes of intellectual property from one computer to another in cyberspace.

Plus, the actors in cyberspace are also digital. To rob a bank in the real world, a person must put himself in physical danger. Bank robbers have to worry about getting shot or attacked. This visceral sense of danger is an effective deterrent because it forces criminals to risk life and limb to commit crime. Robbing an online bank, however, contains no physical danger. When a person steals cyberspace assets, they are not physically present at the scene of the crime. For this reason, the personas of cyber criminals are much more daring than their physical world counterparts.

Therefore, the fact that cyberspace is a digital world makes it easier to commit crime, and it makes the bad guys more fearless, further emboldening them.

1.3 Dynamic

Cyberspace is also a dynamic world. The terrain of physical space is mostly static. In most cases the locations of landmarks today are the same as they were hundreds of years ago. Maps are accurate and reliable. But cyber terrain is shifting constantly and maps do not exist. Computers come online and go offline all of the time, they change where they connect to the Internet, and even the same computer might have multiple different human operators. If a computer has been hacked, even its rightful owner may not know who all of those operators are!

This means that it is relatively easy for a cyber attacker to cover his tracks and remain hidden in the shadows of cyberspace. It is much easier to become a cyber ninja than a real life ninja. Attackers just need to learn a few tricks of the hacking trade, and they can become anonymous online, or at least maintain plausible deniability.

The fact that cyberspace is a dynamic world makes it difficult to answer basic investigative questions like who did what, when, where, and how. This allows attackers to act with a sense of impunity, making them even more audacious.

1.4 Conclusion

"Sunlight is said to be the best of disinfectants; electric light the most efficient policeman." - United States Supreme Court Justice Louis Brandeis

These three Ds paint a dark picture because they demonstrate how in cyberspace temptation and opportunity walk hand-in-hand, and knowing the heart of man, that is a dangerous combination. They also turn cyber threat actors into a kind of supervillain: they can get anywhere at the speed of light (distanceless), they have superhuman strength and are invincible (digital), and they can make themselves invisible (dynamic). It is not a level playing field making us all vulnerable in cyberspace, from the computer illiterate to the tech-savvy, digital natives. In order to combat cyber adversaries and to protect the vulnerable, we need cyber superheroes. These forces for good need to understand cyberspace, what the attackers are capable of and how they think, how to manage cyber risks and tradeoffs, and the tools, principles, and best practices at our disposal. In Sun Tzu's words, these are the "picked soldiers" (*i.e.*, elite troops) we need on the front rank.

But cyber defenders will face temptations of their own in this domain. Therefore, being a force for good involves more than just skilling up to level the playing field in the fight against the forces of evil. *Cybersecurity is the practice of protecting and respecting the rights of every individual and organization in cyberspace.* In order to not just protect but also respect the rights of the vulnerable, our cyber defenders need to embrace the nobility of the cause. Being part of a bigger story makes possible the self-sacrifice and self-denial needed to be a worthy protector. We love our superheroes because we know that no matter what, they will always choose right over wrong. It is their character that makes them the good guys—not just their opposition to evil forces. This is the reason why we cheer for them and know that they will triumph in the end.

This book is an invitation to begin this journey and join the good guys in the fight. My hope is that you will take up this noble and worthy challenge. In our increasingly technical age, our way of life depends on it. We are counting on you!

 **Chapter 2**

2. The Context of Cybersecurity: Cyberspace

"It is not the Internet that is unnatural, nor our feast for information, but a refusal to consider what their origins are, how and why they are here, where they sit in the flow of our history, and what kinds of men and women brought them about. We think there is something of an obligation in beginning to learn these things…"
- A Mind at Play *by Soni and Goodman*

At its most basic level, practicing cybersecurity means securing *cyberspace*. Therefore, to understand cybersecurity, one must understand what cyberspace is and how it works.

Cyberspace is an artificial world that came into existence when the first computer packet was sent over telephone lines in October 1969, three months after man landed on the moon. Only since around 1990, with the birth of the Internet, has it become a substantial world that impacts our daily lives, so it is a very new world. But what is cyberspace exactly? For our purposes, we can define it as *an electronic world composed of computer devices that transmit, receive, and process data (digital information).* Computer devices include anything and everything that connects to the internet, such as laptops and smartphones and also things we do not usually think of as computers, like printers, smart home devices, cars, etc.

This chapter provides a brief overview of how computers and the Internet work and is the most technical chapter in the book. It provides the necessary foundation for understanding cybersecurity.

2.1 How Computers Work

"The course through which I arrived at it was the most entangled and perplexed which probably ever occupied the human mind." - Charles Babbage on his envisioning of the Analytical Engine, the world's first description of a general purpose computer

Computers are probably humanity's greatest scientific achievement. Their ability to out-do humans in mathematical calculations and to outplay us in games like chess and Go is awe-inspiring. Because of these feats, sometimes we think of them as having a more powerful "brain" than we have. But computers, in reality, are very limited machines, and it is our intelligence that makes them useful by contextualizing their input and output.

2.1.1 Boolean Logic

> "'Contrariwise,' continued Tweedledee, 'if it was so, it might be; and if it were so, it would be; but as it isn't, it ain't. That's logic.'" - Through the Looking-Glass by Lewis Carroll

Computers are only capable of doing one basic kind of thing: performing simple *Boolean logic* operations. Boolean logic, devised by George Boole in 1847, is *a system that accepts true and false inputs and produces a true or false output.* In the 1930s, Claude Shannon, a pioneer of computing, discovered the relationship between Boolean logic and digital circuits. He proved a *circuit* can be created to model (and, therefore, solve) any Boolean logic equation. A circuit is *a path through which electricity flows.* What this means is a computer can accept binary inputs, represented by the symbols 1 for true and 0 for false, and can produce binary outputs (see Figure 2.1). That is it. That is fundamentally what computers can do. It is even a misnomer to say computers can *do* anything—as if they had a mind and will of their own. Essentially, they are an encapsulation of wires and switches that are configured to either admit or deny an electrical charge (1 and 0 in binary, respectively). So more accurately, they are passive—they are a machine through which electricity flows.

Figure 2.1

The binary input and output combinations for three foundational Boolean logic operations: AND, Inclusive OR, and Exclusive OR (XOR).

Boolean AND	0	1
0	0	0
1	0	1

Boolean OR	0	1
0	0	1
1	1	1

Boolean XOR	0	1
0	0	1
1	1	0

The twin marvels of computers that enable them to produce extraordinary results are how *incredibly tiny* their wiring and switches are, and how *incredibly quickly* electricity flows through them. Both of these phenomena are literally incomprehensible to us because our human senses operate on much larger and slower scales. The wiring of a computer is encapsulated in cleverly organized *transistors* that form circuits that perform all of the necessary Boolean logic operations and store the results. A transistor is *a device that can control the flow of electricity.* Transistors are so small that millions of them can fit inside the period at the end of this sentence. Read the previous sentence again—incomprehensible!

With every electric pulse of a computer, electricity flows through the circuits based on the state of the transistors and creates a new state—the transistors are set to either 1 or 0. This state then acts as the input that determines the next state, and so on. A computer's clock speed determines how quickly it moves between states. The ticks of the clock that trigger the state changes take place on the order of *gigahertz (GHz)*, or *billions per second*. To help put this type of speed in perspective, think about a digital stopwatch where the last two digits represent tenths and hundredths of a second. When we look at such a stopwatch, we see the seconds go by clearly, and each tenth of second also registers, but just barely (try it!). The hundredths of a second, on the other hand, cycle through so quickly we only see a blur of numbers. This demonstrates that at best, our minds operate at the timescale of tenths of a second. This makes sense because we use the expression "in the blink of an eye" to mean instantaneously, but in fact, an eye blink takes around three tenths of a second. But computers operate on the timescale of billionths of a second, so to a computer the miniscule span of time between each blurry hundredth of a second on the stopwatch, in "human time," feels like a span of weeks. To a computer each tenth of a second seems like months, and each second, years. We can get a lot of things done over the course of weeks—just like a computer can accomplish a lot of work between every seemingly instantaneous hundredth of a second on a stopwatch. Again, incomprehensible!

You may be wondering, what is the big deal about Boolean logic? What can possibly be accomplished by such simple operations, even given the ability to do billions of them per second? Well, remarkably, any computable problem can be solved with Boolean logic. We will not get into the theory of computability and the difference between tractable and intractable problems here, but suffice it to say, the class of tractable computable problems with real-world applications is huge—it includes much of mathematics and all kinds of information processing tasks.

2.1.2 Data Encoding

"Computers are just 1s and 0s." - Common saying

Claude Shannon also invented information theory paving the way for computer processing. He proved that all types of information can be represented using just a pair of differentiable signals or symbols. This means binary strings, composed of the base two number symbols 1 and 0, can be made to represent any kind of information whatsoever, including colors, letters, images, and sounds. This is called *data encoding*. Needless to say, pairing blazing clock speeds and miniscule transistors with the power of Boolean logic and data encoding gives rise to an incredible phenomenon—and we are witnesses to that every time we use a computer.

Table 2.1

Example data encoding of the twelve most common letters of the English alphabet.

Alphabetic Character	Binary Representation	Alphabetic Character	Binary Representation
E	0000	R	1000
T	0001	D	1001
A	0010	L	1010
O	0011	U	1011
I	0100	[not used]	1100
N	0101	. [period]	1101
S	0110	[not used]	1110
H	0111	[space character]	1111

To see how data encoding works, we can do a simple illustration using the twelve most common letters of the English alphabet: E T A O I N S H R D L U.[1] To uniquely represent these twelve letters, we need a string of four 1s and 0s because a binary string of length four gives us sixteen distinct possibilities:

$$2 \times 2 \times 2 \times 2 = 2^4 = 16$$

Table 2.1 shows an example mapping of these letters to binary strings. Since there are more string combinations than we need, we can use the extra ones for the space character and punctuation marks or just leave them unused. Using this catalog of strings, we can unambiguously encode the twelve letters to communicate words and sentences. For example, to encode the phrase *TASTE AND SEE* we would write:

000100100110000100001111001001011001111011000000000

While not nearly as efficient and easy to read as our alphabetic characters, it works! If you know the encoding, then you can read the information. What if we wanted to encode not just twelve letters but the entire alphabet? This would require twenty-six distinct binary strings. This range can be covered by a string of five 1s and 0s because this provides thirty-two possibilities, six more than are needed:

$$2 \times 2 \times 2 \times 2 \times 2 = 2^5 = 32$$

---

[1] This pronounceable string of letters is ordered by approximate letter frequencies in typical English text descending —E is by far the most frequently used letter.

Each 1 or 0 is called a *bit* which is short for *binary digit*. By adding another bit to the string length, we double the number of possibilities. The formula for calculating the total number of possibilities for a given bit string length is straightforward:

$$2^{[\text{BIT STRING LENGTH}]} = \text{Total Number of Possibilities}$$

These powers of two grow quickly. For example, it would require just sixty-three bits to encode every single grain of sand on Earth with its own unique bit string!

One problem with data encoding schemes is that they are arbitrary. They cannot be used to communicate with others unless the mappings are shared, therefore, it is helpful to create standard data encodings that everybody can agree to use. The first standard binary encoding of the English character set was *ASCII* (*American Standard Code for Information Interchange*). It originally used binary strings of length seven. This makes available $2^7 = 128$ representations. 128 seems like way too many for the twenty-six letters of the alphabet, but every unique symbol needs its own string including each uppercase and lowercase letter. To encode the uppercase and lowercase letters as well as the ten numerical digits, sixty-two combinations are needed. Add to this the space character, comma, etc., and other common symbols, plus control signals like backspace, and there is a need for well beyond sixty-four combinations. The total combinations always need to be rounded up to the nearest power of two, so even if only sixty-five distinct encodings were needed, seven bits would have to be used. 7-bit ASCII was created in the 1960s, and since that time, the *byte* has emerged as the standard unit of data in computing. A byte is *eight bits*. The original ASCII encoding is still used, but a leading zero bit has been added to make each character into a byte length instead of just seven bits.

Figure 2.2

A shade of orangish brown created with parts 213 red, 111 green and 56 blue.

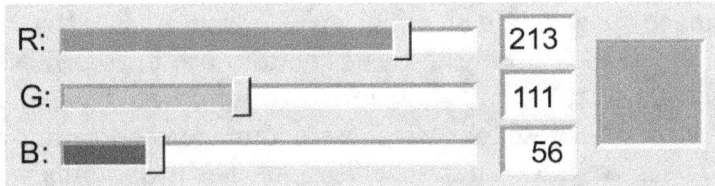

In addition to text, binary strings can also be used to encode images. One standard encoding used for colors is called the *RGB (red, green, blue) color model*. It is a 24-bit system that uses a byte for each part red, green, and blue. By varying the proportions of each color, RGB can represent 2^{24} different color combinations—more than sixteen million. Using this scheme, computer screens can "read" these binary strings to illuminate thousands of extremely tiny *pixels* (*picture elements*) to display detailed color images. To a computer the orangish brown color in Figure 2.2 looks like this:

110101010110111100111000

You may have heard the expression, "Computers are just 1s and 0s." People say this because all the information processed by computers is binary, even the commands that computers execute. Because letters, colors, commands, etc., are all binary strings, without context, there is no way to determine how a binary string should be handled—it is all just 1s and 0s! For example, if the orangish brown color in Figure 2.2 was rendered as text using our custom 4-bit letter encoding scheme, it would read: *.NS OR*. Taken out of context, the results are nonsense...or worse. More on this in future chapters, but hackers have figured out ways to trick computers into confusing user-inputted data for commands causing them to execute hacker-supplied programs!

Because strings of 1s and 0s are difficult for us to read, they are usually written in a base sixteen number system called *hexadecimal*. This is convenient because each group of four bits can be represented by one symbol. A group of four bits is called a *nibble* for half a byte (pun intended!). A pair of hexadecimal digits makes one byte. Hexadecimal uses sixteen symbols: the first ten are the same as the decimal system [0-9], and the last six are the first six letters of the English alphabet [a-f].[2] The prefix 0x is sometimes used with hexadecimal numbers to avoid ambiguity with base ten numbers. For example, 0x10 is not equal to the number of fingers on two hands—it is sixteen in decimal (see Table 2.2). Below is the *TASTE AND SEE* string of bits from before rendered again in binary and then hexadecimal—still difficult to read, but much more manageable. For example, since the encoding used a nibble for each letter, each hexadecimal digit represents a single character (e.g., 1=T, 0=E, f=space, etc.):

000100100110000100001110010010110011111011000000000
12610f259f600

A hex editor is a program that can be used to view and edit the raw bytes of a file. Rather than interpreting the bytes it renders them as hexadecimal values. Figure 2.3 shows a bitmap image opened with an image viewer program and the same file opened with a hex editor. If you look closely at the hex editor output, you will see that the first two bytes of the file are 0x42 and 0x4d, which when interpreted as ASCII, are the letters "B" and "M" for "bitmap." These two bytes instruct image viewer programs to examine this file through the lens of the bitmap specification. The programmers that wrote the image viewer software had to read the bitmap specification so that they could code their software to correctly interpret bitmap bytes. Remember, the "meaning" of the bytes is imposed by humans, and the context dictates how they are handled. In Figure 2.3, the image viewer decoded the bytes the way they were intended, faithfully reproducing an old photograph on the computer screen. The hex editor reveals the reality—it is all just 1s and 0s.

---

[2] The six hexadecimal letters can be written using either lowercase [a-f] or uppercase [A-F] letters—this text follows the convention of using lowercase letters.

Table 2.2

A conversion chart for the binary, decimal, and hexadecimal number systems.

Binary (Base two)	Decimal (Base ten)	Hexadecimal (Base sixteen)
0	0	0
1	1	1
10	2	2
11	3	3
100	4	4
101	5	5
110	6	6
111	7	7
1000	8	8
1001	9	9
1010	10	a
1011	11	b
1100	12	c
1101	13	d
1110	14	e
1111	15	f
10000	16	10

Figure 2.3

A picture of computer pioneers Ken Thompson and Dennis Ritchie (pic.bmp) opened with an image viewer (left) and a hex editor (right).

2.1.3 Binary

"There are only 10 types of people in the world: those who understand binary and those who don't." - A computer geek joke

Because digital computers operate only on binary signals, they use a binary, or base two, number system. We are more familiar with our decimal, or base ten, number system that uses ten distinct symbols [0-9], but the two systems are mathematically equivalent. In base ten, each decimal place is a power of ten, so there is the ones place (10^0), tens place (10^1), hundreds place (10^2), thousands place (10^3), etc. In base two, each binary place is a power of two, so there is the ones place (2^0), twos place (2^1), fours place (2^2), eights place (2^3), etc. To distinguish binary numbers from base ten numbers the prefix 0b is sometimes used. The binary value 0b10 has 1 two and 0 ones so its decimal value is two:

$$0b10 = (1 \times 2^1) + (0 \times 2^0) = 2 + 0 = 2$$

Here again 0b10 is not equal to the number of fingers on two hands—it is two in decimal! Taking a bigger example, the byte 0b01100100 has 0 one hundred twenty-eights, 1 sixty-four, 1 thirty-two, 0 sixteens, 0 eights, 1 four, 0 twos and 0 ones. Add up all these values, and its decimal value is 100:

$$0b01100100 = (0 \times 2^7) + (1 \times 2^6) + (1 \times 2^5) + (0 \times 2^4) + (0 \times 2^3) + (1 \times 2^2) + (0 \times 2^1) + (0 \times 2^0) =$$
$$0 + 64 + 32 + 0 + 0 + 4 + 0 + 0 = 100$$

A byte can range between 0b00000000 (0 in decimal) to 0b11111111 (255 in decimal). Any base two number can be converted to a base ten number using this approach. Naturally, base ten numbers can also be converted to base two numbers. However, base ten numbers with decimal points can be tricky—these are called floating point numbers in computing and are too advanced for this text!

2^{10} bytes are a *kilobyte (KB)*, 2^{20} are a *megabyte (MB)*, and 2^{30} are a *gigabyte (GB)*.[3] The Greek prefixes are used because these numbers are similar in magnitude to thousands (*kilo*), millions (*mega*), and billions (*giga*) in the base ten number system. This simple mapping between powers of two and powers of ten holds and can be used to approximate the order of magnitude of base two numbers (see Table 2.3). *The base two-base ten conversion rule states that ten binary places are approximately equal to three decimal places.*

---

[3] Unfortunately, these terms are ambiguous and in some contexts do refer to powers of ten instead of powers of two—you have to examine the fine print to determine which system is being used.

Table 2.3

Ten binary places are approximately equal to three decimal places.

Power of Two	Actual Decimal Value	Power of Ten Approximation
2^{10}	1,024	10^3
2^{20}	1,048,576	10^6
2^{30}	1,073,741,824	10^9
2^{40}	1,099,511,627,776	10^{12}
2^{50}	1,125,899,906,842,624	10^{15}
2^{60}	1,152,921,504,606,846,976	10^{18}

As an example, if we wanted to assign every human being their own unique binary string, how many bits would we need? The population of Earth is around eight billion people:

$$8,000,000,000 = 8 \times 10^9$$

This number can be converted to a power of two by first converting eight to a power of two:

$$8 = 2^3$$

(Note: when using this rule, it does not always work out for the power of two to equal the leading digits—that is fine because this is just an approximation anyway. In that case, just go with the nearest power of two.) Then 10^9 can be converted to a power of two by converting the three groups of three decimal places ($9 \div 3 = 3$) to three groups of ten binary places ($3 \times 10 = 30$):

$$10^9 \approx 2^{30}$$

The final approximation is:

$$2^3 \times 2^{30} = 2^{33} = 8,589,934,592$$

The quick approximation yields the same order of magnitude and is close enough to the actual number for most purposes. This calculation shows that we would need thirty-three bits to assign every person a unique binary string. If we wanted to be on the safe side for the foreseeable future, we could add another bit which would double the number of possibilities to over seventeen billion—that many should hold up for quite a while!

Using the base two-base ten conversion rule, we can also go the other way from powers of two to powers of ten. Earlier we said we would need sixty-three bits to encode every grain of sand on Earth. To get a better idea of how big of a number this is, we can convert it to a power of ten. To isolate the groups of ten zeros we can write 2^{63} as:

$$2^{63} = 2^3 \times 2^{60}$$

2^{60} has six groups of ten binary places ($60 \div 10 = 6$). Therefore, we need six groups of three decimal places for a total of eighteen ($6 \times 3 = 18$).

$$2^{60} \approx 10^{18}$$

Then we can calculate 2^3:

$$2^3 = 8$$

Therefore, the final approximation is:

$$8 \times 10^{18} = 8,000,000,000,000,000,000$$

This is eight quintillion, an enormous number approximately equal to the number of the grains of sand on Earth. Note that the number of people and the grains of sand, 2^{33} and 2^{63}, are different by a factor of 2^{30}. This means that there are a billion grains of sand for every person on Earth.

2.1.4 Computer Illustration

"'And what is the use of a book,' thought Alice, 'without pictures or conversations?'"
- Alice's Adventures in Wonderland by Lewis Carroll

To illustrate how computers work at the most basic level, and to show how it is human intelligence that extracts the meaning from their calculations, suppose someone needs help deciding where to go on his next vacation. He has determined he wants to go someplace where he can see exotic animals in their natural habitat, or someplace where he can eat at nice restaurants and collect sea shells. Figure 2.4 shows a very simple computer that helps solve this problem.

Figure 2.4

A simple computer.

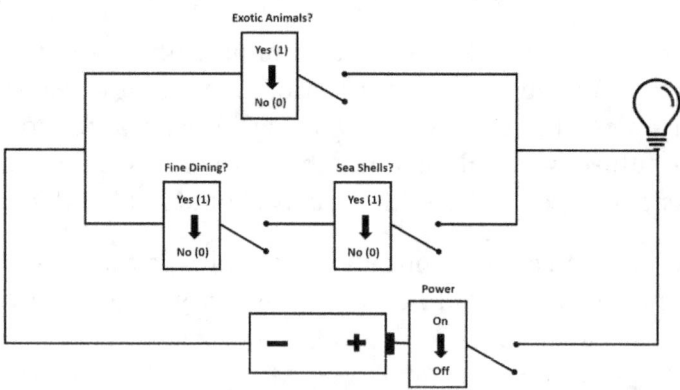

To start, all three of the switches, labeled, "Exotic Animals?", "Fine Dining?", and "Sea Shells?", are set to "No" (0 in binary), and the light bulb is off because the wires connecting it to the battery do not form a closed circuit. For a given candidate destination, imagine flipping the three switches appropriately and then turning the power switch on. If an acceptable destination is identified, the wires will complete the circuit, and the lightbulb will light up indicating a winner. Table 2.4 below shows the values of the switches and their effect on the lightbulb for some candidate destinations.

Table 2.4

Sample results of the simple computer's computations.

Destination	Exotic Animals?	Fine Dining?	Sea Shells?	Winner?
African Safari	Yes	No	No	💡 (on)
New York City	No	Yes	No	💡 (off)
Coast of Greenland	No	No	Yes	💡 (off)
Surfer's Paradise Beach	No	Yes	Yes	💡 (on)

It is easy to see that the "calculation" happens quickly—flip the appropriate switches, turn on the power switch, and instantly, the lightbulb either comes on or not. In this simple case, our brains can solve the problem just as fast. To help us think more about the power and potential of computers, imagine instead of just one person deciding where he wants to go on vacation, there is a room full of people each with their own vacation preferences. For each person, a circuit can be constructed similar to the one in Figure 2.4. If we needed to consider vacation preferences that worked not just for one person, but for all the people in the room, it would be possible to take all of the separate circuits and wire them together in a circuit with a single battery and a single lightbulb (see Figure 2.5). This large circuit would be a complicated and tangled mess of switches and wires, but it could encode the logic of the vacation destination choice for the entire group. Just like the simple computer, this larger computer could be programmed with all of the different preferences for all of the different people by flipping the individual switches. And just like the simple computer, once the programming is done, the power can be turned on, and the lightbulb would either come on or not, revealing the answer.

Electricity flows through wires at rates close to the speed of light. This is incomprehensibly fast. For example, even if this more complex computer used enough wire to stretch all the way from New York to Los Angeles, when the power switch is flipped, if a circuit is formed the light bulb would come on in about one hundredth of a second—the time it takes electricity to travel 2,500 miles. To our senses, this is indistinguishable from the amount of time it would take for electricity to travel the few inches in Figure 2.4. Therefore, the more advanced computer is doing a significant amount of "calculating," but it still returns the result instantly. In this scenario, it would take a human being much more time than the computer to process all of the data and determine the answer. This begins to show the potential of computers.

Figure 2.5

A computer for calculating the vacation preferences for a group of people (this graphic is missing many of the wires and gates necessary to solve the problem).

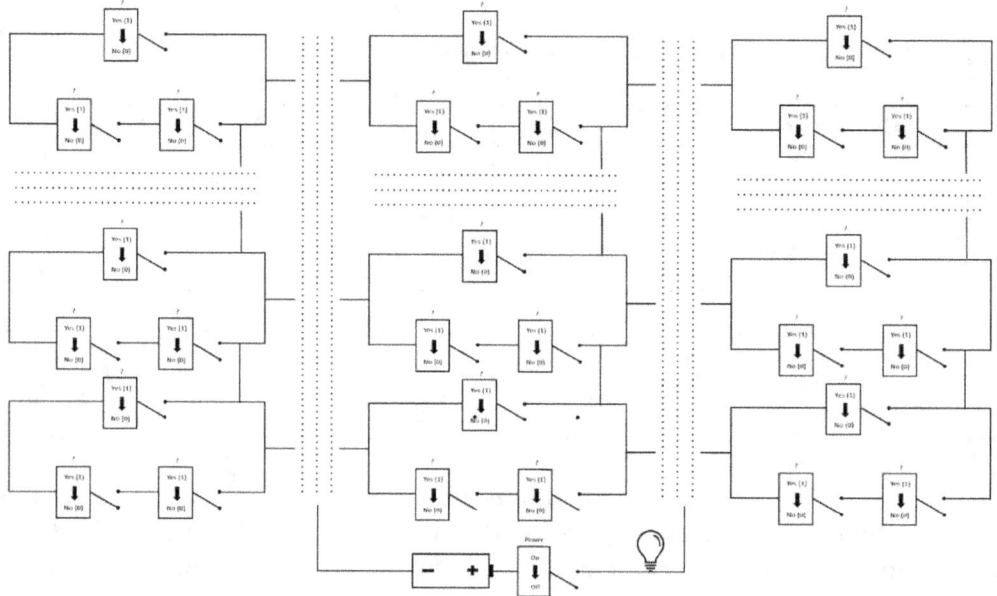

One important feature that these figures do not illustrate well is that circuits also output binary data that can then become input into another circuit. By stringing these inputs and outputs together in general-purpose circuits, instead of just calculating a single state and then stopping, a computer would emerge capable of doing much more than deciding the suitability of vacation destinations. It could in principle solve any logical problem at warp speed.

2.1.5 Encapsulation and Abstraction

> *"Software, we've seen, is a thing of layers, with each layer translating information and processes for the layers above and below. At the bottom of this stack of layers sit the machine with its pure binary ones and zeros. At the top are the human beings, building and using these layers."* - Dreaming in Code *by Scott Rosenberg*

The first computer, believe it not, was probably about as simple as what is illustrated in Figure 2.4. Clearly we have come a long way since the first digital computers were created in the 1940s. How did we get from something so simple to today's computers that allow us to conduct video calls across the world (among many other incredible things)?

The secret to building computers and software is *encapsulation* and *abstraction*. Encapsulation is *the process of grouping functionality into a single simple unit*, and an abstraction is *a high-level summary that retains the essential elements*. The human mind can only comprehend so much complexity, which means we are limited to relatively simple designs. The simple computer in the illustration shows a few components, but imagine a similar design with a little more complexity—maybe a few dozen gates that control the flow of an electric current and a few more inputs and outputs (i.e., lightbulbs). It would still be possible for us to take this in, maybe not at a glance, but with a little study, and understand what is going on and exactly how the circuit works. That would probably look more like the very first primitive computers, as humans started making progress in the field. There would be no mystery, but it could still do something more useful than helping to determine the suitability of vacation destinations.

This is where encapsulation comes in. Once we understood how the simple design worked, and how certain inputs produced certain outputs, we could draw a proverbial black box around that simple design, and refer to it as a single component that performs function *A*. In doing so, we have abstracted the logic from the detailed implementation into a high-level understanding. Then we could make several other simple designs that perform similarly simple functions *B*, *C*, and *D*, and we could do the same process with them—draw black boxes around and abstract them. Now here is the big step: we could then take these simple black box components, and combine them into a new simple design. The new design would still have a manageable number of components, each performing a function that we could understand, but because each component is itself made up of several circuits, the overall level of complexity would be taken up a notch, and could start getting beyond our ability to truly comprehend at the detail level. Maybe function *A* performs Boolean AND logic on several bits, and B performs Boolean OR logic, and by combining several of them in clever ways, we could produce function *SUM* which performs binary addition. Then we could draw a black box around *SUM*, and use it in its own design which is composed of similar more complex black boxes that could perform multiplication. Now imagine doing this building upon building multiple times, each time adding a layer of abstraction—this starts to approach a modern day computer. At its basic level it is still the extremely simple processing of 1s and 0s, but the way we think about how it works is at a highly abstracted level, far above the level of wires and switches. Add to this the inhuman speed and scale of computers, and it all seems like magic, which is generally how we all think of computers. They are truly an engineering marvel, and represent the peak of humanity's technological prowess.

This is relevant for cybersecurity because it shows that computers at their base are highly complex. It takes hundreds of billions of circuits, and millions of black box components to create a modern computer. There is an old saying in cybersecurity you will see again and again in this text: *complexity is the worst enemy of security*. Light, whether that be physical light leading to visibility, or metaphorical light leading to understanding, is vital to security, and computers are full of darkness (i.e., complexity). This is why understanding how computers and the Internet works is vital for practitioners of cybersecurity—the better we understand, the more "light" we are able to shine in the darkness to see what is really going on.

2.2 Computer Hardware

"The considerations which follow deal with the structure of a very high speed automatic digital computing system…" - "First Draft of a Report on the EDVAC" by John von Neumann, a pioneer of computer architecture

Computers are made up of two basic components: *hardware* and *software*. Hardware is *the physical computational components of the computer*. Hardware is generic. It is designed to carry out a basic set of instructions and calculations. Software is *the instructions that the computer hardware executes*. Software is specific. It is designed to accomplish a task. Hardware is configured by software to carry out specific computational tasks.

In a draft report in 1945 John von Neumann described the hardware architecture of a general-purpose computer that became the standard model in computing and is now known as the *von Neumann architecture*. The von Neumann architecture has four main hardware components: *memory*, *storage* (or disk space), a *central processing unit* (*CPU*), and *input/output (I/O) devices*. Computer programs and data are saved in storage. Programs and data are loaded into memory to be executed. I/O devices are what enable people to interact with computers through feeding them data (input) and observing their calculations (output).

2.2.1 Memory

"When I think back on all the crap I learned in high school / It's a wonder I can think at all / And though my lack of education hasn't hurt me none / I can read the writing on the wall." - "Kodachrome" lyrics by Paul Simon

Computer memory can be pictured like a wall of post office boxes with box numbers—each box holds a value. Each block of memory typically holds a byte of data and is directly accessible via extremely tiny wiring at a memory address. Memory in modern computers is known as *random-access memory* (*RAM*) because values stored in memory can be accessed directly and in any order. The amount of addressable memory in a computer is based on a computer's *word size*. The word size is *the number of bits that can be processed at a time by the CPU*. Figure 2.6 shows an 8-bit address scheme with memory locations ranging from 0x00 to 0xff. Each of the 256 blocks stores a byte, but in the figure only memory location 0x42 is shown holding a value (0x10). Modern computers have word sizes of either thirty-two or sixty-four bits, referred to as *32-bit* or *64-bit* architectures. Data in memory changes frequently, and it is normal for memory locations to hold outdated or random values known as *garbage data*. If due to bad programming or hacking, a program accesses one of these garbage values, the result may be an error or a program *crash—the abrupt termination of a running program.* As alluded to earlier, if a hacker is able to input valid machine instructions into memory and then prompt the CPU to execute them, he can hijack the operation of a computer!

Figure 2.6

Computer memory is similar to post office boxes—the only cell with a value shown is at address 0x42.

	0x0	0x1	0x2	0x3	0x4	0x5	0x6	0x7	0x8	0x9	0xa	0xb	0xc	0xd	0xe	0xf
0x0																
0x1																
0x2																
0x3																
0x4			0x10													
0x5																
0x6																
0x7																
0x8																
0x9																
0xa																
0xb																
0xc																
0xd																
0xe																
0xf																

In order for a program to be run, its instructions and data must be loaded into memory. Memory is constantly being written, read, and re-written. Computers work off of data in memory because unlike storage, it is close to the CPU and is able to be accessed quickly. Memory is *volatile*, meaning that when a computer is turned off, all the contents of memory are lost. Therefore, data in memory must be written to storage in order for it to be saved. A typical laptop computer might come with anywhere from four to thirty-two GBs of memory (i.e., RAM).

As we have seen, all of the information processed by a computer is a string of 1s and 0s. This includes memory addresses, instructions that "tell the CPU what to do," and user data, such as documents, videos, and pictures. The general purpose nature of a computer is due to its ability to interpret strings of 1s and 0s differently based on context. Sometimes the 1s and 0s select wires to read or write a value to a memory address, sometimes they cause the CPU to perform an operation, sometimes they instruct a monitor how to render pixels on a computer screen, etc.

2.2.2 Storage

Storage is where computers store data persistently. All of a computer's programs and data, including the operating system, reside in storage. Storage is inexpensive compared to memory and is also much slower. It is needed because data held in storage is saved even when the computer is powered off. In most computers storage takes the form of either a *hard disk drive* (*HDD*) or a *solid state drive* (*SSD*). HDDs store data on spinning metal platters that look similar to CDs and DVDs. SDDs store data in cells and have no moving parts, therefore, they are much faster than HDDs but are also more expensive. A typical laptop computer contains hundreds of GBs of storage.

2.2.3 CPU

The CPU performs all of the calculations in a computer. At every clock tick it performs a *fetch-decode-execute cycle*. It fetches a block of data from memory, decodes that data as a machine instruction, and then executes the instruction. Then the CPU fetches the next instruction, decodes and executes it, and so on. Executing an instruction may involve storing data in memory, loading data from memory, or performing a calculation on values stored in *registers*. Registers are *a type of extremely fast memory that act as a scratchpad for CPU calculations*. They are even closer to the CPU than memory and are constantly updated. Modern laptops may have between sixteen 32-bit registers and thirty-two 64-bit registers.

A CPU on a modern computer may also have multiple *cores*. A core is *a processing unit controlled by the CPU capable of executing instructions*. Having multiple cores allows tasks to be parallelized, improving the performance of a computer.

2.2.4 I/O Devices

In order to be useful, computers also must have I/O devices. These devices are the mechanisms for interacting with a computer. Common input devices are keyboards, mouses, and microphones. Common output devices are computer monitors and speakers. Many of the devices we plug into USB ports on our computer perform both input and output such as memory sticks. The operating system mediates access to the I/O devices in a user-friendly way. It is only through I/O devices that we have visibility into what is going on inside a computer. Through I/O devices we are able to input data, instruct a computer what to do, and see the results.

2.3 Computer Software

> "The Analytical Engine has no pretensions whatever to originate anything. It can do whatever we know how to order it to perform. It can follow analysis; but it has no power of anticipating any analytical relations or truths. Its province is to assist us to making available what we are already acquainted with." - Ada Lovelace, the world's first programmer

Software is the instructions that the computer hardware executes. It is specified and fed into the computer. Unlike hardware it is malleable—it can be easily shaped to perform any computational task. Ultimately, all software runs on hardware—on the billions of tiny circuits and switches inside the computer. By modifying the 1s and 0s in memory and combining the simple instructions that CPUs are hard-wired to execute on binary strings, programs can be written to do everything from playing chess to solving math problems to rendering video games.

2.3.1 Algorithms

"An algorithm must be seen to be believed." - Donald Knuth, a pioneer of algorithmic analysis

Computer programs are the software of a computer and are what gives computers their general-purpose behavior. Generally speaking, computer programs accept input and produce output, and are composed of *algorithms*. An algorithm is *an abstract, step-by-step recipe for solving a well-defined problem*. Abstract in this context means that it is designed to process general, not specific inputs. It is the difference between adding numbers and doing algebra (i.e., 1 + 2 vs. x + y). Algebra is abstract in the same way that algorithms are abstract. For example, an algorithm can be written to sort a list of words alphabetically, and the algorithm would work whether it is fed a few words or millions of words. The only difference would be the length of time it would take for the algorithm to run to completion. What makes algorithms abstract is their use of *variables*. A variable is *a placeholder for data that is provided when the program is run*.

The limits of what kinds of problems an algorithm can solve are the limits of computing. Alan Turing, a pioneering computer scientist, famously boiled down computers to their essence in research he conducted in the 1930s, and he proved that well-defined problems exist that computers will never be able to solve. Besides these problems, called *non-computable* problems, there are computable problems that algorithms can solve, but for even modest size inputs, it would take a computer a prohibitively long time to solve them. These are called *intractable* problems.

While algorithms are abstract, they are in a different way highly specific. They solve well-defined problems with well-defined solutions, not problems that require judgment, interpretation, and commonsense.

2.3.2 Computer Programming

"The programmer, like the poet, works only slightly removed from pure thought-stuff. He builds his castles in the air, from air, creating by exertion of the imagination...Yet the program construct, unlike the poet's words, is real in the sense that it moves and works, producing visible outputs...The magic of myth and legend has come true in our time. One types the correct incantation on a keyboard and a display screen comes to life, showing things that never were nor could be." - The Mythical Man-Month by Frederic Brooks, a pioneer of software engineering

Algorithms are typically implemented in a high-level programming language such as C++, JavaScript, or Python. These languages allow computer programmers to compose algorithms at a much higher level of abstraction than the 1s and 0s and simple instructions that a CPU can actually execute. Computer programs are made up of just three basic types of statements: *sequential statements, conditional statements*, and *looping statements*. Sequential just means *one after the other*. You read a book sequentially, one word, one sentence, one page, after the other. A conditional statement is like a fork in the road (although they are not limited to just two paths) that can take the program down alternative routes based on its inputs. Some books employ a conditional construct—the "choose your own adventure" stories written for children where the reader encounters sentences like, "If Johnny opens the scary door, go to page 10, but if Johnny runs the other way, go to page 20." Just like this example, the *if-statement* is a typical conditional statement *keyword* employed by high level programming languages. A keyword is *a reserved word in the programming language with a predefined meaning*. Looping statements direct the program to repeat the same sequence of instructions over and over again until a condition is met. Loops are how the same algorithm can sort short lists or long lists of words—the loop will continue until there are no words left to process. For a list of five words the loop will repeat five times, but for a list containing thousands of words, the same loop will repeat thousands of times. The *for-statement* and the *while-statement* are the typical looping keywords in high-level programming languages.

Sequential, conditional, and looping constructs can be combined and nested in an infinite number of ways to form algorithms, and these algorithms can be encapsulated into black boxes called *procedures*, which then become usable by other algorithms, and so on. So in the same way computer hardware can be encapsulated and abstracted, so can computer software. At the highest level of abstraction, a program may be relatively simple to understand as a list of procedures (see Figure 2.7). Procedures can take zero, one, or more *arguments* as input that are indicated within parentheses following the procedure name. Any non-trivial computer program is composed of many procedures, and procedures typically call other procedures to perform sub-tasks, and so on. *Software libraries are groups of related procedures that can be imported into a program.* Rather than reinventing the wheel, programmers can just find the procedure they need in a library and call it. Software libraries that perform common tasks can become very popular. These libraries are used and reused many times over by lots of different programmers.

Figure 2.7

Pseudocode for sorting a list of words made up of three procedures.

```
GetWordsFromFile(filename)
SortWords()
StoreWordsToFile(filename)
```

Programs must be tested to be sure they produce the correct outputs because programming is difficult. In one sense, programs always produce the "correct" outputs—meaning the outputs they were specified to produce. However, the specified outputs do not always match a programmer's intended outputs! Programming mistakes occur frequently and are called bugs. Sometimes bugs lead to cybersecurity issues.

Computer programs need to be unambiguous, so programming languages have a well-defined *syntax*. The syntax is *the rules for writing a program in a given language*. Syntax rules specify how variables and procedures are defined and how statements are constructed and delimited. For example, the C++ programming language dictates that statements are delimited with a semicolon and procedure definitions are enclosed in curly braces. A big part of learning a programming language is learning its keywords and syntax. Once a person learns a couple of different programming languages, additional ones come quickly because the algorithmic thinking process remains the same and learning new programming language syntax becomes intuitive.

Computer hardware cannot directly execute a computer program written in a high level programming language. The program has to be transformed into the instructions that a CPU can execute. These simpler instructions are called *machine code* and are usually rendered in hexadecimal. A more human-readable form of machine code called *assembly language* uses mnemonic names as substitutes for hexadecimal machine code instructions. A few lines of a high-level program might result in hundreds of machine code instructions. A *compiler* is a program that takes as input a text document that contains a high-level program, called *source code*, and produces as output machine code that can be executed by a computer. If the source code does not have proper syntax, the compiler will fail and return a syntax error. A compiled program is called an *executable* and can be run by the operating system. Some programming languages do not require that source code be compiled before it can be run. These are called *interpreted languages*. Instead of going through a compiler, these programs are run by an execution engine program called an *interpreter* that transforms the source code into machine instructions line-by-line at execution time. Python is a well-known interpreted language. Interpreted programs are called *scripts* to differentiate them from executables.

Traditionally the first program someone writes in a new language is called the *Hello, World!* program. This is a simple program that just prints "Hello, World!" as output. Table 2.5 shows some *Hello, World!* programs in different languages. The JavaScript and Python code is much shorter because they are interpreted languages.

Table 2.5

Hello, World! programs in different languages.[4]

Programming Language	Hello, World! Program

Assembly Language (Intel x86 architecture)	```section .text```
	```extern printf```
	```global _start```
	```_start:```
	``` mov edx,len```
	``` mov ecx,msg```
	``` mov ebx,1```
	``` mov eax,4```
	``` int 0x80```
	``` mov eax,1```
	``` int 0x80```
	```section .data```
	``` msg db 'Hello, World!',10```
	``` len equ $ - msg```

C++	```#include <iostream>```
	```int main() {```
	``` std::cout << "Hello, World!";```
	``` return 0;```
	```}```

JavaScript	```console.log("Hello, World!");```
Python	```print("Hello, World!")```

2.3.3 The Operating System

"UNIX is basically a simple operating system, but you have to be a genius to understand the simplicity." - Dennis Ritchie, co-creator of UNIX

All general purpose computers come with one very important piece of software that runs continuously called the *operating system* (*OS*). This computer program underlies all of the other computer programs that run on a computer. *UNIX* is one of the most influential OSs ever created. It was designed by *Ken Thompson* and *Dennis Ritchie* in the 1970s and became the model for many modern OSs. The three main OSs in use in computers today are Microsoft's *Windows*, Apple's *macOS*, and the free and open-source operating system *Linux*. These OSs are enormously complex and include tens of millions of lines of source code that took tens of thousands of man hours to produce. Some of the most well-known names in computing became famous because they are behind these OSs which together underpin most of the computers in the world. Bill Gates and Paul Allen co-founded Microsoft in 1975, Steve Jobs and Steve Wozniak co-founded Apple in 1976, and Linus Torvalds developed Linux which is based on UNIX and was first released in 1991 (its name comes from the combination of Linus and UNIX).

Linux is elegant, reliable, and unlike Windows and macOS, it is free! Because it is open source, its source code can be vetted and modified to suit different purposes. Thousands of programmers have contributed to its development over the years, and have packaged

---

[4] The source for the Hello, World! assembly language program is: talaatmagdyx.gitbook.io website. Displaying *"Hello World!"* in x86 Assembly. Retrieved June 2025.

the Linux kernel (the core low-level OS component) with software, tools, and user interfaces called distributions (distros). The two main distributions are Debian and Ubuntu. Both of these distros are widely used among cybersecurity professionals and enthusiasts, and navigating them is not that much different from using Windows and macOS. Probably the most well-known hacker OS is called Kali Linux. As the Kali website states, "Kali Linux is an open-source, Debian-based Linux distribution geared towards various information security tasks, such as Penetration Testing, Security Research, Computer Forensics and Reverse Engineering."[5]

All computers that humans interface with run an OS, and this certainly includes smartphones. The two main smartphone OSs (i.e., mobile operating systems) are Google's Android (first released in 2008) and Apple's iOS (first released in 2007). Android is based on Linux, but the Android OS that runs on most smartphones is Google's proprietary version.

When a computer is powered on, the operating system is the first program loaded from storage into memory. This is called *booting* because the computer has to lift itself up by its own bootstraps in order to get going! The OS runs continuously while the computer is on and presents a user interface so that people can interact with the computer. There are two main types of operating system user interfaces: the *command-line interface* (*CLI*), and the *graphical user interface* (*GUI*) (see Figure 2.8). CLIs are purely text-based and enable users to type in commands and read typed output. GUIs have graphics such as buttons, icons, and text fields that allow the user to point and click as well as type. GUIs are more popular with everyday users because they are intuitive to use, but CLIs are more efficient for adept users.

Figure 2.8

Windows OS CLI (left) and GUI (right) for browsing files.

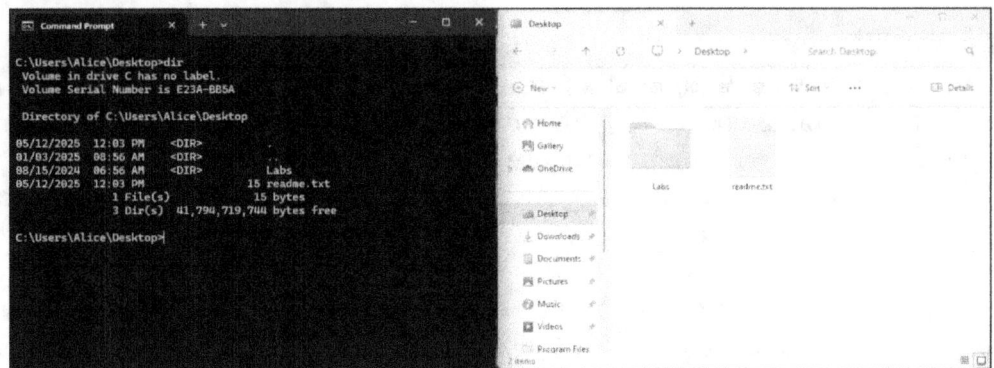

When a program is executed on a computer, the OS loads it from storage into memory where the CPU can access its machine instructions. At this point, the program becomes a *process*. A process is *a running program*. Computers typically have many processes

---

[5] kali.org website. The most advanced Penetration Testing Distribution. Retrieved June 2025.

running at once. The OS switches between them so quickly that they all appear to be running at the same time, but in reality, they all must share the CPU, so technically only one can be run at a time. Managing processes in an OS is kind of like the plate spinning acts in a circus, where the juggler keeps multiple plates spinning by tending to each plate for a moment and then moving from plate to plate to plate. OSs are constantly switching between processes to keep them all going. There is a limit to how many processes can be kept spinning by the OS—if too many programs are started, the computer will start to become sluggish.

Operating systems also manage I/O devices, mediating their access to the CPU. They provide *device drivers* so that *peripheral devices* (i.e., *outside devices that are plugged into a computer*) can have efficient access to the CPU. A device driver is *a low-level program that manages communication between a peripheral device and the CPU*. For example, every time a character is typed on a keyboard, the keyboard device driver triggers a *processor interrupt*. A processor interrupt is *a signal sent to the CPU to prompt it to handle a new action*. In this case, the CPU needs to read the character and take the appropriate action, such as outputting it to the monitor. It is mind-boggling that a computer can do so many operations so quickly amidst such constant interruptions!

Operating systems also make available *system calls* (*syscalls*) to programs. System calls are *OS-defined procedures that allow user programs to exercise some control over OS functionality*. They may prompt the OS to carry out an action on processes, files, devices, etc., and they are typically executed with special privileges, so they have well-defined boundaries.

Operating systems also perform many basic cybersecurity functions, including managing user logons, protecting user data, and isolating processes from one another. Section 8.2.2 covers access control in OSs.

2.3.4 The Web Browser

The *web browser*, next to the operating system, is the second most important user-facing program that runs on a modern day computer. Web browsers are the program through which we run all the other programs we use online such as email, social media, and apps like Google's Docs, Sheets, and Slides. It is the main program we use to gain access to *the cloud*. The cloud is *an expression for the computer servers we access online*. It is a fitting expression because we do not actually need to know the details of where we are connecting and how it works—we can just consider it a black box. The cloud is yet another example of abstraction that hides enormous complexity. *A server is a specialized computer optimized to rapidly process requests for data*. Cloud servers are meant to be accessible at all times from any Internet-connected computer. We spend most of our time online using web browsers to access applications and data on servers. A few popular web browsers are Microsoft's *Edge*, Mozilla's *Firefox*, Google's *Chrome*, and Apple's *Safari* browsers.

Before the high-performing and ubiquitous Internet of today, people worked offline most of the time. They purchased software programs on CDs (and before that, *floppy disks*) and installed programs locally on their computers. These programs were known as *desktop applications*. The Microsoft Office Suite of programs, which includes Word, Excel and PowerPoint, are examples of popular desktop applications still in use today. Desktop applications are similar to the apps that we use on our tablets and smartphones. One major difference, however, is that most apps are dependent on Internet connectivity. In fact, they are essentially specialized web browsers. This is important for computing devices with small, touch-enabled screens because apps can deliver custom content more effectively than smartphone web browsers—that is why the app paradigm is much more popular on smartphones than on laptops.

Web browsers are a desktop application that come pre-installed on our OSs, although many people use their pre-installed browsers to download alternative web browsers that they prefer. Web browsers are the main desktop application we use. They are our window into the data we access and the applications we use in the cloud. Online applications are called *browser-based applications*. Browser-based applications have several advantages over desktop applications. For one, information processing takes place on cloud servers, not on our local computers. This means it is not necessary for our computers to have a lot of processing power and memory. Also, our data is stored online and is accessible from any Internet-connected computer. We do not have to keep track of and synchronize data stored on multiple different local computers (e.g., our home computer and work computer). Nor do we have to worry about backing up our data to protect it from accidental loss—these were major issues before cloud computing. One downside of this data storage model, however, is that we are no longer the sole stewards of our data. We are trusting our data to an outside organization and this raises privacy concerns (more on this in Chapter 9). Also, being online makes our data an attractive and accessible target to hackers all over the world, raising much more serious privacy concerns!

2.3.5 Virtual Machines

"Welcome my son / Welcome to the machine / What did you dream? / It's alright, we told you what to dream." - "Welcome to the Machine" lyrics by Pink Floyd

Traditionally, a one-to-one correspondence has existed between physical computers and operating systems (OSs). A computer has one set of hardware, and the one OS mediates access to that hardware. *Virtual machines (VMs)* break this paradigm. A VM is *an OS that runs as a program on top of a computer's actual OS*. This technology is called *virtual* machines because it enables users to allocate virtual hardware to OSs. This makes it possible to run multiple OSs on a single computer —even different kinds of operating systems. When a computer is running VMs, the actual OS is called the *host OS* and the VMs are called *guest OSs*. There is a limit to how many guest VMs can run on a single computer because virtual CPUs and RAM assigned to VMs ultimately run on physical

CPUs and RAM. Running too many VMs consumes a computer's resources and makes it unusable.

The idea of virtualizing hardware has been around for decades, but it became cost-effective for the average computer user in the 2010s. Since that time, the use of VMs has exploded because they provide many benefits. One of their benefits is that VMs can be saved and copied like any other computer file. This makes it possible to customize a VM and then distribute copies of it to others—it is like "cloning a laptop" and giving the copies to friends. This practice is prevalent in the world of cybersecurity. Experts configure VMs with open-source hacking tools, and then make them freely available for download. One of the most popular of these cyber VMs is *Kali Linux* (see Figure 2.9).

Figure 2.9

A Kali Linux VM running as a browser-based application.

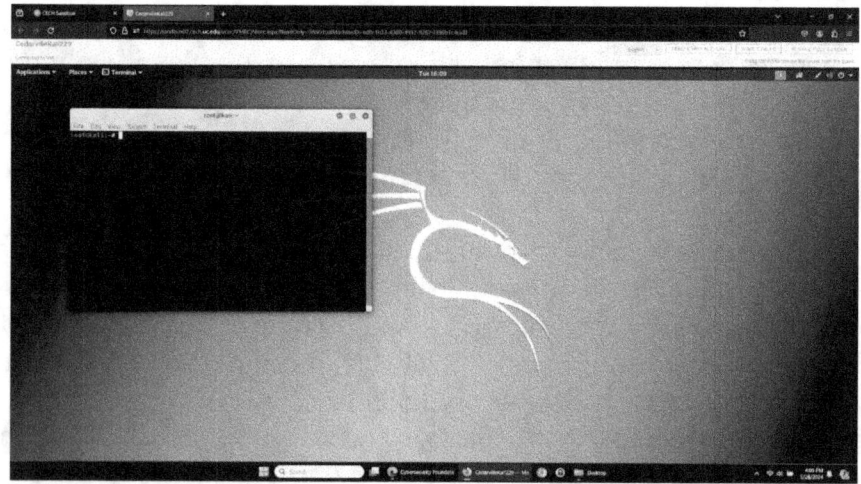

Another benefit of VMs is that machine states can be saved efficiently. This functionality is called taking a *snapshot*. A snapshot is *a capture of the complete state of a VM at a point in time*. Users can take snapshots and revert back to them later. A practical application of this is taking a snapshot before making a series of changes to a VM. If the changes do not work out as expected, it is easy to revert back to the pre-changes snapshot. At that point, it is like the changes never happened, so there is no worry about lingering effects that may be difficult to undo. Reverting to a snapshot is like going back in time.

VMs can also be used as *sandboxes*. A sandbox is *a compartmented safe space for testing and exploring*. Like in the example above, a cyber researcher may want to explore a computer virus but be worried that it may end up infecting his own computer. In that case, he can examine the virus within a guest VM (although some sophisticated types of malware have been known to "escape" from VMs and infect the host OS, so even this is not without risk!). If the guest VM ends up being compromised, no problem—it can be reverted to an earlier snapshot, or if needed, it can even be destroyed. VMs are "cheap" in that way—unlike physical machines, they can be easily destroyed and a new one can be created.

Another use of sandboxes is to create a safe space for practicing hacking. Gaining unauthorized access to a computer is against the law (see Chapter 10), so hacking exercises must be conducted carefully to avoid accidentally overstepping legal boundaries. Multiple VMs can be networked together and isolated from the Internet in a sandboxed computer network called a *VM pod*. Attacks within VM pods can be safely launched from one VM to another without risk of breaking any laws.

VMs can also be accessed through the cloud and run as browser-based applications. When accessed through the cloud, web browsers provide a window to a VM running on a cloud server. This makes it possible for users to control VMs through their web browsers. Because the VMs are cloud-based, they do not consume computing resources on the user's computer. Most laptops are not powerful enough to run multiple VMs at the same time, but they can easily run multiple VMs in different browser tabs.

Hacking labs are a vital part of a cybersecurity education. VM pods can be created and distributed to students through the cloud via a *cyber range* infrastructure (see Figure 2.10). A cyber range is *a safe online space for practicing cybersecurity*, just like a gun range is a safe space for practicing with firearms. Cyber ranges are powered by multiple servers that host several VMs. Students access cyber ranges from their web browsers and work through cyber lab exercises on VMs in the range. The students do not have to install any software locally, avoiding annoying installation issues, and it is not necessary for the students to buy high-powered laptops. Also, since the VMs run on the cyber range servers, students do not have to worry about accidentally releasing malware into the wild or launching attacks inadvertently from their own computers. Making sure the range is isolated is the responsibility of the cyber range administrators, not the students. Repeatability is another nice benefit of this model. Students have access to their own VM pods, and all of the VMs start from the same exact initial state. Therefore, if students follow the same steps, they can fully expect to see the same results. Also, if during the course of an exercise a VM becomes unstable due to miscues, the VM can easily be reverted to a previous snapshot or recreated, and the student can try again.

Figure 2.10

A cyber range architecture showing a student accessing three VMs through his web browser running on range servers.

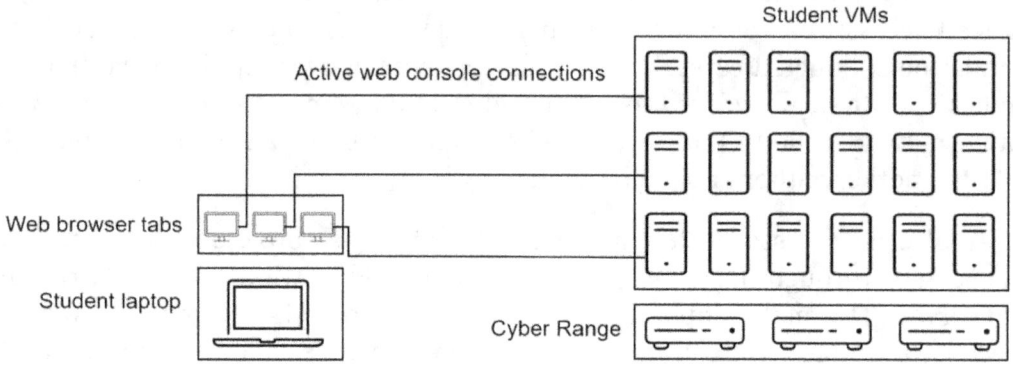

2.4 How the Internet Works

> *"When I helped to develop the open standards that computers use to communicate with one another across the Net, I hoped for but could not predict how it would blossom and how much human ingenuity it would unleash." - Vint Cerf, a pioneer of the Internet*

"What hath God wrought?" was the message sent by Samuel Morse in the world's first telegram. This monumental historical event received much fanfare when it occurred in May 1844. No such publicity accompanied the world's first computer network transmission in October 1969. The first packets were sent between UCLA and Stanford on the *ARPANET, a United States' Department of Defense-sponsored project to create a computer communications network that evolved into the Internet.* The message that was received at Stanford before the connection crashed was "LO" for the first two characters of the LOGIN command. However, it is much more fitting to take a revisionist approach to history and reinterpret it as "LO" as in "LO AND BEHOLD!"

A standard I/O device on every modern computer is the *network interface card (NIC)*. NICs enable computers to send and receive data. These communication links are the basis of a computer network. The Internet is one example of a computer network—it is by far the biggest one and spans the entire globe. The term *Internet* was coined in the 1980s when several isolated computer networks were connected to form a single large computer network—this became a network of networks or an internetwork. The Internet has expanded continuously and rapidly since its inception. We derive increased benefits from technology when it is networked together, and this has driven us to connect more and more "things" to the Internet. This has led to today's *Internet of Things* that includes everything from smart home devices to cars and much more. There are billions of devices connected to today's Internet.

The computing devices on the Internet form a *connected graph* made up of *nodes* and *links*. A connected graph means that there exists at least one *path* between any two nodes. In this case, the nodes are computing devices. This means it is possible for any two devices on the Internet to communicate with one another—at least in principle. In practice, firewalls and other technologies prevent certain paths. This is similar to the way the roadway system works in the continental United States. Our roadway system forms a connected graph made up of homes and intersections (nodes) and roads (links). A car can get from any home A to any other home B by traveling along a series of roads. Some roads are small and quiet and others are major highways full of traffic. The Internet has an analogous system of network links that carry data across the country and, via undersea cables and satellite links, all over the world. And just like the roadways, links vary in *bandwidth*. Bandwidth is *the amount of data that a link can carry*, and is comparable to the number of lanes on a road. Highways have high bandwidth—many cars can travel on them at the same time. The backbone links on the *information superhighway* (i.e., the Internet) also have high bandwidth and carry huge amounts of data traffic.

2.4.1 The Five Layer Model

"We believe in rough consensus and running code." - Internet Engineering Task Force slogan

Many different *protocols* are involved in computer network communication. A protocol is *a specification for communicating over a network*. Protocols are like programming languages in that they define keywords and a syntax. They also define actions and the proper order of messages. When two computers "speak" the same protocol, they can communicate information to one another. The Internet protocols are managed by the *Internet Engineering Task Force (IETF)*—a group of mostly volunteers who since 1986 has aspired to "make the Internet work better."[6] The standards documents used by the IETF are called *RFCs (requests for comments)*. The vast majority of RFCs are serious and professional, but not all. As an example particularly fitting for cybersecurity, RFC 3514 was released April 1, 2003 (yes, April Fools' Day), and it creates a standard declaring that hackers must set an "Evil Bit" field in their attack traffic so that cyber defenders can more easily recognize malicious activity and defend against it!

Computers communicate by sending signals to one another over links. The communication is coordinated through a layered network architecture. The Internet architecture is a *five layer model* (see Table 2.6). Each layer handles different aspects of the communication, and neighboring layers pass data up and down to one another. Messages start their journey at the application layer and are passed down through the other layers before being sent onto the network by the physical layer. When the message reaches the destination computer, it is received first by the physical layer, and then it is passed back up to the application layer.

---

[6] ietf.org website. Introduction to the IETF - Mission. Retrieved June 2025.

The physical layer manages the transmission and reception of physical signals over links. These signals include radio transmissions over the air (e.g., wireless networks), electromagnetic transmissions over wires, and light pulse transmissions over fiber optic cables. The link layer manages the interpretation of these physical signals as 1s and 0s on a per link basis. The most well-known link layer protocol is *Ethernet* invented by Robert Metcalfe in 1973. Most computers use either wired or wireless Ethernet to connect to the Internet. Ethernet addresses are forty-eight bits long and are called *MAC* (*Media Access Control*) *addresses*. NICs implement both the physical and link layers, and every NIC has a unique MAC address. There is no chance of running out since there are approximately 256 trillion of them!

Table 2.6

The Internet's five layer model.

Layer	Layer Name	Example Protocol	Address Scheme
5	Application	HTTP	N/A
4	Transport	TCP	Port numbers
3	Network	IP	IP addresses
2	Link	Ethernet	MAC addresses
1	Physical	802.11	N/A

The OS handles the network and transport layers. These layers work together to manage the end-to-end connections between computers. The Transmission Control Protocol (TCP) and Internet Protocol (IP) were invented by Vint Cerf and Bob Kahn in the 1970s, and are the foundation of the Internet. They define how computer network data is addressed. By definition, every computer on the Internet has an IP address. IP addresses are like phone numbers. A phone cannot make or receive calls without a phone number, and a computer cannot send or receive data on the Internet without an IP address. IP addresses, like everything else in computers, are binary. They are thirty-two bits long and for convenience are written in dotted decimal notation as four numbers separated by periods ranging from [0-255] because each decimal number represents one byte of the IP address.[7] Because they are thirty-two bits, there are approximately four billion IP addresses. Fortunately, this is not a limit on the number of devices that can connect to the Internet because multiple computers on a local network can share the same public IP address. For example, Internet service providers (ISPs) typically assign only one IP address to each home, but most households have multiple online devices. The one public IP address is assigned to the home's router, and the router manages all the devices connected to it using private IP addresses. Private IP addresses are not globally unique, so mil-

---

[7] This describes IP version 4 (IPv4) addresses. IPv6 is an updated version of the Internet Protocol, but IPv4 is still in common use.

lions of computers are able to use the same IPs at the same time, greatly multiplying the number of devices that are able to connect to the Internet to way beyond four billion.

The IP address of Cedarville University's website (www.cedarville.edu) written in dotted decimal notation is 163.11.75.44. Written in binary it is:

    10100011000010110100101100101100

As a decimal number, this value is 2,735,426,348, and as a hex number, it is 0xa30b4b2c. Most web browsers can use multiple different IP formats to "dial the web server's number." For example, all of the following links go to the same webpage (type them into your web browser's address bar see!):

    http://www.cedarville.edu

    http://163.11.75.44

    http://2735426348

    http://0xa30b4b2c

2.4.2 Packet Switching

"Go West, young man." - Horace Greeley

Sending data over the Internet is analogous to mailing a letter via the United States Postal Service (USPS). If Alice wants to send a letter to Bob, she writes a message on a piece of paper, sticks the paper in an envelope, writes Bob's name and mailing address on the front of the envelope, and drops it in the mail (let's forget about the stamp—Internet traffic requires no postage!). From there, the USPS takes over, and the address is dissected first by state and zip, then by city, road, and eventually house number as the envelope makes its way through the postal service from Alice's house to Bob's house. Once it reaches Bob's house, the name on the envelope is the last part of the address needed to make sure that Bob opens the letter and not another member of his household. When computers send messages over the Internet, they address them with something like an envelope called a *header*. On the Internet, a header is *a prefix added to a packet that specifies delivery-related information*. The non-header part of the message data is called the *payload*. The IP header contains a source and destination IP address as well as some other fields. The protocols at each layer add their own header to communicate with their peer protocol at the destination computer. When the packet arrives at the destination and is processed, the header is stripped off and the payload is passed up the stack to the next layer—just like the envelope is thrown in the trash after a letter has reached its destination.

Data is sent over the Internet in *packets*, and the computing devices that process and route packets towards their destinations are called *routers*. Routers make up the core of the Internet and are the basis of the network layer. The network layer is in charge of routing packets from the source computer to the destination computer. Routers examine the destination IP address in the IP header. IP addresses are hierarchical similar to mailing addresses. This makes it possible for a router to send a packet to a neighboring router closer to the destination without needing to know the precise location of all the world's IP addresses. This is similar to the way a person in Ohio can drive to a specific home in California without a GPS or even a map. He does not need to know the location of the home in California in order to get started in the right direction, he just needs to know that California is west of Ohio, and I-80 is a major United States highway that runs east-west. He may need someone to point him in the direction of I-80 to get started, but once he hits that intersection, he goes west! When he eventually gets to California, he can ask someone to point him in the direction of the Californian city he is trying to find. This might involve taking a California state highway angling north or south. Once he gets to the correct city, a local person can direct him to the proper street, and once on the street, a neighbor perhaps walking his dog can help him find the correct house number. The point is that his path does not need to be fully mapped out before he can begin his journey—he just sets out in the general direction, and he finds more specific guidance when he needs it. Due to local knowledge, more specific guidance is available as he gets closer to his destination. Of course, traveling by this method does not result in an optimal route, but a decent route is good enough, especially for blazingly fast Internet packets.

The hierarchical nature of IP addresses is due to them being grouped by leading bits. For example, all IP addresses beginning with the sixteen bits 1010001100001011 are reserved for Cedarville University. In dotted decimal notation, this is the IP address range 163.11.0.0 through 163.11.255.255. Therefore, all packets addressed to any of the 65,536 IP addresses in this range will be sent to Cedarville's router, and from there, Cedarville's router is responsible for getting the packets to the correct computer on Cedarville's campus network. (In case you are wondering, the answer is no, most of these IP addresses are not needed and are not used!)

On the Internet, this process of directing packets one link at a time towards their destination is called *packet switching* (see Figure 2.11). A major benefit of packet switching is that it is dynamic—when links go down unexpectedly, new paths can be created on the fly. This makes the Internet robust to failures. A downside of this approach is that packets must set out without knowing whether their destination is reachable or even if it exists. This leads to *dropped packets*. A dropped packet is *a packet that never reaches its destination*. Because dropped packets are always a possibility, TCP implements a reliable packet delivery protocol involving acknowledgements, timers, and retransmissions.

Packets travel from the source computer through the network core and eventually reach the destination computer. Once packets reach the destination, they still need to be directed to the correct process because computers run multiple processes at once. Processes are identified by port numbers. Port numbers are Layer 4 addresses. They are like Bob's name on the envelope in the mailing address analogy—they only come into play after the packet has reached the destination computer. Port numbers are sixteen bits long and are written as decimal numbers ranging from [0-65,535].

Figure 2.11

Packet switching with routers shown as discs.

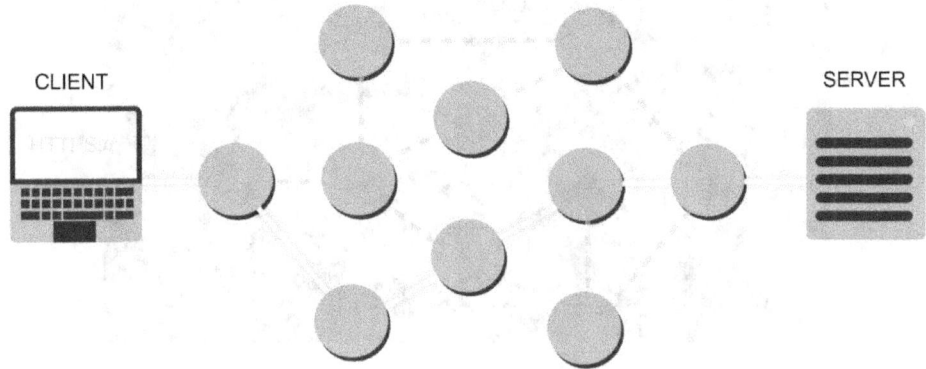

All computers that connect to the Internet have a *default gateway*. The default gateway is *the first hop for all of a computer's Internet traffic.* A *hop* is *shorthand for a communication link*, and when computers communicate with one another over the Internet, typically many hops are involved. *Traceroute is a command line utility that shows the hops to a destination on the Internet* (implemented as *tracert* in Windows). Traceroute works by sending three packets to each hop to determine the router name or IP address at that hop and calculate the roundtrip time. Figure 2.12 shows a traceroute from a client in Cedarville, Ohio, to a server located in Los Angeles, California. It shows that there are sixteen hops to the destination server, which means that the packets are received and processed by fifteen different routers along the way. The default gateway, i.e., the first hop, is 10.40.3.254. From trying to decipher the router names, it would appear the packets bounce around Ohio and then to Virginia, Georgia, and Texas before finally making it to California. This is not the most direct route, but, despite this, the last row of numbers show that the roundtrip time for the complete journey to Los Angeles and back is around eighty-three milliseconds (ms)—this is .083 seconds! This delay is called *network latency* and includes *propagation delay* (i.e., *the amount of time it takes a signal to travel a distance*) and *processing delay* (i.e., *the amount of time it takes a router to receive, process, and resend a packet*). To put this speed in perspective, it would take around seven hours to make a roundtrip flight from Cedarville to Los Angeles.

Figure 2.12

A traceroute from Cedarville to Los Angeles.

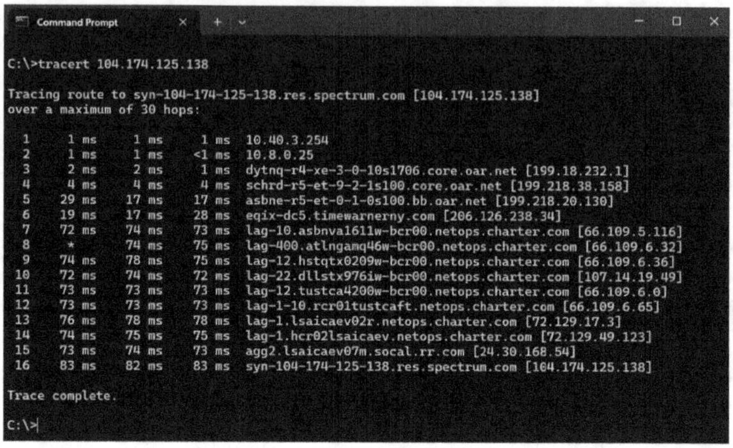

```
Command Prompt              ×   +  ∨                                        —   □   ×

C:\>tracert 104.174.125.138

Tracing route to syn-104-174-125-138.res.spectrum.com [104.174.125.138]
over a maximum of 30 hops:

  1    1 ms     1 ms     1 ms  10.40.3.254
  2    1 ms     1 ms    <1 ms  10.8.0.25
  3    2 ms     2 ms     1 ms  dytnq-r4-xe-3-0-10s1706.core.oar.net [199.18.232.1]
  4    4 ms     4 ms     4 ms  schrd-r5-et-9-2-1s100.core.oar.net [199.218.38.158]
  5   29 ms    17 ms    17 ms  asbne-r5-et-0-1-0s100.bb.oar.net [199.218.20.130]
  6   19 ms    17 ms    28 ms  eqix-dc5.timewarnerny.com [206.126.238.34]
  7   72 ms    74 ms    73 ms  lag-10.asbnva1611w-bcr00.netops.charter.com [66.109.5.116]
  8    *        74 ms    75 ms  lag-400.atlngamq46w-bcr00.netops.charter.com [66.109.6.32]
  9   74 ms    78 ms    75 ms  lag-12.hstqtx0209w-bcr00.netops.charter.com [66.109.6.36]
 10   72 ms    74 ms    72 ms  lag-22.dllstx976iw-bcr00.netops.charter.com [107.14.19.49]
 11   73 ms    73 ms    73 ms  lag-12.tustca4200w-bcr00.netops.charter.com [66.109.6.0]
 12   73 ms    73 ms    73 ms  lag-1-10.rcr01tustcaft.netops.charter.com [66.109.6.65]
 13   76 ms    78 ms    78 ms  lag-1.lsaicaev02r.netops.charter.com [72.129.17.3]
 14   74 ms    75 ms    75 ms  lag-1.hcr02lsaicaev.netops.charter.com [72.129.49.123]
 15   73 ms    74 ms    73 ms  agg2.lsaicaev07m.socal.rr.com [24.30.168.54]
 16   83 ms    82 ms    83 ms  syn-104-174-125-138.res.spectrum.com [104.174.125.138]

Trace complete.

C:\>|
```

2.4.3 The Network Edge

> *"It was really hard explaining the Web before people just got used to it because they didn't even have words like 'click' and 'jump' and 'page.'"* - Tim Berners-Lee, the inventor of the World Wide Web

Programs implement the application layer and interface with the transport layer to send and receive messages. Due to the Internet's layered design, programmers do not need to worry about sending electronic pulses over wires (Layer 1), communicating bits over a link (Layer 2), routing packets through the network core (Layer 3), or creating a communication channel with the destination process (Layer 4)—all this happens through services provided "under the hood." This frees the programmer to concentrate on sending and receiving application-level messages.

Programming for communicating over a network is called *socket programming*. The word *socket* was coined because it is a helpful word picture, as in plugging a cable into a socket like in the old-fashioned telephone switchboards. Table 2.7 shows two minimal socket programs written in Python: a *server program* and a *client program*. A server program is *a program that listens for incoming network connections* and a client program is *a program that initiates network connections to servers*. Both programs import the socket software library so they can call its procedures. The client connects to the server at 163.11.13.37:2600 (IP address:port number) and sends the message "LO" which is then received and printed by the server, recreating the first ever computer network message. Much work has to happen in order for this high-level message to be delivered, but the programmer can invoke it with just a few lines of code and be happily oblivious to all the underlying details.

Table 2.7

A minimal Python server and client that recreates the first "Internet" message.

Server	Client
```import socket``` ``s = socket.socket()`` ``s.bind(("163.11.13.37", 2600))`` ``s.listen()`` ``c, a = s.accept()`` ``print(c.recv(1024).decode())`` ``c.close()``	``import socket`` ``s = socket.socket()`` ``s.connect(("163.11.13.37", 2600))`` ``s.send("LO".encode())``

We (Internet users) operate at the edge of the network using applications written by programmers. All of the Internet processing and routing takes place in the core of the network and is transparent to us. Unlike programmers, we do not even need to know about IP addresses or port numbers in order to use the Internet. As stated in Section 2.3.4, the main Internet application we use is the web browser. We use web browsers to access the *World Wide Web* (*the Web*). Tim Berners-Lee invented the Web in 1989 when he defined the *URL* (*Uniform Resource Locator*), *HTTP* (*Hypertext Transfer Protocol*), and *HTML* (*Hypertext Markup Language*). These three technologies work together to enable web browsing. *Hypertext is formatted text that enables linking to URLs.* This is a powerful concept and is the reason why it is called "the Web" because links are like spider web strands that tie Internet resources to one another. HTML defines *the syntax for web pages* and HTTP is *the network protocol web browsers and web servers use to communicate.* HTTP has evolved considerably since Berners-Lee invented it. Today a secure version is used called *HTTPS*.

When browsing the web, we access resources using URLs. A URL is *a unique name for a resource on the world wide web.* URLs have a well-defined hierarchical structure:

[protocol]://[subdomain].[primary domain].[top level domain]/[resource path]

Not all URLs include a forward slash and resource path, and resource paths can include multiple forward slashes and at most one period. Subdomains are optional and there can be multiple sub-subdomains. The most common subdomain is *www* which stands for *world wide web*. Figure 2.13 shows a labeled URL.

The most important part of the URL is the *primary domain*—it ties a real-life entity to the resource because domains are leased and publicly registered. The primary domain is the section of the URL immediately before the *top level domain* (*TLD*). Some common TLDs are *.com*, *.org*, *.edu*, *.net*, and *.gov*, but numerous other TLDs exist. You can always identify the TLD because it comes immediately before the first forward slash if any are present, and if no forward slashes are present, it is the last part of the URL.

Figure 2.13

An example URL with sections labeled.

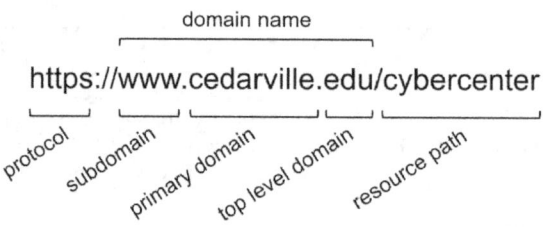

The first part of the URL after the protocol and up to and including the TLD is called the *domain name*. Domain names are tied to IP addresses. This is similar to how smartphones map our contacts' names to phone numbers. We do not need to remember our friends' phone numbers, we just click on their names and our phone makes the connection to the correct phone number. Like contact names, domain names act as memorable names for IP addresses. The *Domain Name System (DNS) is a public distributed database that maps domain names to IP addresses*. When we click on a URL our web browser uses DNS to find the IP address of the domain name in the URL. Once the IP address has been identified, the web browser makes a network connection to the web server at that IP address. The protocol field in the URL determines the port number to use. HTTP uses port 80 and HTTPS uses port 443. Once connected, the web server examines the resource path of the URL to connect the browser to the requested resource. There are many websites that provide DNS services and there are also command-line utilities such as *dig* for Linux and *nslookup* for Windows (see Figure 2.14).

Figure 2.14

The dig and nslookup utilities showing that the IP address for the domain www.cedarville.edu is 163.11.75.44.

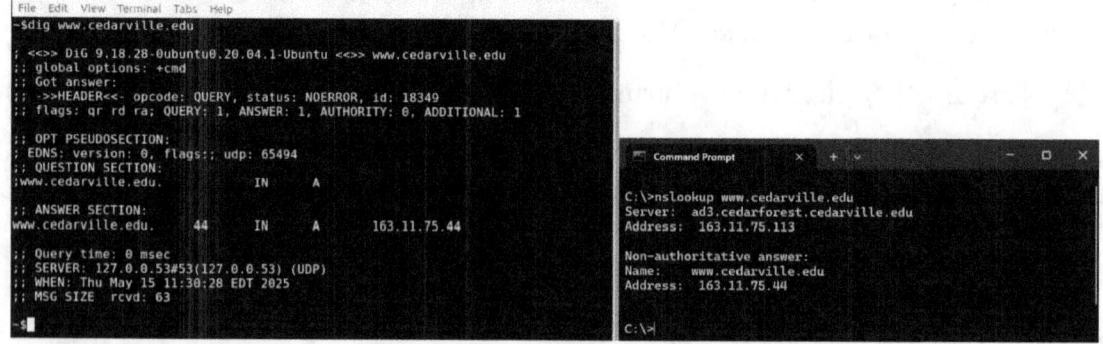

When a computer is online, its NIC is constantly sending and receiving data. Applications called packet sniffers log these packets and allow users to inspect them. Wireshark is a free and open-source packet sniffer. It provides visibility into all of the packets that a computer is sending and receiving. It helpfully formats the packets so that they can be inspected by layer. It even provides visibility all the way down to the 1s and 0s (displayed as hexadecimal digits) to help remove as much mystery as possible—at the end of the day, it really is just 1s and 0s! Figure 2.15 shows a packet capture for the network traffic involved in navigating to a simple website. The user types the URL in the address bar of the browser, hits enter, and instantly the webpage appears. Meanwhile, at least six packets had to be sent and received (three roundtrips) in order for the web page to be downloaded.

Figure 2.15

A Wireshark packet capture.

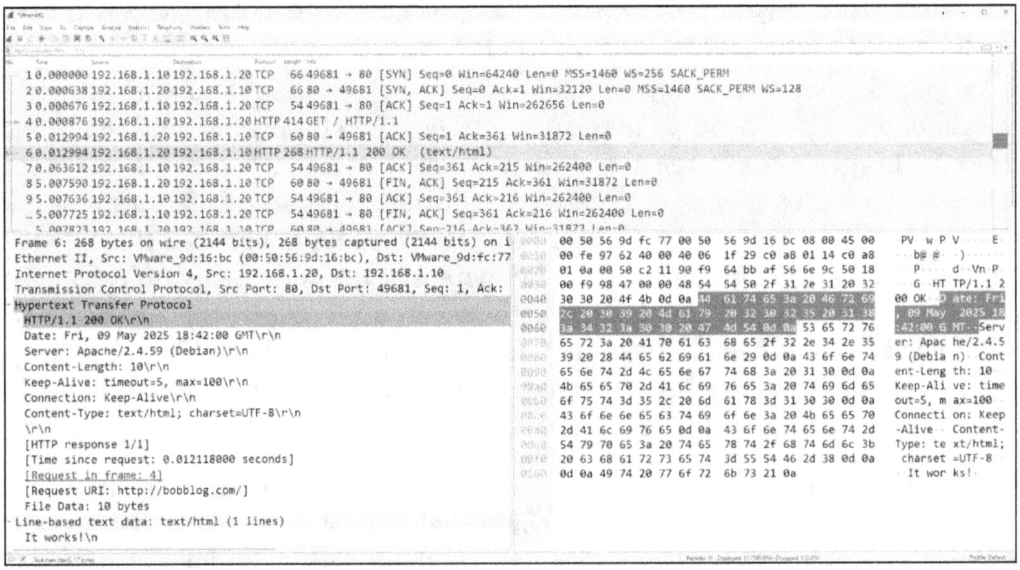

2.5 The Frontiers of Computing

Artificial intelligence and quantum computing are at the forefront of computing technology and have ramifications for cybersecurity. In keeping with its foundational emphasis, these topics are not addressed at length in this textbook, but this section includes a short primer on both technologies and some summary thoughts on their impact on cybersecurity.

2.5.1 Artificial Intelligence (AI)

> *"THE HEAD AND IN FRONTAL ATTACK ON AN ENGLISH WRITER THAT THE CHARACTER OF THIS POINT..." - Statistically chosen English words using a "second-order word approximation" in Claude Shannon's, "A Mathematical Theory of Communication," anticipating the development of generative AI*

Recent advances in *artificial intelligence* (*AI*) might seem to contradict this chapter's claim that the appearance of intelligence in computers is actually entirely due to human intelligence that extracts meaning from their computations. AI is *cutting edge computing technology that performs at human-like levels or better*. In every era of computing, AI represents the forefront of capabilities and the height of complexity. As of this writing, AI is synonymous with machine learning and large language models that have given rise to *generative AI* technologies. Generative AI is *computer technology capable of generating coherent media such as text, audio, images, and video.* For example, intelligent chatbots speak English fluently and can provide expert answers to questions in seemingly every domain of human knowledge. They can also write computer programs, fiction, poetry, and much more. While not perfect, many believe it is inevitable that they will improve over time and will someday be practically flawless.

Regardless of how good computers get at generative AI (and at any other type of AI), foundationally, they are still using Boolean logic to process binary input signals to produce binary output signals. Due to layers upon layers of abstraction and encapsulation and extreme amounts of processing, the relationship between inputs and outputs is impossible to trace. This gives rise to the appearance of true intelligence, but it is an illusion. In 1948 Claude Shannon anticipated generative AI when in his study of information theory he proved that languages have statistical properties. He demonstrated this by manually selecting random letters from a book. As he increased the complexity of his calculations he was able to randomly produce semi-coherent sentences. Due to the escalating amount of work involved in each step, he was not able to proceed further. What Shannon could not do at that time, computers today can due to the enormous amount of data available to them and their incredible processing power.

The AI technologies that have produced generative AI are based on the statistical nature of information. While it is incomprehensible that coherent language can be produced based sheerly on statistical probabilities, that is their underlying program. This is not how humans process information, and this is easily verified by comparing the minimal amount of energy human brains consume to the vast amount of energy generative AI models consume.

In short, AI in all of its forms, is still just a computer program. Human intelligence is needed to write it, feed it inputs, and evaluate its outputs. Its outputs are deemed good or bad based on human evaluation—that is the ultimate criteria proving that computers serve humans, not the other way around. We need to embrace AI as a tool that augments human intelligence. We are able to do more with AI technologies than we can without them—they multiply productivity. As for cybersecurity, AI is another technology that needs to be secured, the goal of *secure AI*, and it is also a technology that can help enhance cybersecurity, the goal of *AI for cyber*. Because AI will assist both cyber attackers and cyber defenders, it is not clear if it will change the overall balance of power in cybersecurity. Chapter 9 covers some ramifications of AI for cybersecurity.

2.5.2 Quantum Computing

"I think I can safely say that nobody understands quantum mechanics." - Richard Feynman, a pioneer of quantum physics

Quantum computing is a paradigm of computation based on qubits. A qubit is *a quantum bit*, and unlike the discrete bits in classical computing that are equal to either 1 or 0, qubits can contain the entire range of real values between 1 and 0 simultaneously—this is called *quantum superposition*. This property is due to *quantum mechanics*, the behavior of atoms and subatomic particles. Scientists have demonstrated that these particles can become *entangled*, meaning, they can communicate literally instantaneously (not just apparently instantaneously as in the speed of electronic communication) despite being separated by any amount of distance.

The goal of quantum computing is to harness these mysterious properties to perform certain types of calculations quickly that are intractable with classical computers. Algorithms have been devised on paper that show how these qubit computations would work. These calculations have tremendous value, therefore, much research is being performed in this field. To date, while progress has been made, no practical applications have yet been achieved. There is ongoing debate over how long it will take to produce a quantum computer of the scale needed to produce real-world value, and some maintain that it is a fool's errand and will never happen.

Even if quantum computers are produced, they will not replace classical computers. Different types of algorithmic problems are better suited to one computing paradigm than the other. In terms of cybersecurity, quantum computing has important ramifications for cryptography, the bedrock of cybersecurity (see Chapter 7). The main encryption algorithm responsible for securing Internet communication can theoretically be broken with a quantum computer. Without secure encryption, the Internet as we know it cannot exist. For this reason, research into quantum-resistant cryptography is ongoing, and the hope is that new algorithms will be developed and implemented before quantum computing is realized.

2.6 Conclusion

This short primer has covered a massive amount of cyberspace terrain! It is important because cyberspace is the context of cybersecurity. We need to know the basics of how computers and the Internet work if we have any hope of securing cyberspace.

The biggest takeaway from this chapter is that computers leverage Boolean logic and binary encoding at mind-boggling speeds and scale together with highly engineered layers upon layers of encapsulation and abstraction to accomplish feats of seeming magic. They are incredibly simple but at the same time amazingly complex. In this complexity lurk innumerable cybersecurity weaknesses.

The Internet is a connected graph of machines that send signals to and from one another across distances, again, at incomprehensible speeds. Computers communicate using well-defined protocols. Fundamental communication problems, such as signal processing, addressing, and transmission reliability, are encapsulated within neatly defined layers. Application programmers and Internet users exist at the network edge and can remain oblivious to most of the work that takes place. In this chapter we uncovered just some of the magic involved in computer networks as we examined packet switching, IP addressing, and packet sniffing.

This chapter provides enough depth to set the stage for an introduction to the main concerns of cybersecurity covered in the remainder of this book. However, students desiring to go into cyber operations and the more technical aspects of cybersecurity will need a much deeper level of familiarity with cyberspace—this is just the start!

 **Chapter 3**

3. The Adversary of Cybersecurity: Hackers

"If you know the enemy and know yourself, you need not fear the result of a hundred battles. If you know yourself but not the enemy, for every victory gained you will also suffer a defeat. If you know neither the enemy nor yourself, you will succumb in every battle." - The Art of War *by Sun Tzu*

Cybersecurity is predicated on the existence of cyber threat actors. These people are commonly called *hackers*. This term predates cybersecurity, and it was originally a compliment, referring to students working with the earliest computers on college campuses in the 1960s. It meant they were good at finding creative ways to make computers accomplish different tasks. Later, the media used the word hacker to describe a few early cyber-criminals, and the term stuck. Today, it is a good catch-all term for cyber adversaries. A hacker is *a person who attempts to gain unauthorized access to a computer system or deny access to authorized users.* Hackers may work alone or in groups. In this text the term hacker will frequently be used as a short-hand for cyber attackers.

The previous chapter outlined the context of cybersecurity, cyberspace, from a technical perspective. Because of the centrality of hackers to cybersecurity, this chapter covers who these people are and what motivates them. The next chapter explores how they accomplish their objectives.

3.1 Hackers

Hackers exploit vulnerabilities. A vulnerability is a weakness and an exploit is an action that takes advantage of a vulnerability to compromise security. While exploiting vulnerabilities sounds nefarious, it is simplistic to say that hackers are bad guys. For example, good guys may attempt to gain unauthorized access to computers belonging to terrorists to thwart an attack. So while the term hacker generally has a negative connotation, the hacker's motivation and goals must be considered before passing judgment. Not all hacking is against the law or even unethical.

Hacking hat colors are used to help distinguish legal hacking from criminal hacking. *White hat hackers are individuals who hack legally and ethically.* This means, ironically, that they gain unauthorized access to computers with authorization! *Black hat hackers*, on the other hand, are *individuals who engage in illegal hacking.* Most of this text is concerned with black hat hackers, so the qualifier *black hat* is usually omitted.

There is also a third hat color that is sometimes used: *gray hat hackers.* Gray hat hackers are *individuals who blur the lines between legal and illegal hacking and engage in both kinds.* Hacking can be highly stimulating due to the combination of anonymity, social distance, and the pursuit of "forbidden fruit." This overstimulation can cause hackers to get into a flow state and experience euphoria. This feeling can lead to addiction where hackers keep coming back and pushing boundaries seeking a rush. This desire can ultimately impair their sound judgment and lead to illegal activities. Gray hat hackers may also purposely disregard the law to accomplish what they feel are worthy cybersecurity ends. An example of this is *hacking back*, which is the practice of trying to hack somebody who has hacked you. It may sound legitimate, but gaining unauthorized access to a computer is illegal, even if the other person hacked you first. Some have tried to argue that hacking back is an act of self-defense, but self-defense does not carry the same weight in cyberspace as it does in physical space. In cyberspace, hackers pose no imminent threat of physical danger.

Hacking hat colors are derived from old western films. In those movies, it was often the case that the good, law-abiding, and heroic cowboy, like the famous actor Roy Rogers, would wear a white hat, whereas the villainous cowboys would wear black hats (see Figure 3.1).

Figure 3.1

Roy Rogers wearing a white hat in an old western movie poster.

3.1.1 Ethical Hacking

Before focusing on black hat hackers, we will quickly examine some ways that white hat hackers hack for good, otherwise known as *ethical hacking*.

There are three principles of ethical hacking: *behaving ethically at all times, respecting the rights of all citizens, and obeying all applicable laws and authorities*. This is a high bar for those who would engage in ethical hacking—the bar is high because the stakes are high. Ethical hacking is serious business. Ethical hackers must know the law and understand the legal authorities so that they can operate within the appropriate legal boundaries at all times. There is overlap between these three principles, but they have different emphases. Chapter 10 explores each area in detail. This section provides an overview of the four main types of ethical hackers: penetration testers, cyber warriors, law enforcement officials, and cybersecurity researchers.

3.1.1.1 Pentesters

Penetration testing (pentesting) is the active probing of the cybersecurity defenses of an organization for the purpose of improving security. Companies pay professional pentesters to hack them so that they can discover security weaknesses before the bad guys do. Pentesters must be vigilant in how they maneuver within an organization because they need to avoid creating harmful side-effects from their testing. This is not easy because hacking sometimes involves uncertainty and trial and error. Therefore, it is important that pentesters be highly competent hackers. They need to minimize risks such as deleting important data or crashing computers.

Pentesters must also be of high moral character. During their pentesting, they may observe sensitive information that needs to be kept confidential. Also, they need to be trustworthy enough to do their job thoroughly and to report their findings accurately. Pentesters have opportunities to cross the line into unethical areas, and if they do, it is likely nobody would ever find out. Their character must be strong enough to safeguard them from these types of temptations.

Pentesters need to abide by important conditions that are codified in official documents that have legal standing. First, they must gain authorization. Gaining authorization is achieved by having an authorized party sign a *permission memo* before the pentest begins. Colloquially this document is known as a "get out of jail free card." A permission memo explicitly grants the pentesters authorization to hack the organization. If you ever consider engaging in pentesting, always remember: procuring proper permission prior to performing pentesting is paramount! Second, they must define the boundaries of the pentest. Boundaries are defined in agreements called the *scope of work* (*SOW*) and the *rules of engagement* (*ROE*). Boundaries include things like specific IP addresses, workstations, servers, and routers that are eligible to be pentested; the times, dates, and duration of the pentest; who will be performing the pentesting; and the types of attacks that are permitted. Lastly, they must maintain confidentiality. Pentesters sign documents such as *non-disclosure agreements* (*NDAs*) that state that they will keep the results of the pentest and any information they discover confidential.

After the penetration testing is completed, pentesters deliver a pentest report to the client. This report lists the activities undertaken by the pentesters and the detailed results. It also includes a list of the vulnerabilities discovered, their severity, and prioritized recommendations.

3.1.1.2 Cyber Warriors

> "USCYBERCOM plans, coordinates, integrates, synchronizes and conducts activities to: direct the operations and defense of specified Department of Defense information networks and; prepare to, and when directed, conduct full spectrum military cyberspace operations in order to enable actions in all domains, ensure US/Allied freedom of action in cyberspace and deny the same to our adversaries."
> *- USCYBERCOM original mission statement*

Conflict between nation states has taken place in different domains across time. First there was land, sometime shortly after followed sea, many years later, after the Wright brothers invented flight in the 1900s, came air, and in the 1950s came rockets and outer space. The newest domain since the 1980s is cyberspace. The United States Department of Defense (DOD) has referred to these settings as the *five domains of warfare*.

Cyber operations are intelligence and military operations that take place in and through cyberspace. Cyber operations include activities like espionage (e.g., spying), sabotage, and subversion (e.g., disinformation campaigns). The United States sometimes refers to cyber operations activities as the *five Ds: disrupting, degrading, denying, destroying, and deceiving our adversary's capabilities in and through cyberspace.*

Cyber warriors are individuals that hack with the authorization of the government. In the United States, cyber warriors work for the DOD, either in the military or for an intelligence agency. It is important for ethical hackers, and in particular cyber operators, to understand the authorities applicable to international scenarios. Their work is protected by Titles 10 and 50, respectively, of the United States Code. They are explicitly authorized to obtain unauthorized access to computer systems owned and operated by our adversaries. This type of hacking happens both in peacetime and wartime. The rules describing what is in and out of bounds are carefully defined and lawyers are consulted when needed to determine whether or not a specific activity is permitted.

Title 50: War and National Defense, outlines national security, including all foreign intelligence and counterintelligence activities. *Signals intelligence* (*SIGINT*) focuses on electronic communications and includes covert cyber operations to gather foreign intelligence—this is a primary function of the United States *National Security Agency* (*NSA*). These cyber operations activities are governed under Title 50 and must comply with the United States Constitution, federal law, and executive orders. For example, Title 50 outlines the *Foreign Intelligence Surveillance Act* (*FISA*). FISA explicitly allows the United

States government to collect foreign intelligence signals on domestic soil under specific circumstances. The NSA relies heavily on FISA to accomplish its national security mission. FISA and the *Electronic Communications Privacy Act* (*ECPA*) are occasionally seen to be in conflict because the ECPA protects the right to privacy for United States citizens (ECPA is explained in more detail in Chapter 10). It is vital that cyber warriors understand these laws so that they can properly protect the rights of all citizens.

Title 10: Armed Forces of the United States Code, outlines the role of the United States Armed Forces (the Army, Marine Corps, Navy, Air Force, Space Force, and Coast Guard) and the *Unified Combatant Commands* (*UCCs*). The UCCs have broad, continuing missions and coordinate military activity either by geography or by specialty domain area. The *United States Cyber Command* (*USCYBERCOM*) was established in 2010 to unify the cyberspace operations of the military and has several components, including Army Cyber Command and Air Forces Cyber. Cyber operators acting under the authority of Title 10 carry out military missions in cyberspace, which have much broader goals than just SIGINT. USCYBERCOM's seal contains the MD5 hash of their original mission statement (see Figure 3.2 and look closely around the inner ring):

9ec4c12949a4f31474f299058ce2b22a

Section 7.2.3 of this text covers hashes and their significance to cybersecurity.

Because technology underpins all five domains of warfare and intelligence operations, the DOD is a pioneer of cybersecurity in the United States and remains a foremost influence. *Challenge coins* are an important part of cybersecurity culture that has arisen out of the influence of the NSA, USCYBERCOM, and the military. A challenge coin is *a custom coin presented by a leader as a commendation for a job well done.* Challenge coins originated in the military as a way of recognizing outstanding effort and accomplishments, and soldiers would carry them around in their pockets as a treasured keepsake. They became known as challenge coins because a group of soldiers could be challenged to present a coin and any that failed to produce one would have to buy drinks for those that did! Different organizations and units within an organization create their own coin designs bearing a seal or motto or other phrases and symbols of special significance, and leaders within those organizations hand them out at their discretion. The challenge coin tradition has been integrated into the cybersecurity industry and cybersecurity education in the academy, and collecting coins is a badge of honor for cyber professionals and students. An NSA challenge coin is shown in Figure 3.3.

The motivation of cyber warriors is to improve their nation's cyber defense and advance their national interests.

Figure 3.2

USCYBERCOM seal.

Figure 3.3

An NSA challenge coin—see Figure 7.2 for the key to decoding the ciphertext on the inner ring of the reverse side.

3.1.1.3 Law Enforcement Officials

> *"Malicious cyber activity threatens the public's safety and our national and economic security. The FBI's cyber strategy is to impose risk and consequences on cyber adversaries." - United States Federal Bureau of Investigation (FBI)[1]*

Law enforcement officials, like those that work for the United States Federal Bureau of Investigation (FBI) and state and local police forces, investigate crimes and make arrests. They gather evidence and make a case to prosecute criminals. Increasingly, much of the evidence is digital and must be carefully analyzed with high-tech tools and high levels of expertise. *Cyber forensics is the process of collecting, analyzing, and preserving cyberspace evidence.* It encompasses activities like determining how a cyber intrusion occurred. What was the initial vulnerability exploited by the attackers? What were their

---

[1] fbi.gov website. *What We Investigate - Cybercrime - The Cyber Threat.* Retrieved June 2025.

post-exploit activities? This involves the analysis of evidence left behind on computer systems, including operating system and other types of log files. Forensics experts examine log files and other data artifacts to find evidence and to reconstruct events. Sometimes this is done after a person has deliberately tried to destroy evidence by deleting data, modifying logs, or damaging physical hardware. Some cyber forensics experts work in specialized labs to reconstruct damaged hardware to recover data (see Figure 3.4).

Individual hackers or hacker teams may develop a signature of a certain sequence of steps and types of tools. These are called *techniques, tactics, and procedures (TTPs)*. They can function almost like a fingerprint—if a group uses the same TTPs in different attacks, investigators may be able to attribute the attacks to the same group. Hackers may do some of the same things as a habit, or because they trust the TTPs because they have been effective in the past. In general, as we saw in Chapter 1, because of the fundamental features of cyberspace, it is difficult to determine who did what when. To make things even more difficult, a hacking group may deliberately plant subtle clues to try and pin the crime on somebody else. The difficulty of determining who is responsible for a cyber attack is called the *attribution problem*. Cyber forensics specialists, like detectives, study past attacks, look at all the evidence of a crime scene, and try to solve the crime by tracking cyber attacks to their source.

Figure 3.4

A cyber forensics lab.

Cyber forensics also includes hacking into computer systems to assist in criminal investigations. Unlike cyber warriors, this type of hacking can be performed against United States citizens, but, again, only with the proper authorization. For example, if an encrypted smartphone is discovered in the course of a criminal investigation that might contain evidence, and a judge issues a warrant, then law enforcement officials may hack the device to obtain access to its data. Chapter 10 provides more details on the rights of citizens.

Law enforcement officials are motivated by the pursuit of justice.

3.1.1.4 Cybersecurity Researchers

"Our Bug Bounty Program encourages collaboration with the research community and incentivize [sic] researchers to report vulnerabilities in Intel products. Through the Bug Bounty program, Intel invites researchers to test specific targets, submit vulnerabilities, and get paid for their work." - Intel Bug Bounty Program[2]

Cybersecurity researchers perform vulnerability assessments on websites, software, and hardware products. Like pentesters, their goal is to find and disclose vulnerabilities to improve cybersecurity. Some work for large organizations. Others are academics—professors, graduate students, and fellows at a university. Still others are independent cybersecurity researchers. One way that cybersecurity researchers earn money is through *bug bounty* programs. A bug bounty is *a payment made by an organization for finding a vulnerability in one of their products.* Companies promote these programs, in effect inviting cybersecurity researchers to probe their products, as a way to "hire" outside researchers to help make their products more secure. They would prefer to know that a vulnerability exists before cyber attackers discover it and cause fall-out for their customers. Bug bounty programs outline the ROE, how to contact the company with findings, and how much money they will pay for different classes of vulnerabilities.

Cybersecurity researchers might also probe for vulnerabilities even when there are no bug bounty programs. Here, they must be careful if they are proceeding without explicit permission from the company. Copyright-related laws, including the *Digital Millennium Copyright Act* (*DMCA*), provide *intellectual property* (*IP*) protections for companies. IP is *proprietary information that provides a competitive advantage to an organization.* The DMCA prohibits the dissecting of hardware and software and the undermining of security controls. There are carve outs in the DMCA for cybersecurity researchers who are acting in good faith. However, engaging in reverse engineering of products without explicit permission is a gray area. There have been incidents where researchers have discovered vulnerabilities and informed the company of their findings, and instead of thanking them for helping to improve cybersecurity, the company promptly sued the researchers for violating their intellectual property rights.

Another major target of cybersecurity researchers is *free and open-source software* (*FOSS*). FOSS is *software that is free to use and whose source code is publicly available.* The computing industry and the Internet are heavily dependent on FOSS. Occasionally, researchers find major vulnerabilities in FOSS that impact systems across the world, and these vulnerabilities become frontpage news. One such vulnerability was discovered in the OpenSSL library in 2014. The vulnerability could be exploited to allow attackers to eavesdrop on encrypted communications, among other things. Around this era the

---

[2] intel.com website. *Intel Bug Bounty Program - Collaborating with the Research Community.* Retrieved June 2025.

practice of "marketing" vulnerabilities emerged. Researchers created catchy names, logos, and official websites to make their findings more accessible and for public relations purposes—discovering an important vulnerability can make a cyber researcher's career (see Figure 3.5). The OpenSSL vulnerability was dubbed *Heartbleed* because it exploited OpenSSL's heartbeat protocol (i.e., a protocol to determine whether a connection was still "alive"). Other examples of famous vulnerabilities in FOSS include *ShellShock* and *Log4j*

Figure 3.5

The Heartbleed Bug's website and logo.

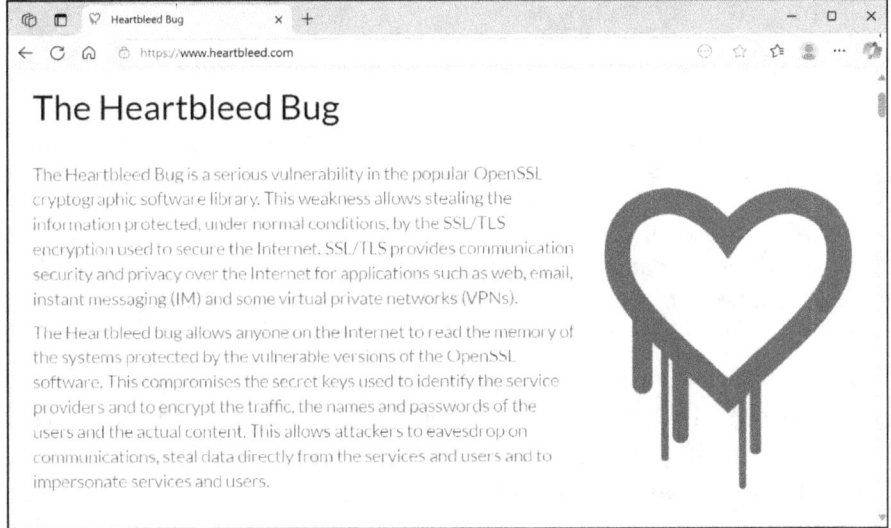

Over time a de facto process has emerged for finding and reporting vulnerabilities called *responsible disclosure*. Responsible disclosure is *the steps taken to report a discovered vulnerability to an organization*. In the early era of cybersecurity research, individuals would quietly notify companies that they had found a security flaw in the company's product or website that could be exploited by bad actors, and then the researchers would assume they had done their good duty and move on. However, the researchers were surprised to discover that companies either did not believe them, did not understand the risk, or did not care, and in many instances, the vulnerabilities were never fixed. This frustrated cybersecurity researchers who understood the risks the companies were exposing to their customers. To force companies to take them more seriously, researchers began informing companies that after some reasonable period of time based on the perceived effort to address the vulnerability (e.g., 30 days), the researchers would publicly announce the vulnerability and provide a proof of concept for exploiting it. This subtle "threat" forces companies to engage with the researchers and take them seriously instead of just ignoring them. Providing a proof of concept makes it easy for others (including bad actors) to exploit the vulnerability and is somewhat controversial. However, this removes any doubt that the vulnerability is legitimate. Many cybersecurity researchers found that the proof of concept was necessary because sometimes companies would deny that the vulnerability existed or they would say it is overblown and not a practical concern. Researchers provide companies advanced notice so the companies will have time to fix the vulnerability and release a patch if necessary before bad actors find out about it. The idea is that by the time of the public disclosure, the vulnerability would be a moot point, but the cybersecurity researcher would still receive credit for his effort.

Cybersecurity researchers, like all ethical hackers, must be careful to act ethically at all times, respect the rights of all citizens, and obey all applicable laws. The responsible disclosure process in particular is a good case study for an ethical analysis like the ones outlined in Chapter 10 of this textbook.

3.1.2 Illegal Hacking

> *"I have a lot of regrets, but I think my essential failing was that I lost touch with the accountability and responsibility that comes with being a member of society. A friend of mine once told me to behave as though everyone could see what I was doing all the time. A sure way to avoid engaging in illegal conduct, but I guess I wasn't a believer because when I was invisible, I forgot all about this advice. I know now that we can't be invisible, and that it's dangerous thinking." - notorious hacker Max Butler at his sentencing*

Unfortunately, there are far more black hat hackers than white hats. Illegal hacking is the main concern of this textbook, and it is helpful to distinguish among several different categories of illegal hacking. The first main distinction is the difference between internal and external hackers relative to the organization that is being hacked. An internal hacker is called an *insider threat*. An insider threat is *a person that works for the organization they attack*. Insider threats are difficult to prevent because they are trusted and have at least some legitimate access to computer systems and data in order to do their jobs. Where they cross the line into illegal hacking is when they exceed their authorized access—a form of obtaining unauthorized access. The threat posed by external versus insider threats are comparable to the difference a random bank robber poses versus the manager of the bank. The manager has legitimate access to the assets of the bank, and knows all the security protocols, including the lock combination to the safe. Also, insider threats are difficult to detect because it is easier for them to cover up their crimes—in the bank example, the manager can tamper with logs and other official records. Organizations have to put many controls in place to deter, prevent, and detect insider threats (more on controls in Chapter 5).

Any hacker who is not an insider threat is an external threat. All the categories of hacking in the following subsections apply to both insider and external threats. Most hackers are external threats since the outsiders of any organization greatly outnumber the insiders. Whether an insider or an external threat, illegal hacking can be categorized by the motivations of the hackers.

3.1.2.1 Hacking for Cash

The main motivation of most illegal hacking is to make money. This category of hacking is called *criminal hacking*. Just like in physical space, hackers can make money criminally through activities like stealing, fraud, espionage, sabotage, and extortion. Examples of stealing money include logging into a victim's bank account and transferring money, or logging into a victim's human resources portal at work and changing their paycheck's direct deposit information to a bank account controlled by the attacker. In both of these cases, the attacker impersonates the victim by using their username and password which is a type of fraud. Attackers may also steal cryptocurrency by gaining unauthorized access to another person's crypto wallet and initiating transactions. Recovering stolen cryptocurrency is difficult since all transactions logged in the blockchain are instantly permanent, whereas there is a short grace period for banking transactions that allow them to be cancelled if caught in time.

Besides directly targeting a victim's finances, hackers may also profit from obtaining access to data. When this happens it is called a data breach. The data obtained might be IP, *personally identifiable information (PII)*, or other types of sensitive information such as health and financial records. Criminal hackers may steal, or the equivalent in cyberspace, make a copy of, data that contains valuable information. An example of IP theft would be stealing research and development data from a pharmaceutical company. Attackers might obtain PII or credit card numbers so they can sell them on the black market. PII is *data that can be used to identify a person and commit identity theft*. Some data breaches impact millions of people. Therefore, even if individual records stolen in a data breach sell for a small amount each, the total amount of money made by hackers can be substantial.

In business settings, hackers may gain unauthorized access to a competitor's network for the purposes of spying and learning their plans (*corporate espionage*) or to impair their competitor's ability to operate (*corporate sabotage*). Hackers may also obtain access to sensitive information and then threaten to release it publicly unless they are paid. They may also commandeer computer systems and data and then prevent the owners from accessing them until a payment is made— these are forms of extortion.

These few examples of criminal hacking barely scratch the surface of all the ways that hackers can use their skills to make money. The amount of money cybercrime costs society is measured in trillions of dollars annually, so this is a major cybersecurity threat and a significant national concern.

3.1.2.2 Hacking for a Cause

Some individuals or groups of individuals hack to advance their cause. When they hack for political or ideological reasons this category of hacking is called *hacktivism* (short for hacktivist hacking). Hacktivists may publish private documents belonging to a company the activist group opposes. For example, a hacking group that is concerned about the environment may hack a company they believe is polluting a local stream. After gaining access to their network, the hacktivists may download and disclose evidence to the media or directly to the public. The goal of the hackers is to raise awareness and harm the company. This is a form of vigilante justice, and it is against the law.

Insider threat hacktivists might publish private documents in order to expose what they consider are illegal or unethical practices. This is known as a *data leak*. Sometimes this type of hacktivist is a legitimate whistleblower and is able to obtain legal protection, but other times their actions are illegal and they are charged with crimes. Edward Snowden famously leaked classified documents he obtained from the NSA and claimed whistleblower status, but the United States government disagreed and charged him with serious crimes. *Wikileaks* is a website that publishes documents from data leaks.

There are several well-known groups that call themselves hacktivists. One pioneering group based in Europe is called the *Chaos Computer Club*. Another famous group based in the United States was called the *Cult of the Dead Cow* (abbreviated *cDc*). cDc engaged in gray hat hacking. Some of their early members later formed a group called *Lopht* that pioneered cybersecurity research and famously appeared before Congress using their hacker nicknames. They were called to testify about the insecurity of the Internet. Some of the members became tech industry leaders. *Anonymous* is another notorious, and darker, hacktivist group that has a decentralized leadership structure and has members all over the world. Many hacktivists consider themselves on the side of good versus evil. However, hacking is illegal, and hacktivists are subject to criminal charges regardless of their "good" intent.

Another cause that motivates hackers is national pride. These individuals are called *patriotic hackers*. As civilians they are not technically authorized to hack by their country, but they take it upon themselves to defend their nation by hacking perceived enemy nation states. Russian nationals have been accused of patriotic hacking, particularly in the conflict against Estonia in 2007 (see Section 10.3.1.1). The Russian government benefited from their activities even though they could officially denounce them because they were not sanctioned by the government.

Another cause that motivates hackers is terrorism. Terrorist groups are motivated by hate. Their goal is to inflict maximum damage on their enemies, and they attempt to do so using any means necessary, including hacking. The terrorist group al Qaeda conducted the 9/11 terrorist attacks against the United States on September 11, 2001, using commercial airplanes as missiles. Today, many people worry about a *cyber 9/11*. A cyber 9/11 might target critical infrastructure, like shipping ports, the power grid, or water treatment plants. Disrupting critical infrastructure could harm the economy and could also result in mass casualties and death. A similar idea is a *cyber Pearl Harbor*—a large-scale cyber attack carried out by an enemy nation state. Some hackers may leave a *calling card* to prove that they were the ones that carried out an attack. In some attacks motivated by political or ideological reasons, it can be important for the hackers to be able to take credit for the harm they caused.

3.1.2.3 Hacking for Curiosity, Fun, and Bragging Rights

> *"ELK CLONER: / THE PROGRAM WITH A PERSONALITY / IT WILL GET ON ALL YOUR DISKS / IT WILL INFILTRATE YOUR CHIPS / YES IT'S CLONER! / IT WILL STICK TO YOU LIKE GLUE / IT WILL MODIFY RAM TOO / SEND IN THE CLONER!" - Elk Cloner virus poem*

Other motivations for hacking are simply curiosity, fun, and bragging rights. When curiosity is unchecked, it can result in illegal hacking. This category of hacking is known as *nuisance hacking*. Despite the name, the effects can be damaging beyond causing a mere nuisance. Many nuisance hackers do not believe they are doing anything wrong due to immaturity or lack of a moral compass. They might even blame the victims for being sloppy and inviting an attack. Some students are susceptible to the allure of nuisance hacking. Unfortunately for them, it is illegal and can result in expulsion from school and even criminal prosecution.

Some people hack for the joy of it. Their motivation is to have fun and see if they can accomplish a personal goal. They may be trying to boost their street cred by collecting trophies. They likely brag to their friends about their hacking exploits. A specific example is *website defacing*. Website defacing is in cyberspace what spray painting graffiti on a building is in physical space. In a website defacing attack, a website is taken over and replaced with a defaced version or a new webpage promoting the hacker. Hacktivists engage in website defacing to embarrass victims and promote their cause.

Many of the earliest hackers fell into the category of nuisance hacking. They were not hacking to make money. Rather, they were just trying to pull pranks and test the limits of what they could accomplish. Sometimes their hacking exploits resulted in silly annoyances. The *Elk Cloner virus* was one of the first viruses ever created. It was created as a prank by a fifteen year old named Rich Skrenta in 1982, during the early era of personal computers. When it infected systems, it would randomly pop up a poem on the victim's computer. Other examples of nuisance hacking can be much more damaging, like other early computer viruses that erased all of the files on the victim's machine.

Phone phreakers are another category of early nuisance hackers. Phone phreaking was an early form of nuisance hacking that exploited vulnerabilities in the landline, and later cellular, telephone system. It could be argued that they did benefit monetarily because they made phone calls for free. Kevin Poulsen allegedly hacked the phone system to win multiple radio "9th caller" competitions and later went on to become a journalist and author. So while there were financial benefits, mostly phone phreakers were motivated by curiosity and bragging rights.

3.1.3 Hacking Skill Levels

When categorizing hackers, it is important to not only examine their motivations but also their skill levels. Skill levels vary from *script kiddies* at one end of the spectrum to *elite hackers* at the other.

3.1.3.1 Script kiddies

> *"Today's top-secret programs become tomorrow's PhD theses and the next day's hacker tools." - Bruce Schneier*

Script kiddies are *unskilled individuals who utilize user-friendly tools and scripts developed by others to hack into computer systems.* It is a pejorative term because it connotes a person with little technical skill. Script kiddies can be any age, not just kids or teenagers. They do not have a lot of technical expertise or understanding, but they can follow a step-by-step script to conduct a hack, and they can wield point-and-click hacking tools built by other more highly skilled hackers. Script kiddies can cause damage even if they do not understand the technical details of what they are doing or how their attacks work. Script kiddies are significant threats because as time passes, once complex hacks known to only an elite few become scripted and are made available to the masses in the form of hacker tools. Script kiddies do not create novel attacks, but recycled attacks often remain effective for a long time.

The Social Engineering Toolkit (*SET*) is a command line tool that provides a step-by-step wizard for configuring and deploying social engineering attacks that can be used to steal user credentials and gain access to a victim's machine. *OpenVAS* is a vulnerability scanner that can be used to identify a weakness in a target system that could lead to unauthorized access. *Metasploit* is a popular command line tool that makes gaining access easy. It steps users through the process of pairing an exploit for a vulnerability with a payload that will be executed on the target—a typical payload provides the hacker command line access to the victim. Once the "cyber weapon" is configured with the parameters of the target machine (e.g., the IP address and port number), it can be fired while the hacker sits back and waits for a call-back from the victim machine. *Armitage* is a GUI wrapper for Metasploit that makes its functionality even more accessible. *Mimikatz* is a simple hacking program that can be run after a hacker has gained access to a victim to extract plaintext passwords from memory—the passwords can then be used to gain even deeper access. All of these tools (plus many more) are free and come bundled with Kali Linux and other hacking distros. While they may make hacking easy, it would be unfair to say that they are script-kiddie tools because they are used by professional pentesters and advanced hackers as well. They are sophisticated tools built by expert developers and elite hackers that perform complex operations. Advanced hackers actually understand how they work and can wield them in even more powerful ways. When script kiddies use them, they typically make a lot of "noise," jeopardizing their chances of success and making it more likely they will be caught.

3.1.3.2 Elite hackers

> *"Kaspersky Lab's experts can confirm they have discovered a threat actor that surpasses anything known in terms of complexity and sophistication of techniques, and that has been active for almost two decades – The Equation Group...they use tools that are very complicated and expensive to develop, in order to infect victims, retrieve data and hide activity in an outstandingly professional way, and utilize classic spying techniques to deliver malicious payloads to the victims." - from "Equation Group: The Crown Creator of Cyber-Espionage" by Kaspersky Labs*

On the other end of the spectrum from script kiddies are elite hackers. Elite hackers possess an enormous amount of technical understanding and expertise. They are expert programmers and can write code in low-level programming languages such as C and assembly. They understand the minute details of computer network protocols. They are adept systems administrators and can write and modify operating systems. They are also skilled at *software reverse engineering*. Software reverse engineering is *the skill of dissecting an executable program to determine how it works and where it may have vulnerabilities*. This takes a significant amount of skill because executable programs are compiled, and machine code is not easy for humans to read and understand (see Chapter 2), although there are tools called *disassemblers* that help.

Elite hackers discover new vulnerabilities and develop novel techniques for gaining unauthorized access to target machines. They study the technologies used by the target and figure out how to exploit weaknesses to gain access. They build tools like the ones mentioned in the prior section and use them in advanced ways. Unlike script kiddies, they use stealth so that their activities are not noticed, allowing them to obtain access and then maintain it for as long as needed to accomplish their objectives (more on this in the next chapter).

When they work in teams, elite hackers are called *Advanced Persistent Threats* (*APTs*). History has shown that if an APT targets an organization, that organization is almost sure to fall victim to the attackers. One of the more well-known APT groups is called *The Equation Group*. This group was discovered in 2015 by Kaspersky Labs, a Russian cybersecurity research firm and antivirus provider. Kaspersky described them as the most elite cyber force in the world. While the exact identity of this group is unknown, many people believe that it is the NSA or USCYBERCOM.

3.1.4 Hacker Profile

> *"My misdeeds were motivated by curiosity. I wanted to know as much as I could about how phone networks worked and the ins-and-outs of computer security." -* The Art of Deception *by Kevin Mitnick*

What type of person is drawn to hacking? Studies show that there are some general personality characteristics shared by many hackers. For example, hackers are typically curious and intelligent. Their curiosity leads them to start probing and exploring computers. They may start by taking computing devices apart physically to try and see how they work. Later, they may want to learn programming and explore how operating systems and the Internet work. Coupled with intelligence and persistence, they may derive an emotional high from the experience of discovery, motivating them to continue pushing boundaries.

Hackers may also express some obsessive compulsive behaviors. Hacking can be tedious and frustrating. It takes a lot of trial and error and a persistent effort to find a vulnerability. Many hackers possess a drive that pushes them to keep trying long after others would have given up.

Black hat hackers especially may have a strong dislike of authority. This could be due to a traumatic childhood and the lack of positive role models. They may think of the Internet as a meritocracy where real-world systems of status and acceptance are irrelevant. As hackers, they can compete with others on a level playing field. They may hack into systems as a way to seek power or control over others.

3.2 Hacking Culture

Part of "knowing thy enemy" is exploring the history and culture of hacking. Studies of culture typically examine facets of a people such as language, art, literature, commerce, folklore, and traditions. This section provides a brief overview of hacking culture in the United States. Within hacking culture, especially in the realm of cybercrime, there's a distinct lack of trust. Nobody knows who can be trusted, making everybody paranoid and suspicious. Somebody that seems to be trustworthy one day might be revealed to be an undercover law enforcement officer the next. It is actually possible they really were a cybercriminal one day and work for law enforcement the next because they were caught and agreed to turn on their fellow hackers in hopes of a more lenient sentence! This paranoia creates isolation in the hacking community. Even though some aspects of hacking culture may seem glamorous, keep in mind the darker and more sinister forms of hacking that wreak devastation in the lives of victims.

3.2.1 Language and Art

Leet speak arose early in hacking culture as a way to circumvent online bulletin board (i.e., web forum) word filters. People developed a simple system where numbers substituted for letters. For example, the number 1 replaced the letter *L*, 3 replaced *E*, 7 replaced *T*, 5 replaced *S*, and so on. This became known as leet speak which is short for *elite speak* and is sometimes written *1337* for leet.

The world of hacking overlaps with both computer programming and video gaming culture, especially when it comes to acronyms and lingo. In the gaming realm, *pwn* means "to own." It expresses the idea of dominating the other player. It has a similar meaning for hackers. When hackers take over a victim's computer, the victim has been *pwned*. Table 3.1 lists some popular hacking terms.

Table 3.1

A selection of hacking-related terms.

Term	Definition
root or *rooted*	gaining root level, or administrative access, to a computer through hacking
warez	pirated software (pronounced "wares" as in goods)
ng	next generation, often used in software names
trojan	trojan horse program
bot	a pwned computer that can be remotely controlled
botnet	a collection of bots that can remotely controlled as a group by a hacker
carders	cybercriminals who sell stolen credit card information

Hacker culture also includes art. *ASCII art* creates images out of ASCII characters (see Figure 3.6). Hackers use command-line interfaces (CLIs) extensively. Since CLIs can only display ASCII characters, ASCII art allows for creative expressions.

Figure 3.6

ASCII art used in the hacking tool Metasploit.

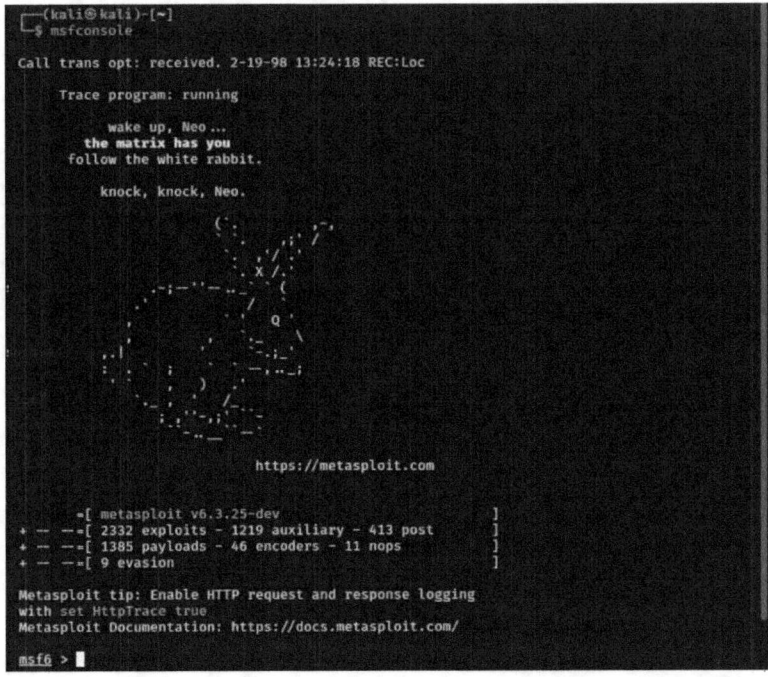

3.2.2 Nicks

For obvious reasons, hackers like to keep their real-life identities secret. To conceal their identities, they are known online by their *nicks* (short for nicknames), also known as *handles*. *MasterSplyntr* is a nick that was inspired by the sensei character, Splinter, from the Teenage Mutant Ninja Turtles comic. Splinter was a rat, and ironically, the person who used this nick was an FBI agent who infiltrated the cybercrime underground and ended up "ratting out" many cyber criminals! Table 3.2 lists a few notorious nicks.

Table 3.2

A selection of notorious hacker nicks.

Nick	Real-life Identity
Iceman	Max Butler, AKA Max Vision, the hacker whose story is told in *Kingpin: How One Hacker Took Over the Billion-Dollar Cybercrime Underground* by Kevin Poulsen
Aleph One	Elias Levy, the hacker wrote the famous paper, "Smashing the Stack for Fun and Profit"
Condor	Kevin Mitnick, a phone phreaker and famous social engineer
MafiaBoy	Michael Calce, as a fifteen year old he made national headlines for taking down yahoo.com
Mudge	Peiter Zatko, a member of L0pht that testified before Congress
Dark Dante	Kevin Poulsen, a phone phreaker turned hacker journalist

Hackers may have multiple nicks or change them over time making the job of tracking their real-life identities difficult. Because of the seeming anonymity of online interactions, many times hackers act with a sense of impunity. Impunity does unfortunately exist sometimes due to the difficulty in prosecuting cybercrime, and this emboldens many hackers and cybercriminals. However, cybercriminals have to be extremely careful not to get their real-life identities tied in any way to their nicks since that can lead to evidence that may be used against them in a court of law. The investigative journalist Brian Krebs is famous for tracking cybercriminals across platforms and over time and tying them to real-life identities in posts he titles, "Breadcrumbs." Many of these posts start with information about a notorious hacker known only by his nick and then end with his real-life name and photo. Some of these exposés have resulted in arrests!

3.2.3 Literature and Films

> *"Another one got caught today, it's all over the papers." - the first line of "The Conscience of a Hacker" by The Mentor in Phrack*

Early hacking literature was captured on *bulletin board systems* (*BBSs*) popular in the 1980s before the World Wide Web and the modern Internet. Hackers frequented BBSs to trade secrets and share information. They also created publications devoted to hacking such as *Phrack* and *2600: The Hacker Quarterly*. Phrack's name comes from combining the terms *phreak* (for phone phreaking) and *hack*. 2600's name is derived from 2600 Hz, the frequency that when "whistled" into the old landline telephone system would trigger operator mode—the gateway to phone phreaking (see Figure 3.7).

Figure 3.7

A portion of the title page from an issue of Phrack.

```
            Volume Three, Issue 26, File 10 of 11

PWN PWN PWN PWN PWN PWN PWN PWN PWN PWN PWN PWN PWN PWN
PWN                                                 PWN
PWN          P h r a c k   W o r l d   N e w s       PWN
PWN          %%%%%%%%%%%%   %%%%%%%%%   %%%%%%%       PWN
PWN                   Issue XXVI/Part 2              PWN
PWN                                                 PWN
PWN                   April 25, 1989                 PWN
PWN                                                 PWN
PWN          Created, Written, and Edited            PWN
PWN                 by Knight Lightning              PWN
PWN                                                 PWN
PWN PWN PWN PWN PWN PWN PWN PWN PWN PWN PWN PWN PWN PWN


And TAP Someone                              April 3, 1989
%%%%%%%%%%%%%%%
```

"The Conscience of a Hacker" is an essay by The Mentor that appeared in *Phrack* in 1986. It became known as "The Hacker Manifesto" and is famous in hacker culture for its portrayal of the hacker psychological profile. "Smashing the Stack for Fun and Profit" by Aleph One appeared in *Phrack* in 1996. It is an academic-quality paper that describes buffer overflow attacks, and is a great example of hacker tradecraft being shared with other hackers.

Table 3.3 lists some films that are also part of hacking culture for their portrayal of hackers, technology, and cyberspace.

Table 3.3

Cult-classic films that illustrate hacking culture.

Film	Synopsis
WarGames (1983)	Seattle high schooler David Lightman (Matthew Broderick) impresses his classmate Jennifer Mack (Alley Sheely) by hacking into a computer he believes belongs to a video game company but is actually an AI-driven Department of Defense computer and almost triggers a nuclear war with the Soviet Union.
Sneakers (1992)	Martin Bishop (Robert Redford) leads a team of pentesters including Darren Roskow (Dan Aykroyd), AKA Mother, who work with the NSA to recover a Russian code-cracking device.
Hackers (1995)	Prodigy hacker Dade Murphy (Jonny Lee Miller), AKA Crash Override, leads a group of teenage hackers including Kate Libby (Angelina Jolie), AKA Acid Burn, in a battle against law enforcement and other hackers.

3.2.4 Commerce

The medium of exchange of hacking culture is online currencies and money orders. These types of transactions are popular because they are difficult to trace to real-life identities and are irreversible. An early cyberspace currency was called *e-gold*, and it ended up being so abused by cybercriminals that eventually law enforcement seized it and shut it down. Bitcoin is a modern day cryptocurrency that has come under criticism for the role it plays in cybercrime because hacker extortion demands are often paid in Bitcoin. Of course, there are many other cryptocurrencies with new ones being invented all the time, and any one of them is fair game for the hacking community who seek anonymity and ease of use. Money orders are a physical space alternative to cryptocurrencies. They are reliable but lack convenience and carry physical space risks because they must be purchased in-person.

While the cybercrime market is illegal and therefore unregulated, it does sometimes exhibit ethical business practices. If a hacker buys stolen credit card numbers and then discovers they are not valid, he might actually get his money back. Even black market businesses are worried about reputational damage. Also, some cyber black markets do enforce some level of morality. They may ban certain "products" like human trafficking because of their moral repugnancy.

Generally speaking, however, financial dealings in the cybercrime market place come with risks. Some hacking groups have been accused of faking law enforcement takeovers to avoid paying their partners. They take their business offline and either eventually reopen under a new name, or retire with their ill-gotten gains.

3.2.5 Folk Stories

"I was hooked in before hacking was even illegal." – Kevin Mitnick

The Discovery Channel released a documentary in 2001 called *A Secret History of Hacking* that highlights the role of phone phreakers and folk heroes in hacking culture. The documentary provides vignettes of famous hackers such as Apple co-founders *Steve Jobs* and *Steve Wozniak* and the famous social engineer *Kevin Mitnick*. The most famous phone phreaker highlighted is *John Draper*. He discovered that the toy plastic whistle that came in Cap'n Crunch cereal boxes was tuned to 2600 Hz—the tone that triggered operator mode when whistled into a phone (see Figure 3.8). Draper became known as "Captain Crunch" because of his discovery.

Figure 3.8

The Cap'n Crunch toy whistle that triggered operator mode.

Robert Tappan Morris, AKA *RTM*, unleashed the world's first crippling malware which became known as the *Morris Worm* on the Internet in 1988. It was the product of an experiment gone awry that was mostly motivated by curiosity. Unfortunately, it crashed many computers and cost organizations a substantial amount of time and money in recovery costs. He was twenty-two, had just graduated from Harvard, and was starting his PhD program at Cornell University. His dad, Robert Morris, held a high-level position at the NSA. He was the first hacker convicted of the Computer Fraud and Abuse Act and was sentenced to a fine and community service. Even though he broke the law and was charged with a crime, he was a nuisance hacker and not a cyber criminal. More on RTM in Chapter 10.

Figure 3.9

Gray hat and later white hat hackers Kevin Mitnick (left) and Kevin Poulsen (right).

Max Butler, AKA *Max Vision*, whose online nick was Iceman, was a gray hat hacker who was sentenced to prison in 2010. He may have designed the first ever spear phishing attack which he used to steal login credentials from employees of an organization he later hacked. He had enormous success in stealing and selling credit card credentials (i.e., carding), and was the subject of the book *Kingpin* by Kevin Poulsen—the same Kevin Poulsen of phone phreaking fame (see Figure 3.9).

Tsutomu Shimomura was an early white hat hacker who became the nemesis of Kevin Mitnick after Mitnick hacked into his computer. Mitnick was on the lam from the FBI for several months trying to avoid capture, always managing to stay one step ahead of law enforcement as he moved from city to city, changing his identity multiple times. Eventually, the FBI hired Shimomura as a consultant, and he helped them locate Mitnick in a matter of days, proving the adage, "it takes a hacker to catch a hacker."

3.2.6 Traditions

There are two large hacker conferences every summer in Las Vegas that draw thousands of hackers. *Black Hat* is held one week and is focused on industry and vendors, and it is followed the next week by *DEF CON* which focuses more on hackers. The conference badges at DEF CON are an important part of the conference culture and sometimes contain electronic components with built-in hacking challenges (see Figure 3.10). DEF CON also features a popular scavenger hunt with hacker-themed clues and challenges.

Figure 3.10

An electronic DEF CON conference badge.

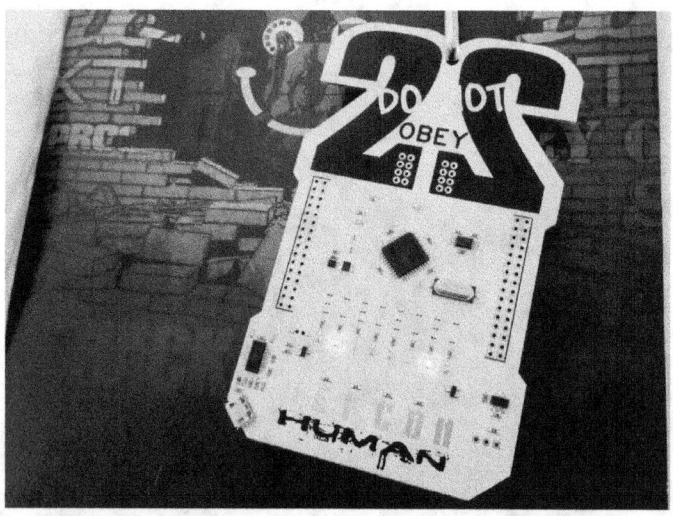

DEF CON holds a famous *capture-the-flag* (*CTF*) contest that draws elite hackers from all over the world. CTFs provide a legal way for hackers to hack and compete with one another, and are an important part of hacking culture. There are two main types of CTFs: *Jeopardy-style* and *attack and defend*. In a Jeopardy-style CTF, organizers set up a sandboxed cyber infrastructure with embedded flags (usually text strings) and also pose challenges in different categories worth various point values that the competitors can select among to try and solve (like in the Jeopardy game show). Competitors earn points by finding flags and solving challenges. In an attack and defend CTF, the competitors are tasked with either defending or attacking a sandboxed cyber infrastructure. They earn points based on the amount of time that systems and services remain online. In a CTF, the *red team* is the attackers, and the *blue team* is the defenders. Some CTFs focus on blue teaming, some focus on red teaming, and in some contests, the competitors defend their systems while attacking their competitors' systems.

3.3 Conclusion

Cybersecurity would not exist without hackers—humans that deliberately attack computer systems. As we saw in this chapter, not all hacking is illegal—there are good hackers, too! We explored the motivations for hacking whether for good or bad and also the fact that hackers differ based on skill level. There are script kiddies and elite hackers and everything in between.

We also took a brief look at the hacker profile and hacking culture. We focused on some of the history of hacking and traditions to understand how the world of hacking works. This is important because cybersecurity and hackers are inextricably linked. Many cybersecurity professionals are white hat hackers and share a lot in common with their foes. The most important distinction between them is that hackers engage in illegal behavior because they abuse systems without proper authorization. The more cybersecurity professionals understand hackers, the better they will be able to defend against them. The next chapter explores the TTPs of hackers in more depth.

Chapter 4

4. The Need for Cybersecurity: Cyber Attacks

"It's about time someone wrote a book to teach the good guys what the bad guys already know." - Bruce Schneier's endorsement for Exploiting Software

The previous chapter on hackers previewed some of the content in this chapter: the activities of hackers. Learning how cyber adversaries go about their hacking and the tricks of the hacking trade make clear the risk and the corresponding need for cybersecurity. At the end of the chapter, after we review the basic approaches to hacking, we examine the overarching goals of cybersecurity. Understanding cyber attackers and their activities is prerequisite knowledge for cyber defenders seeking to thwart attacks. The more the defense understands the offense, the better chance they have of competing on a level playing field.

4.1 Cyber Attacks

Not all cyberattacks are carefully planned. Some are conducted by unsophisticated script kiddies attacking targets at random. Other attackers look for low-hanging fruit—any victim that is an easy target. The *Shodan Search Engine is a website that continually scans the Internet and catalogs accessible software and hardware devices.* If a hacker knows of a vulnerability for a particular hardware device, he may be able to use Shodan to find a list of targets. For example, if he discovers a vulnerability in a specific webcam make and model, Shodan may be able to return a list of those webcams that he can start attacking.

There are also sophisticated attacks that are carefully planned against deliberately chosen targets. These types of attacks usually follow certain steps. One general model for the phases of a cyber attack was defined by white hat hacker, Ed Skoudis. He identifies five phases (see Table 4.1).

Table 4.1

Ed Skoudis' five phase cyber attack model.

Ed Skoudis Phase	Description
Reconnaissance	Gathering information about the target.
Scanning	Identifying the cyberspace perimeter of the target.
Gaining Access	Exploiting a vulnerability to gain unauthorized access to the target.
Maintaining Access	Taking measures to maintain access to the target such as creating alternative ways in.
Covering Tracks and Hiding	Taking measures to conceal the presence of the attacker in the target computer and network.

One of the nation's largest defense contractors, Lockheed Martin, has developed another framework to outline attacks called The Cyber Kill Chain®. The United States military originated the idea of a *kill chain*. A kill chain describes the chain of events leading up to a successful attack. Like a physical chain, a kill chain is dependent on every link in the chain. If any link is broken, the attack cannot succeed. Defenders focus on the links to see where they have opportunities to foil attacks, and they take security measures at every link to bolster their defenses. The Cyber Kill Chain® is similar to Skoudis' model, and contains seven links (see Table 4.2).

Table 4.2

The Cyber Kill Chain® seven link cyber attack model.

Chain Link	Description
Reconnaissance	Gathering information about the target.
Weaponization	Creating malware to infiltrate the target.
Delivery	Delivering the malware to the target.
Exploitation	Exploiting a vulnerability in the target.
Installation	Installing malware on the target.
Command and Control	Executing instructions on the victim machine.
Actions on Objectives	Accomplishing the goals of the cyber attack.

This section provides an overview of the main steps of a cyber attack and borrows from both Skoudis' model and The Cyber Kill Chain®. The four phases we will examine are Reconnaissance, Gaining Unauthorized Access, Post Exploitation, and Actions on Objectives.

4.1.1 Reconnaissance

"To rely on rustics and not prepare is the greatest of crimes; to be prepared beforehand for any contingency is the greatest of virtues." - The Art of War *by Sun Tzu*

Reconnaissance (recon) is the first step of a carefully planned cyber attack. Bank robbers are known to "case the joint." Like scoping out a bank in preparation for a robbery, the recon step of a cyber attack is about gaining as much information about the target as possible. This helps the attacker define the attack surface, including the *cyberspace perimeter*. The cyberspace perimeter includes the domain names and IP addresses that belong to the target and all the software and hardware accessible via those domain names and IP addresses. Other types of information may also be helpful to the attackers, such as the names and contact information of employees, recent job postings, and social media posts. In general, the more information that is gleaned, the better for the attacker. Information provides clues to potential vulnerabilities, and it also helps the attacker to adapt and improvise as needed—this is what Sun Tzu means in the quote above about being prepared for any contingency. Recon does not usually involve any illegal activities, although it could be incriminating evidence if a crime is later committed.

In this phase, attackers conduct both low- and high-tech recon to gain as much information about the target as possible. Some low-tech recon tactics would be snooping around the physical premises of the target, and perhaps even *dumpster diving*. Dumpster diving was popular among phone phreakers. They were notorious for literally crawling into the trash dumpsters belonging to big phone companies looking for unpublished documentation and other discarded information that might help them hack the phone network. Dumpster diving can reveal useful insider information.

An example of high-tech recon is *open source intelligence (OSINT)*. Open source intelligence is *identifying and collecting information that is available to the public* (i.e., information that is open source). Google is a valuable tool for conducting OSINT. Advanced uses of Google to search for vulnerabilities and sensitive information about the target is called *Google hacking*. Complex query strings can be crafted to identify sensitive information that may have been inadvertently posted online including spreadsheets and other files and even passwords. Ready-made query strings are available in online databases for hackers to use. Cyber defenders also may use these queries to find and take sensitive information offline before the hackers find it first!

4.1.2 Gaining Unauthorized Access

Most cyber attacks involve gaining unauthorized access to the victim's network. The information gathered in the recon phase is used to find potential vulnerabilities and other ways into the target network. This section outlines five main ways that hackers obtain their initial unauthorized access: deceiving authorized users, exploiting technical vulnerabilities, credential stealing, supply chain attacks, and obtaining physical access.

4.1.2.1 Deceiving Authorized Users

> *"What's the greatest threat to the security of your business assets? That's easy: the social engineer—an unscrupulous magician who has you watching his left hand while with his right he steals your secrets."* - The Art of Deception *by Kevin Mitnick*

It is commonly said that people are the weakest link in cybersecurity. Cyber attackers look for the easiest way to compromise their target, and frequently this is through the people that work for the organization they are targeting. *Social engineering* is sometimes called "the art of human hacking." Social engineering is *the practice of deceiving people into divulging sensitive information or performing actions that undermine security.* Social engineering can take place over any communication medium, including in person, over the phone, through email, via social media, and any combination of these and more. Dropping a malware-laced USB stick in the parking lot of a target and hoping a curious employee picks it up and plugs it into their work computer is an example of social engineering. More sophisticated types of social engineering typically follow a multi-step process.

The first step is gathering background information. This involves uncovering information to find a contact and gain trust from the contact. During this phase, the attacker learns the vocabulary of the organization, including esoteric acronyms and other insider information like the names of employees and vendors. Speaking the language of the organization helps the attacker gain trust. The famous social engineer Kevin Mitnick had an uncanny ability to quickly gain trust by convincingly sounding like an insider.

The next step is *pretexting*. Pretexting involves creating a believable background story for making contact. Essentially, it is lying about who the social engineer is and why they are contacting the target. An example of pretexting is someone dressing up as a plumber and explaining to the front desk receptionist that he is responding to an urgent water leak and needs to be let inside the building immediately. Another example is calling an employee pretending to be an information technology (IT) staff member. Sometimes pretexting involves lying about an imminent crisis to scare the target and create urgency.

The final step is *influencing*. This is getting the target to divulge sensitive information or take actions that undermine security. In some sophisticated cyber attacks, the attacker works over a long period of time to build the trust of an important gatekeeper before asking them to do anything that would undermine security. By then, the social engineer has put himself beyond suspicion. Table 4.3 provides an example of the three phases in a realistic social engineering attack.

Phishing emails are a type of social engineering. In some phishing attacks, the attacker sends the target an email with an attachment and tries to entice the target into opening it. The attachment might be a document that executes a script that allows the hacker to access the victim's computer. Other phishing attacks are used to steal credentials by getting the victim to login to a spoofed website (more on this below).

Table 4.3

A social engineering attack example.

Phase	Social Engineer Activity
Gathering Background Information	The social engineer identifies the name, title, and contact information of an employee and his manager, the type of computer and operating system he is using, and the name of new software that was recently installed by the organization.
Pretexting	The social engineer calls the employee with a spoofed number to match a company phone number. He claims to be with IT support and reports suspicious behavior on the employee's computer that needs to be investigated immediately. To gain trust, the attacker mentions the manager's name, details information about the employee's computer, and blames the new software installation for the problem. The social engineer says the situation is urgent because a hacker may be on the network. He asks for the employee's assistance in running some diagnostics on the computer.
Influencing	The social engineer asks the employee to open a command prompt and type in commands that the employee does not understand. The commands create an exploitable vulnerability on the employee's computer. At the end of the call, the social engineer happily assures the employee that everything looks good, hoping the victim moves on without thinking too much about the interaction.

Users can also be deceived into running *malware* that provides hackers with direct access to their machine. Malware is an abbreviation for "malicious software," and it encompasses a wide variety of nefarious software used by hackers. *Trojan horse malware is malware that appears to be and functions like a normal program, but it comes bundled with malware that creates a backdoor into the victim's machine. A wrapper program is software that binds two different programs together and is useful for creating trojan horse malware.* For example, a calculator application can be bundled with malware. When the victim uses the calculator in the foreground, the malware is silently activated in the background. Trojan horse malware is named after the famous Greek story of the Trojan attack on the city of Troy.

4.1.2.2 Exploiting Technical Vulnerabilities

> *"On many C implementations it is possible to corrupt the execution stack by writing past the end of an array declared auto in a routine. Code that does this is said to smash the stack, and can cause return from the routine to jump to a random address. This can produce some of the most insidious data-dependent bugs known to mankind." - from "Smashing the Stack for Fun and Profit" by Aleph One in* Phrack

Hackers may also gain unauthorized access to target systems by exploiting technical vulnerabilities. *A hack is an action that is allowed by the system but that undermines the intent of the system.*[1] The people who designed the technology are not aware that the system is capable of such behaviors—otherwise they would have fixed the problem. Fixing vulnerabilities in software is called *software patching*. This is done by installing an updated version of the software.

Hackers work hard to identify vulnerabilities. Finding a new vulnerability that nobody knows about in a widely used software program is a significant coup. Previously unknown vulnerabilities are called *zero-day vulnerabilities* because they have been known about for zero days. *Zero days* are exploits that target zero-day vulnerabilities. Zero days are valuable because they have a high probability of succeeding—nobody knows about them so they have not been patched. Once a zero day has been used "in the wild," the underlying vulnerability is likely to be exposed, publicized, and patched. At that point, the zero day vulnerability becomes an *n-day vulnerability* because it has now been known about for some n greater than zero number of days. N-days are not as effective as zero days because updates may have been installed, and many security conscious organizations work diligently to do so as quickly as possible. However, many systems are not patched in a timely manner, meaning n-days are still effective. A surprising number of major hacks are actually due to n-days, not zero days, including the *NotPetya* malware attack of 2017, the most financially devastating cyber attack in history up to that time.

Technical vulnerabilities reside in software including operating systems. These vulnerabilities are usually found by hackers with elite levels of skill. Many of them are due to programming errors or oversights. For example, one of the best known technical exploits is the *buffer overflow attack*. In this type of attack, the attacker is able to send code to the target computer and force it to execute it. Sometimes this provides the attacker with root access, or administrator-level access, to the target system. This is known as gaining *root* and is especially damaging for the victim because the attacker has complete control of the system. The hacker Aleph One popularized buffer overflows in a tutorial he published in *Phrack* he titled, "Smashing the Stack for Fun and Profit."

A buffer overflow is a type of *injection attack*. An injection attack is an attack where hackers provide code as user input and trick the target computer into executing their input. This is possible because as we learned in Chapter 2, all information within a computer system is 1s and 0s including machine instructions. The 1s and 0s must be interpreted within the proper context in order to be processed appropriately. Injection attacks trick computers into mistaking data 1s and 0s for code 1s and 0s. Some injection attacks are *remote code execution (RCE)* attacks. An RCE attack is *an attack where hackers are able to execute their code on a victim's computer from over the network.*

Websites are also vulnerable to attack. This is called *web exploitation*. Historically, many websites have been vulnerable to injection attacks. The most popular web-based injection attacks are *SQL injection* and *cross-site scripting (XSS)* which attack database servers

---

[1] This definition of a hack is from Bruce Schneier.

and JavaScript engines, respectively. In these attacks, the web server or web browser fails to compartmentalize user-inputted data from code that it executes. The result is that the target runs code provided by the attacker, providing the attacker with unauthorized access to data or the machine. Another example of web exploitation is *cross-site request forgery* (*CSRF*). In this type of attack, the attacker tricks the target's web browser into making a website request crafted by the attacker. This could result in making a transaction or modifying data on behalf of but unbeknownst to the victim.

Technical vulnerabilities also exist in computer networking protocols. For example, in the early days of the Internet, the *Ping of Death* attack caused victim computers to crash just by sending them a single packet. The packet did not comply with protocol standards, and the receiving computers did not know how to process it, causing them to crash. The Ping of Death attack is an example of a *denial of service* (*DoS*) attack. In DoS attacks, the attacker does not gain unauthorized access to the victim, but he denies authorized users access. This can be costly for victims because the computing resources are not available to employees or customers, and it takes time (i.e., more resources) to bring the resources back online.

Some attacks can occur without any human user intervention. *A drive-by-download is an attack that exploits vulnerabilities in web browsers and is triggered by just visiting a malicious website.* Smartphones have been known to be compromised by just receiving a text message—this is called a *no-click attack*. When major operating systems and widely used software programs have vulnerabilities that can be exploited without human intervention, then it is possible for hackers to create a *worm*. Worms are a type of computer *virus*. Computer viruses are so-called because like biological viruses, their "infections" are capable of spreading. Worms are programmed to propagate on their own. Once they compromise a victim machine, they use it as a launching point to spread further, infecting other victims, and so on. Due to the potential for exponential growth at Internet speeds, worms can spread very quickly. The *SQL Slammer Worm* infected seventy-five thousand computers in less than ten minutes in 2003. Another famous worm affected millions of machines starting in 2008. That year Microsoft released a notorious security bulletin known as *MS08-067* (the sixty-seventh bulletin released in 2008). It disclosed a major RCE vulnerability in a Windows server service that existed in almost every Windows computer across the globe. After the bulletin was published, threat actors quickly crafted an exploit that became known as the *Conficker Worm*. It infected millions of computers, enlisting them into a *botnet*.

A botnet is *a collection of "slave" computers that respond to the commands from a "master" computer.* Botnets greatly multiply a hacker's ability to wreak havoc. Sometimes botnets are rented out on a short-term basis to other hackers for a fee. Many computer owners do not know that their machines have been enlisted in a botnet because they may continue to operate like normal most of the time. The slaves in some botnets are IoT devices, and the owners have no visibility into their operation at all. Therefore, botnets can persist for a long time. For example, the *Mirai botnet* is made up of mostly IP cameras and home routers. Botnets are sometimes used to send spam email, and they can also be used to overwhelm a website with a deluge of network traffic. This makes the website unreachable because the web server is so occupied with bogus network traffic that it is unable to respond to legitimate requests. This is known as a *distributed denial of service* (*DDoS*) attack. The Mirai botnet was responsible for some large-scale and high-profile DDoS attacks in 2016, including one against the cyber investigative journalist Brian Krebs' website KrebsOnSecurity in retaliation for an exposé he published the week before.

The Conficker worm is an example of how publicizing vulnerabilities for defensive purposes can lead to attackers learning about them and creating n-day exploits. The *time-to-exploit is the time between vulnerability disclosure to exploitation.* Attackers have been getting faster at creating n-day exploits. This puts pressure on cyber defenders to patch vulnerabilities quickly, but patching carries a risk of downtime and care must be taken to do it correctly. Plus, many organizations are under-resourced when it comes to cybersecurity, and they are not always able to patch systems quickly enough to avoid being victims of n-day exploits. This is a major cybersecurity challenge.

Some technical vulnerabilities are not due to bugs in software, but to software misconfigurations. For example, some products are shipped with default administrative usernames and passwords, and there are hacking websites devoted to compiling lists of these defaults. When these products are purchased and installed, if the default password is not changed, then hackers can gain admin access by guessing the password. Misconfigurations also happen when users install products but do not understand the security ramifications of the installation parameters. Software with lots of features can be complex, and it takes knowledge to know how to make the software as secure as possible for the needs of the organization. Some systems administrators are so focused on just getting the software to work that they neglect to investigate all the options to the extent necessary to make it secure. Hackers who know the software better than the people who installed it can take advantage of this ignorance and exploit the misconfiguration vulnerabilities to hack the organization.

4.1.2.3 Credential Stealing

Credential stealing is a common way that hackers obtain unauthorized access to computer systems and data. When hackers obtain valid credentials for users, they can login as those users and are able to use the victims' authorizations to access data, run programs, and initiate transactions. This leads to many different types of damaging attacks. Passwords are the most common authentication mechanism (more on this in Chapter 8), therefore, credential stealing typically involves attacks on passwords. Passwords are a serious vulnerability and are frequently exploited. When passwords are used as the sole form of authentication, password compromises are devastating. Many variations of password attacks are described in this section.

Passwords can be stolen through tricking the target into logging into a *spoofed website*. A spoofed website is *a fake version of a legitimate website*. Because the attacker controls the spoofed website, he can see the information that people submit to it, including usernames and passwords. *Website cloning* is easy to do because as we learned in Chapter 2, digital assets can be perfectly replicated, and websites are no exception. The difficult aspect of the attack is getting the victim to navigate to the spoofed website instead of the real website. One way this can be done is through a phishing email that includes a link to a login page. If the target is not vigilant, he may click on the link and thoughtlessly enter in his username and password on the spoofed website without taking the time to check the address bar to confirm he is on the correct website. Attackers might return users to the actual login landing page after the victim "logs in" to avoid arousing suspicion.

Shoulder surfing is a technique for stealing passwords by observing users entering them. Shoulder surfing can be done in person or via video camera. Video recordings have the advantage of being able to be replayed and played in slow motion so that the keystrokes can be clearly seen.

Keystroke logging (keylogging) is another technique for stealing passwords. Keyloggers record all of a user's keystrokes. Hackers can then sift through the keystrokes to discover passwords typed by the user. (They can also see everything else the user typed, such as emails, search queries, etc.) People are at risk for keylogger attacks whenever they use computers owned by others, whether in public places like a library, or when borrowing a "friend's" computer. In these situations, people should be careful what they type!

Keyloggers can be software-based or hardware-based. Software-based keystroke loggers need to be installed on the target computer and this creates a barrier for hackers—they need access before they can get users' keystrokes. Therefore, keystroke loggers are sometimes installed on public machines, however, admin-level access is required in order to steal other users' keystrokes. Once a machine has been hacked, software-based keyloggers are easy to install to spy on the victim's activities.

Hardware keyloggers are small physical devices (see Figure 4.1). For wired keyboards, they sit on the wire between the keyboard and the computer, and act as a man-in-the-middle, copying all the keystrokes that pass through them. They are inexpensive and require little technical expertise to install and use. They are designed to blend in so that they will go unnoticed, and they do not slow down or affect the target computer in any way. One advantage they have over software-based keyloggers is that they do not require login access to the target machine. However, they do require physical access since they have to be physically installed. For wireless keyboards, hardware keyloggers sniff the air waves for the signals transmitted by the keyboard. Wireless keyloggers need to be close to the target but can work through walls, therefore, they can be hidden easily. Some wireless keyboards create an encrypted channel between the keyboard and the computer— keylogger attacks fail in those cases. Some hardware-based keyloggers, whether for wired or wireless keyboards, can connect to wireless networks and transmit the keystrokes they capture back to the hacker over the Internet. Others need to be physically retrieved in order to obtain the data they captured, creating more risk for the hacker.

Figure 4.1

A hardware-based keylogger.

Passwords can also be compromised through *password cracking* attacks. As we will see in Section 7.2.3.3, computers do not store users' passwords but instead store *password hashes*. Password hashes are scrambled "fingerprints" of passwords, and cannot be used for logging in. Hashes need to be "cracked" in order for them to be useful. Hashes are cracked by generating password guesses, hashing the guesses, and then comparing the resulting hash to the hashes in the *hash dump*. A hash dump is *a file that contains password hashes*. In password cracking attacks, billions of password guesses are tried per second. Many effective password cracking tools are freely available such as *John the Ripper*, a command-line tool that runs on Linux (see Figure 4.2). Hackers obtain password hashes when they gain admin access to servers. Sometimes they post hash dumps online, making them free game for other hackers to try to crack. Some hash dumps contain millions of hashes. A website called *haveibeenpwned.com* hosts a searchable database of stolen credentials that have been posted online—users can visit the site to see if their passwords have been cracked. (Note, there are more details on hashing in Section 7.2.3.3 and password cracking in Section 9.2.1.1 of this text).

Password guessing is another attack on passwords. In this attack, the hacker attempts to login as a valid user by guessing the user's password. It is assumed that the hacker is able to obtain valid usernames but does not know the corresponding passwords. Obtaining valid usernames is not difficult since they are not normally secret—many times email addresses are used as usernames. Password guessing is not the same as password cracking because the hacker is forced to go through the authentication server (e.g., a website login screen), and this is much slower. Even when guesses can be sped up by using hacking programs such as *THC Hydra* (see Figure 4.3) to automate the attempts, servers can mitigate attacks by enforcing time delays between attempts or by locking access to accounts after too many failed login attempts. Hackers are forced to work around these mitigations by adjusting the frequency of their guesses. For example, if the authentication server triggers some action after five bad attempts over a certain period of time, hackers will try four guesses and then wait until the count resets before trying again. Password guessing is most effective when a hacker performs recon and uncovers personal information about the target that might have been incorporated into their password, such as family names, pet names, school mascots, important dates, etc.

Figure 4.2

John the Ripper showing the cracked passwords for bob (batman) and alice (catwoman).

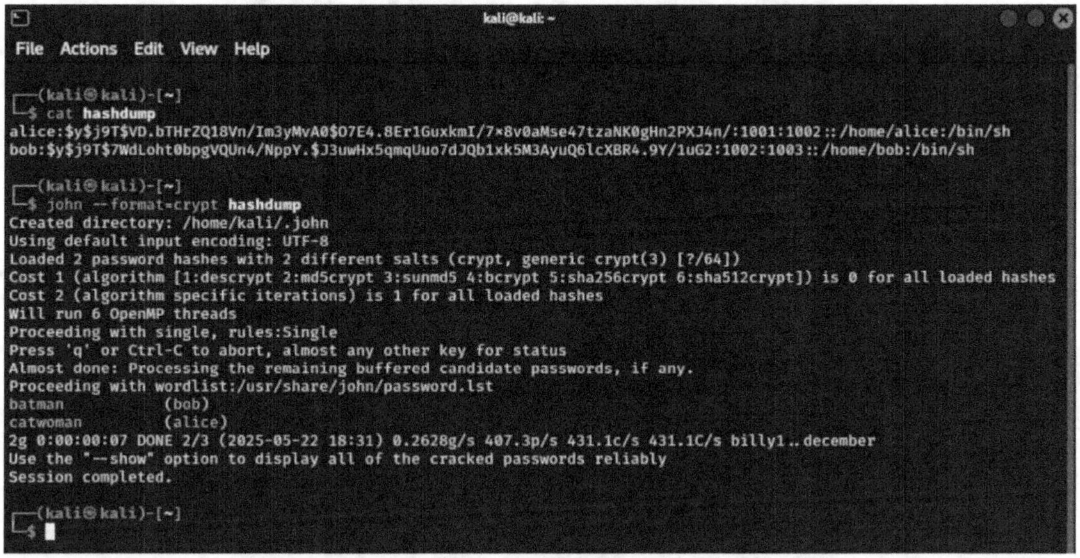

Password spraying is similar to password guessing but instead of targeting specific users, it focuses on a small number of password guesses. The goal of password spraying is to login to *any* user's account. If a hacker is able to obtain a list of valid usernames, he can try a few password guesses for every user. This will not trip the "too many failed attempts" trigger for any one user, and it gives the hacker many opportunities to guess a valid set of login credentials. This attack relies on the probability that at least one user chose an easily guessable password. For example, at a university where the mascot is the Tigers, a password spraying attack might try to login as every student using a small set of passwords based on the word Tiger, such as Tigers, Tigers1!, t1g3r5, etc. It is likely that at least one student has chosen such a password.

Credential stuffing is yet another attack on passwords. It is the main reason why security experts warn people not to reuse the same password on different websites. This attack works by finding a valid username and password combination for one site and trying the same pair of credentials on another site. The valid username and password may have been discovered through a password cracking attack on an organization. A user may not care that his credentials were stolen for some throwaway account he created and used one time, but if he uses those same credentials for his online bank account, he has a problem! In credential stuffing attacks hackers use stolen and known valid credentials to try to login to many different high-value websites such as banks and email accounts hoping to find a match.

Figure 4.3

THC Hydra, a hacking tool for password guessing.

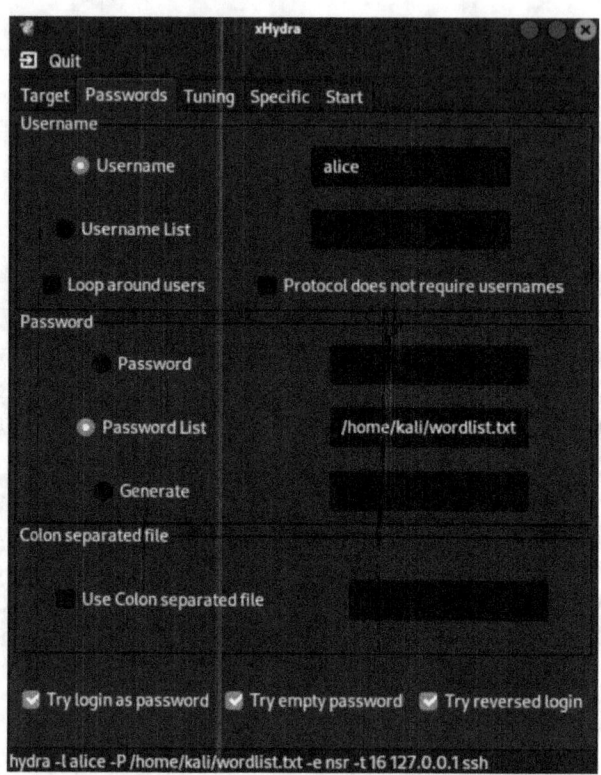

All of the credential stealing attacks mentioned in this section can be mitigated with multi-factor authentication (MFA). MFA does not rely solely on passwords for authenticating users. MFA will be discussed more in Chapters 8 and 9.

4.1.2.4 Supply Chain Attacks

> *"The attackers reportedly first gained access to Target's system by stealing credentials from an HVAC and refrigeration company, Fazio Mechanical Services, based in Sharpsburg, Pennsylvania. This company specializes as a refrigeration contractor for supermarkets in the mid-Atlantic region and had remote access to Target's network for electronic billing, contract submission, and project management purposes." - A "Kill Chain" Analysis of the 2013 Target Data Breach by the United States Senate Committee on Commerce, Science, and Transportation*

Supply chain attacks focus on first compromising a supplier of the target and then using the supplier to gain access to the target. Because a trust relationship exists between an organization and its suppliers, the supplier may have some privileged access to the target, including computer and network access. If the supplier's cybersecurity is weaker than the target's, then it may make sense to use them as a means to compromise the target. Large companies frequently have many suppliers, making them more susceptible to supply chain attacks.

One of the first highly publicized mega data breaches occurred in 2013 against the retail department store Target and compromised over 100 million Target customers (confusingly, in this case, Target was the target!). Target had a trust relationship with a refrigeration vendor and the attackers exploited that trust to gain a foothold in Target's network by first compromising the vendor. Once they had trusted access to Target's networks, they were able to obtain customer records.

Many organizations rely on vendors to supply IT support. These relationships entail a significant level of trust because many IT functions require privileged access. SolarWinds is a software company that provides IT monitoring tools for large organizations. In 2019, hackers obtained unauthorized access to SolarWinds' computers and network, but SolarWinds did not detect the attack. Over a period of time, the hackers used their access to SolarWinds to compromise SolarWinds' customers, many of whom were United States government agencies. This attack was highly effective. From the perspective of the organizations being hacked, the attack came from SolarWinds, but SolarWinds would never attack their own clients and were therefore beyond suspicion! This is how trust relationships can be exploited.

Software is complex, creating opportunities for bad actors to implant malicious code in software supply chain attacks. Ken Thompson won the *Turing Award* (the "Nobel Prize" of computer science) along with Dennis Ritchie in 1983 and in his acceptance lecture titled, "Reflections on Trusting Trust," he made the point that trust is inherent in software. He wrote, "You can't trust code that you did not totally create yourself...No amount of source-level verification or scrutiny will protect you from using untrusted code." His point is that organizations have no choice but to trust the companies that developed the software they use. But, even if software companies are honest, this still does not eliminate the threat of a software supply chain attack. As we saw in Chapter 3, software developers rely on software libraries written by other developers. Software libraries are imported into software development projects. They become de facto standards and save software developers from "reinventing the wheel" when they need routine code. It is prohibitively expensive to analyze software libraries for security vulnerabilities or malicious code, so developers typically just trust the code to be secure when they incorporate it into their products. When software libraries are compromised, all the customers using those libraries become vulnerable. The Heartbleed vulnerability in OpenSSL affected hundreds of thousands of systems around the globe because it is such a popular software library. In this case, it is not known for sure whether the vulnerability was accidentally introduced or if it was malicious implanted. Either way, it demonstrates the potential power of software supply chain attacks.

Another type of supply chain attack is known as a *watering hole attack*. In this attack, hackers go after a website that their target frequents. The hackers gain unauthorized access to the third party's web server and implant malicious code on it. The website then becomes the point of attack. When the target visits the website, the malicious code can potentially exploit the target. Drive-by-download exploits are useful in watering hole attacks. One downside of watering hole attacks is that they are coarse-grained. Everybody that visits the compromised website is subject to attack, not just the target.

Computer hardware can also be compromised in a supply chain attack. The United States government has issued directives prohibiting federal agencies from purchasing certain computer hardware products made in China. The concern was that the hardware might be used for espionage and possibly other malicious purposes. China has denied the allegations, but it is impossible for them to completely validate their claims. As Ken Thompson noted, if you do not trust the source, no amount of reassurance from them will be able to convince you.

In a sophisticated form of a hardware supply chain attack known as *interdiction*, attackers intercept otherwise secure hardware on the way to the target and compromise it before delivery. Attackers need to do this quickly and undetectably to avoid raising suspicion. Products like smartphones and smart home devices, if intercepted and compromised, could provide attackers with extraordinary access to spy on their targets.

4.1.2.5 Obtaining Physical Access

Obtaining physical access to a computer system is a powerful way for an attacker to gain unauthorized access. *An evil maid attack is an attack where a hacker gains physical access to an unattended computer and compromises it.* It is so-called because people often leave their laptops in their hotel rooms when they go out, creating an opportunity for the hotel housekeeping service to access their devices. Even if the housekeeping service does not have a resident hacker, a hacker could gain access to the room by social engineering hotel staff or through some other way.

If the unattended device is not protected with a password, or if the user is still logged in when the attacker starts using it, then the attacker has easy access. If the user is an administrator, then the attacker can view and change anything and everything on the device. This would include viewing stored wireless network and web browser passwords. The attacker can also use the web browser to navigate to websites. If the user is still logged in or if the browser automatically populates usernames and passwords, then the attacker would have control over the victim's cloud accounts. The attacker could also add malware to the computer for ongoing access, and could change security settings such as firewall, user accounts, and certificate settings. The attacker could also exfiltrate data from the computer either over the network or onto a portable hard drive or USB stick, or take pictures of some of the data on the device. If the logged in user is not an administrator, the attacker would have more limited access but could still accomplish some of these objectives. The attacker could also try to exploit a vulnerability to escalate his privileges (more on this below).

In some evil maid attacks, the attacker may have only a short window of time with the target device. This would be the case if a person is using a computer in a public place and briefly walks away to take a phone call, get a drink, etc. *A rubber ducky attack is an attack that uses a special-purpose USB stick to open a command prompt and quickly execute a series of commands by "typing" at computer speeds.* The attacker plugs the device into a USB port of the unattended computer, waits for a few seconds, and then unplugs it and walks away. In that time, the rubber ducky could create a connection back to the attacker's device, giving him access over the network. This attack only works if the user is still logged in when the rubber ducky is plugged in. This is why it is important for users to activate the lock screen whenever they step away from their computers.

If an attacker gains physical access to a locked device, he could try a password guessing attack. If he is unable to login, the attacker can restart the device and boot to an alternative OS stored on a USB stick. This may allow the attacker to see the contents of the hard drive and copy files and potentially even discover the login credentials to the installed OS. Using full disk encryption would thwart this attack (see Section 9.2.2.1).

An evil maid attacker could easily perform a denial attack on the device by damaging it in some way physically or by formatting or encrypting the hard drive. An attacker could also use his unattended access to install an inconspicuous hardware-based keystroke logger on the device as explained above.

For all these reasons, maintaining the physical security of computing devices is a vital component of cybersecurity.

4.1.3 Post Exploitation

> *"By reprogramming the hard drive firmware (i.e. rewriting the hard drive's operating system), the group achieves...an extreme level of persistence that helps to survive disk formatting and OS reinstallation...[and] the ability to create an invisible, persistent area hidden inside the hard drive." - from "Equation Group: The Crown Creator of Cyber-Espionage" by Kaspersky Labs*

After attackers gain initial access to the target, the post exploitation phase begins. In this phase, many times the attacker gains *command and control (C2)* access to the victim machine. C2 allows the attacker the ability to remotely issue commands. It usually begins with the victim machine "calling home" and creating a connection back out to the attacker's machine—this is called a *reverse shell*. At this point, the attacker may be able to quickly accomplish his actions on objectives (see next section) and move on. But in many attacks, the entry point acts as merely an initial foothold. From there, hackers may attempt to do three things: maintain access, escalate their privileges, and pivot to other computers.

4.1.3.1 Maintaining Access

"The fire consumes the town, the foe commands; / And armed hosts, an unexpected force, / Break from the bowels of the fatal horse."- The Trojan horse attack in The Aeneid *by Virgil (translated by Dryden)*

Usually hackers want to maintain access to the victim for as long as possible. This goal is known as *persistence,* and it allows them to spy on the victim and search for valuable information over a period of time. Maintaining access gives the hackers the ability to increase their access to the victim's computers and network. It is analogous to how in physical space, before an intruder can gain access to the interior rooms of a large building complex, he must first breach the perimeter. Once the attacker gains initial access, it is much easier to gain additional access because they are in the position of an insider.

The first rule of maintaining access is to remain undetected. This is accomplished through covering your tracks and remaining stealthy. Cyber defenders are always on the lookout for *indicators of compromise (IOCs).* IOCs are *detectable evidence that indicate a device has been compromised.* Therefore, from a hacker's perspective, covering tracks involves hiding as many IOCs as possible. It may include deleting or altering log files and deactivating antivirus software. Stealth also involves minimizing IOCs in the first place to avoid tripping alarms. Some computer users have discovered they had been hacked when they noticed their mouse cursor moving by itself as they looked on in disbelief at their computer screen! This is an obvious IOC and poor hacking tradecraft. In general, the better cybersecurity the victim has the more difficult it is for the hacker to remain undetected. Hackers can afford to be sloppy when the victim is not vigilant. Unfortunately, many households and organizations around the world are not as vigilant as they need to be. We will discuss how cybersecurity principles and best practices can help detect unauthorized access in Chapter 9.

A good way for hackers to maintain access is to create one or more alternative ways back into the victim's network. The initial method of gaining access may not be available in the future. In some cases the hacker might even fix the original vulnerability he used to gain access himself! This is done to prevent other hackers from exploiting the same vulnerability and potentially blowing both hackers' cover. *A backdoor is an unauthorized access point.* To gain persistence, hackers may create multiple backdoors. This way, even if the victim detects the hacker's presence and deletes an access point, as long as at least one foothold still remains, the hacker is able to maintain access. This fact causes consternation for cyber defenders because it can be difficult to be certain that a hacker has been completely eradicated from a machine or network. How do defenders know when they have identified and eliminated every backdoor? It is difficult to prove that they have—there always could be one more that they just have not found yet. The SolarWinds hack mentioned above resulted in such deep access to the victim networks that some cybersecurity experts advised that the victims should just replace the devices on the network with brand new machines!

In many hacks, malware is installed on the victim's machine. A *remote access trojan* (*RAT*) is a powerful type of malware. RATs are also known euphemistically as *remote administration tools* because they provide the ability to remotely administer a computer (see Figure 4.4). RATs make it easy to connect back into the victim machine at will. They also provide user-friendly ways to accomplish many common hacker tasks, such as keystroke logging, turning on the victim's webcam or microphone, viewing the web browsing history of the victim, browsing and downloading files from the victim's machine, etc. Antivirus companies catalog RATs, so hackers must be careful when deploying them. The presence of a file associated with a RAT on a machine is an obvious IOC.

Figure 4.4

A RAT called Quasar showing file browsing on a victim machine.

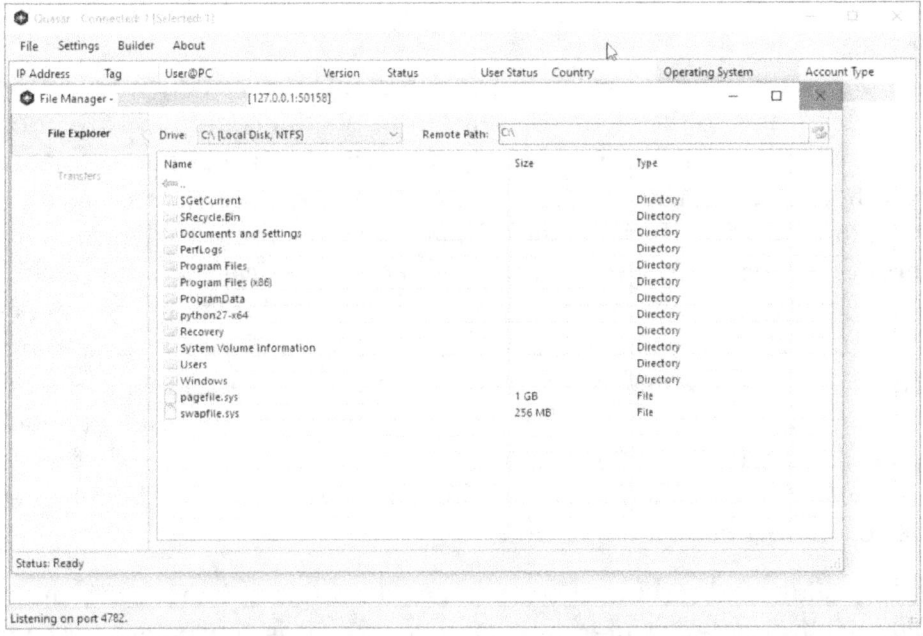

A *rootkit* is a sophisticated form of malware that exploits the operating system of the victim. As we learned in Chapter 2, the OS is the most important software running on a computer. It mediates access to all the software and hardware on the computer. It handles all the basic security tasks and is the source of truth for what processes are running, what network connections are occurring, and what files exist. Rootkits compromise the OS, so they can undermine all of the above trusted functions. If a user searches for malware-related processes, network connections, or files, the rootkit can return all of the valid information *except* what is related to the malware. Because OSs are implicitly trusted, rootkits provide an extreme level of persistence. They conceal many IOCs. A hacker running a rootkit has more privileges on a machine than even the owner himself!

4.1.3.2 Escalating Privileges

When hackers gain their initial access to a victim machine, they inherit the privileges of the victim user or process that was exploited. For hackers, the deeper the access the better. It is preferable for hackers to exploit administrator users and processes so they have root access from the beginning. However, hackers will take whatever initial access they can get, and this sometimes means a lower level of permissions that might not be enough for them to accomplish their actions on objectives. For example, a cyber attacker may want to copy certain files that his initial victim does not have access to. In this case, the hacker must *escalate privileges*. The goal is to work their way up into administrator access so they can gain root. A hacker can escalate privileges by exploiting vulnerabilities in the OS or in a higher-level process.

4.1.3.3 Pivoting

Escalating privileges allows a hacker deeper access into one computer; *pivoting* gives the hacker broader access to the victim organization. Victim computers are often part of a network of computers. The initial foothold provides the hacker the opportunity to gain access to other systems. The initially compromised computer is seen as a jumping off point to explore further. Hackers perform *network scanning* to determine what other computers are accessible from the victim machine. In physical space this is similar to getting the lay of the land in an unfamiliar setting by looking for landmarks and referencing a map. *Nmap* is a popular command line tool used for scanning networks and computers (see Figure 4.5). From his new vantage point on the victim machine, the hacker is likely to uncover new targets that he could not have accessed from his original attack position. In other words, the attack surface has expanded, and hackers have further territory to explore. Once they pivot to a new victim machine, the process can be repeated, because they have yet another new vantage point.

Hacking into additional machines on the network may not be that difficult because of trust relationships that exist internally on networks. It is convenient for computers on the same network to have easy access to one another, and hackers can take advantage of that trust. One famous attack called *passing the hash* exploits a trust-relationship vulnerability by allowing a hacker to log in to other machines without knowing the password. Shared drives may also be easily accessible as well as internal servers. There may be little-to-no security in place for devices that are not Internet facing because it is assumed that only trusted users can ever access them.

Some actions on objectives can only be accomplished by navigating through the network. One of the most famous and sophisticated attacks on record is known as *Stuxnet*. The victim was an Iranian nuclear enrichment facility in Natanz, and the attackers are alleged to be Israeli and United States working in a joint operation. Because the victim network was not Internet-connected, the attackers had no C2 access. Therefore, the malware was programmed to pivot autonomously from its initial access point, and it succeeded in its goal to pivot to and infect the computers that controlled the uranium centrifuges. From there, the malware caused the centrifuges to self-destruct by forcing them to operate outside of safe operating conditions (more on Stuxnet and nation state cyber conflict in Section 10.3.3).

Figure 4.5

Nmap showing that four devices were discovered on the local area network.

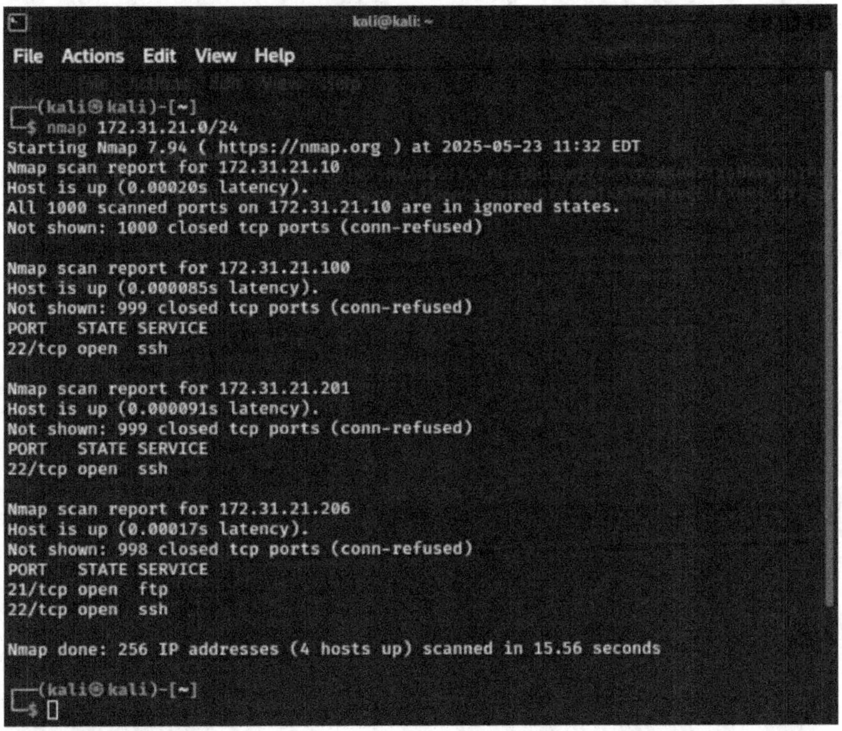

4.1.4 Actions on Objectives

By this phase the attacker has gained unauthorized access and positioned himself to accomplish his goals. In cyber attacks, goals are sometimes known as *actions on objectives*. They represent the reason for the attack, and once the hacker accomplishes this step, he has met his primary objective. Actions on objectives fall into three broad categories. They are easy to remember with the acronym *DAD*: *disclosure*, *alteration*, and *denial*.

4.1.4.1 Disclosure

A major objective of hackers is to disclose data. In this context, disclosing means that the hackers obtain access to data—it does not necessarily entail sharing data with others or even copying data. In some cases, viewing sensitive documents, pictures, and videos may be an end in and of itself. Some cyber attacks involve surreptitiously turning on the webcam or microphone on a victim's computer so the attacker can spy on them. Surreptitious spying can also be performed for espionage purposes, whether by nation states or rival corporations.

In most cases, the hacker will not only view data but will also want to obtain a copy of it. This is called *data exfiltration* or *exfil* for short. It is essentially stealing data, but unlike in physical space where stealing entails taking it away from the victim, in cyberspace exfiltration means *copying data from a victim to the attacker*. This usually occurs over a computer network—in other words, the attacker downloads the data. It could also occur locally—the attacker copies the data onto a USB stick. Once this happens, the data is outside of the victim's control. In some cases the data may be used to extort the victim. The attacker may threaten to publish the data unless the victim pays him money. The scary thing for the victim is that even if he gives into the hacker's demands, he can never really be sure that the threat will go away—there is no way to ensure that the hacker's copy (or more likely, copies) of the data will be permanently deleted. The hacker could potentially come back later with another extortion demand. Hackers may also want to exfil data to sell it on the black market. As we saw in Chapter 3, personally identifiable information (PII) and intellectual property (IP) have market value. In May 2026 a hacker group called ShinyHunters gained unauthorized admin access to Canvas, a popular learning management system used by thousands of educational institutions, and published an extortion note on course websites across the country. This happened right at the end of the semester at many universities—imagine students' surprise when they went to submit their final exams and projects and found an extortion demand instead (see Figure 4.6)! The note threatened to publish private data unless a settlement was negotiated with Instructure (the parent company of Canvas), or with the affected schools directly. Incredibly, Instructure paid the cyber criminals in return for their assurance that the exfiltrated data would be permanently deleted—so problem solved! (or not)

To avoid detection, hackers may try to exfil data using a *covert channel*. A covert channel is *a hidden communication path*. Covert channels usually involve *steganography—the art and science of hiding information in plain sight* (covered in more depth in Section 7.3). An example of a covert channel for exfiltrating data is using flag bits and other fields in network protocol headers. These headers are only supposed to contain metadata, not payload data, and therefore can potentially pass out of a network undetected. The data may also be broken up between several messages and spread out over time to avoid arousing suspicion. What using covert channels costs in efficiency it gains in maintaining cover.

In a *doxxing attack* the attacker publishes exfiltrated data to embarrass or otherwise harm the victim. The *Sony Pictures hack* in 2014 was a doxxing attack. Sony Pictures is a major American film studio. The hackers gained unauthorized access to Sony's network and exfiltrated as much data as they could and then published it on the Internet. The published data included everything from not-yet-released films to employee PII to internal emails. The emails exposed sensitive business dealings and private communications that embarrassed Sony executives. The United States government attributed the attack to North Korea, and the motive was to financially harm Sony Pictures for producing a satirical film making fun of North Korea.

Figure 4.6

The extortion demand students saw when trying to access their Canvas course websites at the end of the semester in early May 2026.

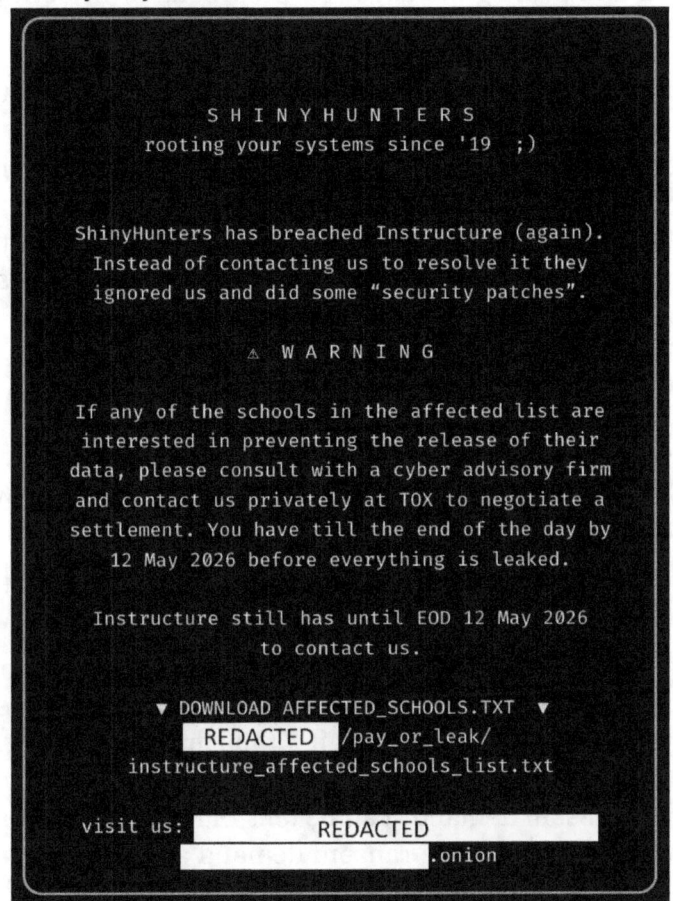

4.1.4.2 Alteration

In many cyber attacks hackers modify data. In this context, modifying encompasses creating, altering, and deleting data. Defacing a website falls into this category. Hackers gain access to a web server and modify the victim's website to advance the cause of the hacker or cause reputational harm to the victim. In watering hole attacks, hackers gain access to web servers so they can install malware that targets website visitors—this is another example of an alteration attack.

Hackers may want to get access to data to modify it in a way that benefits them. In the movie *WarGames*, the main character is a teenage hacker. He steals user credentials to his high school's server, then logs in and changes his grades—an alteration attack. Criminal hackers may try to change the direct deposit information for employees to divert money into their own bank accounts. Hackers may also want to modify official records to make themselves look better or to make others look bad.

In an *email spoofing attack* the hacker changes the "From:" field in an email to make it appear like the email came from someone else. This is similar to hacking into a person's social media account and posting messages on their behalf. In both of these cases, the attackers alter data without authorization.

An especially damaging form of an alteration attack is modifying data with the goal of undermining trust across an entire system. This type of attack could be used against electronic voting systems. The goal is not necessarily to sway the vote in one candidate's favor or another, but to create chaos as citizens realize that the election results are not reliable. This could cause significant harm because democratic governments depend on trustworthy elections. A hacker who gains access to a system and starts changing records at random could cause significant consternation and trigger an expensive investigative process.

4.1.4.3 Denial

Some cyber attacks involve denying legitimate users access to their data and computing systems, like the DoS attacks we covered in Section 4.1.2.2. Stuxnet was a denial attack that destroyed physical equipment as an act of sabotage. The Morris Worm also ended up being a denial attack because it crashed systems by exhausting their resources. Denial attacks harm victims by disrupting their ability to do business and costing them time.

Ransomware is an example of a denial attack. Ransomware is a type of malware that encrypts the data on the victim's computer or network making it inaccessible without the decryption key. The attackers then post a digital ransom note offering to provide the key in exchange for money, usually in the form of cryptocurrency. Depending on the financial capacity of the victim, ransom demands can be tens of millions of dollars. In some cases, victims of ransomware can ignore the ransom demand and recover the data themselves from backups. In other cases, the hackers are able to gain access to the backups as well and either encrypt them, too, or just delete them. Victims of ransomware attacks sometimes pay the ransom because it is cheaper than the costs they would incur to recover the data themselves. When ransoms are paid, cyber attackers usually provide the decryption key to the victims. This is just smart business—if hacking groups gain a reputation for not providing the key after payment, then future victims would not pay ransoms.

Wiperware is another form of malware that falls into the denial category. Wiperware deletes (in other words, *wipes*) data from victim machines. In some cases, wiperware may also cause physical damage to computer systems and render them useless. Saudi Aramco, a huge Saudi Arabian oil company, was hit with a wiperware attack in 2012 that destroyed tens of thousands of machines and hard drives, causing millions of dollars in damage.

Logic bomb attacks are a denial attack often perpetrated by disgruntled insiders. In a logic bomb attack, the hacker plants malware and sets it to "detonate" at a later date. This provides the hacker with an opportunity to distance himself from the attack. When the date arrives, the malware may delete data or tamper with systems in another way. Some systems administrators have planted logic bombs on their employer's network after they learned they were being fired but before they lost their administrative access.

4.2 Cybersecurity Goals

"Genius is making complex ideas simple, not making simple ideas complex." - Albert Einstein

Now that we understand the cyber threat, we are in a position to characterize the goals of cybersecurity. Cybersecurity is difficult to simplify because it is massive, messy, and complicated. Knowing the essence of cybersecurity is helpful for prioritizing what to do and why it needs to be done.

Cybersecurity's primary concern is preventing hackers from achieving their objectives. As we have seen, hackers have different skills and motivations and employ different techniques, tactics, and procedures. Hacktivists might gain access to data to expose it (doxxing). Criminal hackers might encrypt their victim's data so they can extort them (ransomware). Nation state hackers might gain access to spy on their adversaries (espionage). As the previous section shows, all the different types of cyber attacks can be generalized into the three main categories of disclosure, alteration, and denial. If these are the goals of hackers, the opposite of these are the goals of cybersecurity.

The goals of cybersecurity are known as the *CIA triad*. This CIA is not the Central Intelligence Agency, but that is a helpful association. CIA is probably the most well-known acronym in cybersecurity. It stands for *confidentiality*, *integrity*, and *availability*.

4.2.1 Confidentiality

Confidentiality means *preventing the unauthorized reading of data*. "Read" is used here in the technical sense—it means accessing. Reading may mean literally viewing data but oftentimes it involves copying or transmitting it. Cybersecurity entails keeping private data private. Users of the same computer should not be able to see one another's data, and unauthorized users should not be able to see anyone's data. If cyber defenders can ensure confidentiality, then they will avoid numerous types of attacks, including data breaches, data theft, spying, and more. In physical space, privacy is of utmost importance, and it is obtained with physical measures such as walls, locked doors, and safes. In cyberspace, access control and cryptography are the main ways we keep attackers out and "lock up" data to make it inaccessible.

4.2.2 Integrity

Integrity means *preventing, or at least detecting, the unauthorized writing of data*. Like with the term "reading" in the definition for confidentiality, "writing" is used in the technical sense—it means modifying data. This includes creating data, changing data, and deleting data. Data integrity is vital to cybersecurity. When attackers are able to modify data without authorization, they can wreak havoc on an individual or organization. They can create fraudulent records. They can encrypt data so that its owners cannot access it. They can delete data to revise history. They can modify software or install malicious programs.

The definition of integrity is fundamentally different from the definition of confidentiality because it contains the "at least detecting" clause. This clause is necessary because in some cyberspace contexts it is not possible to prevent hackers from modifying data. For example, as we learned in Chapter 2, data passes through many computers as it traverses between endpoints on the Internet. Once the data is "on the wire" it is outside of the data owner's control. Intermediaries have the opportunity and the ability to modify the data—this cannot be prevented. But the I of the CIA triad is still intact if, when this happens, the recipient of the data knows that the data has been tampered with. Integrity does not require that the exact modifications be known—it only requires that data cannot be modified without being detected. The success of some attacks depend on modifying data undetectably. When hackers modify software to make it do something malicious they are counting on nobody noticing.

Confidentiality and integrity usually go together but not always. They are two different concerns. It is possible to violate one while not violating the other. For example, if hackers access encrypted data, and they cannot break the encryption, then the confidentiality of the data is still intact. However, the hackers may still be able to modify encrypted data even though they cannot read it. If the tampering of the data is not detected, then the I of the CIA triad would be violated but not the C.

Also, sometimes it is not important that data remain confidential, but it is important that it does maintain integrity. If a person wants to make an important public statement, they do not need confidentiality—they want people to read their statement. But they may want to ensure that their message is not changed in any way. Their concern is integrity, not confidentiality.

Like with confidentiality, access control and cryptography are the main tools used to ensure integrity. It seems counterintuitive that message tampering can be detected even when the recipient has no way of knowing what the original message was, but we will see in Chapter 7 how this is achieved with cryptography.

4.2.3 Availability

Availability means *ensuring authorized users have access to their data and computer systems*. We saw in the previous section on actions on objectives that some cyber attackers attempt to deny their targets access to data and computer systems. This can have a big impact because many times people cannot perform productive work if they cannot access their computers. Hackers that successfully attack availability can cause substantial financial harm.

Ransomware is an attack on availability because it makes data unavailable. Hospitals are sometimes victimized by ransomware attacks. This causes a major and urgent crisis because if patients' medical records are not available, it may not be possible to treat them. DoS attacks impact the availability of computer resources. Online stores can be taken offline in DoS attacks costing the victim sales.

Availability is a core concern of IT even outside the presence of adversaries. An organization's IT staff must not only prepare for deliberate attacks on availability, but they must also mitigate random acts of nature, accidents, and mistakes. They backup data regularly in case of hardware failures or accidental deletions. They install fire suppression systems to reduce the risk of fires destroying data and computer systems. They invest in generators and battery backups so systems can stay online even when there are power outages. Some of these measures perform double duty and are important for cybersecurity as well.

Availability is achieved through access control and risk management.

4.3 Conclusion

Cybersecurity is about protecting and respecting the rights of every individual and organization in cyberspace. This chapter has shown that a major part of this is preventing unauthorized access to computer systems and data and making sure that authorized users have access. Authorization is a central concept in cybersecurity. We have also seen that the goals of cybersecurity can be simplified into three categories that are the opposite of the goals of hackers: confidentiality, integrity, and availability (see Table 4.4).

Table 4.4

Hacker goals and the corresponding cybersecurity goals.

Hacker Goals: DAD	Cybersecurity Goals: CIA
Disclosure: obtaining unauthorized access to data	*Confidentiality:* preventing the unauthorized reading of data
Alteration: modifying or creating data without authorization	*Integrity:* preventing, or at least detecting, the unauthorized writing of data
Denial: denying authorized users access to their data and computer systems	*Availability:* ensuring authorized users have access to their data and computer systems

The central question answered in Chapters 5 through 10 of this textbook is how we go about accomplishing cybersecurity. Chapter 5 covers the overall approach to cybersecurity: cyber risk management. Chapter 6 covers the fundamental skill that we need to bring to all things concerning cybersecurity: adversarial thinking. Chapter 7 addresses the bedrock of cybersecurity: cryptography. Chapter 8 covers the means of preventing unauthorized access: access control. Chapter 9 covers practical strategies embodied in cybersecurity principles and practices. Finally, Chapter 10 covers the boundaries of cybersecurity: ethics, rights, and laws.

Chapter 5

5. The Approach to Cybersecurity: Cyber Risk Management

"There is no such thing as 100% security." - Cybersecurity axiom

Now that we have seen what hackers are capable of and identified the goals of cybersecurity, how do we go about implementing cybersecurity in practice? When we think of cybersecurity, we normally focus on technical approaches to keeping our computers and data safe (e.g., configuring a firewall, using strong passwords, etc.). The technical approaches are absolutely necessary, but to get the biggest bang for the buck, a prior decision must be made before they are implemented.

Due to the asymmetry in cybersecurity where the attacker needs to only find one vulnerability but the defender needs complete protection, it is clear that the attack surface is vast and the protection resources are limited. Organizations need to figure out how to allocate their resources as efficiently as possible. Therefore, before implementing cyber defenses comes a business decision: how can we most efficiently allocate scarce protection resources across our cyber assets? In this chapter we look at cybersecurity from the business perspective. While some of the insights are applicable to personal cybersecurity as well, the focus is on organizational cybersecurity where cyber risk management must be thoughtfully and diligently performed.

5.1 Cybersecurity Governance

Managing cybersecurity goes beyond any one organization. An organization's cybersecurity posture obviously impacts the organization itself for good or bad, but it also impacts the organization's constituents, and this makes cybersecurity a matter of public concern. Organizational leaders have outside accountability for the decisions they make, and they have external resources at their disposal to help guide them. *Cybersecurity governance is the oversight of the security risks of an organization.* It is performed by leaders who support, define, and direct the security efforts of the organization. Organizational leaders are sometimes referred to as the *C-suite*. The C-suite is *the topmost leaders of an organization*, and it is so-called because their titles begin with the letter *C* for the word *Chief*. For example, many large organizations are led by a Chief Executive Officer (CEO), Chief Operations Officer (COO), Chief Financial Officer (CFO), and a *Chief Information Security Officer* (*CISO*). The CISO (pronounced see so) is *the officer in charge of cybersecurity*. Organizational leaders may also be accountable to third parties. *Third-party governance is oversight imposed on an organization by an outside organization*. The third-party might be the local, state, or federal government, or some other body that sets operating standards in the name of public safety and security.

The United States *National Institute of Standards and Technology* (*NIST*) produces standards and provides guidance to organizations. NIST has introduced several important standards related to cyber risk management. Some organizations that do business with the federal government are required by law to comply with some of NIST's standards, but compliance is voluntary for most organizations. Because they define best practices, familiarity with these standards is a vital first step for practicing cybersecurity governance. Two of NIST's most important cybersecurity standards are the *Cybersecurity Framework* (*CSF*) and the *Risk Management Framework* (*RMF*). These and numerous supporting documents are freely available online and provide guidance on how to manage cyber risks. They define benchmarks for appropriate cybersecurity. The CSF focuses on the functions of cyber risk management: *govern, identify, protect, detect, respond,* and *recover* (see Table 5.1). Each function has multiple categories and subcategories. The CSF walks organizations through the process of understanding and managing their cyber risks.

Table 5.1

NIST Cybersecurity Framework (CSF)[1]

Functions	Summary
Govern	The organization's cybersecurity risk management strategy, expectations, and policy are established, communicated, and monitored.
Identify	The organization's current cybersecurity risks are understood.
Protect	Safeguards to manage the organization's cybersecurity risks are used.
Detect	Possible cybersecurity attacks and compromises are found and analyzed.
Respond	Actions regarding a detected cybersecurity incident are taken.
Recover	Assets and operations affected by a cybersecurity incident are restored.

The RMF defines the steps that organizations need to take to implement cyber risk management: *prepare, categorize, select, implement, assess, authorize,* and *monitor* (see Table 5.2). There are several tasks and outcomes defined for each step. The RMF zooms

---

[1] Source: NIST Cybersecurity Framework (CSF) 2.0

in on the details of putting controls in place to manage cyber risks. A control is a measure taken to reduce risk (explained in more detail later in the chapter).

Table 5.2

NIST Risk Management Framework (RMF)[2]

Steps	Summary
Prepare	Prepare to execute the RMF from an organization- and a system-level perspective by establishing a context and priorities for managing security and privacy risk.
Categorize	Categorize the system and the information processed, stored, and transmitted by the system based on an analysis of the impact of loss.
Select	Select an initial set of controls for the system and tailor the controls as needed to reduce risk to an acceptable level based on an assessment of risk.
Implement	Implement the controls and describe how the controls are employed within the system and its environment of operation.
Assess	Assess the controls to determine if the controls are implemented correctly, operating as intended, and producing the desired outcomes with respect to satisfying the security and privacy requirements.
Authorize	Authorize the system or common controls based on a determination that the risk to organizational operations and assets, individuals, other organizations, and the Nation is acceptable.
Monitor	Monitor the system and the associated controls on an ongoing basis to include assessing control effectiveness, documenting changes to the system and environment of operation, conducting risk assessments and impact analyses, and reporting the security and privacy posture of the system.

Both the CSF and the RMF are general in form so that they can apply to any organization. It is up to organizations to determine how to apply the guidance to their unique circumstances.

A *cybersecurity audit* may be necessary to demonstrate that an organization is in compliance with standards. A cybersecurity audit is *an accounting of how an organization's cybersecurity complies with a standard*. Standards may need to be met in order for an organization to reduce their liability in the case of an incident. An outside consultancy is typically hired to perform the audit, catalog their findings, and make recommendations. Auditors come onsite and go through a checklist of items. They conduct interviews with key personnel, make observations, and examine evidence. Each item is rated on a scale such as 1 (no awareness) to 5 (fully implemented). This documentation acts as an official record that the organization can use to prove they are in compliance. Table 5.3 shows an example of a small portion of a CSF audit checklist that examines the *Asset Management* category under the *Identify* function.

---

[2] Source: NIST Risk Management Framework for Information Systems and Organizations

Table 5.3

Partial NIST CSF audit checklist[3]

Function	Category	Subcategory	Examples	Score
Identify (ID): The organization's current cybersecurity risks are understood				
	Asset Management (ID.AM): Assets (e.g., data, hardware, software, systems, facilities, services, people) that enable the organization to achieve business purposes are identified and managed consistent with their relative importance to organizational objectives and the organization's risk strategy			
		ID.AM-01: Inventories of hardware managed by the organization are maintained	Ex1: Maintain inventories for all types of hardware, including IT, IoT, OT, and mobile devices Ex2: Constantly monitor networks to detect new hardware and automatically update inventories	
		ID.AM-02: Inventories of software, services, and systems managed by the organization are maintained	Ex1: Maintain inventories for all types of software and services, including commercial-off-the-shelf, open-source, custom applications, API services, and cloud-based applications and services Ex2: Constantly monitor all platforms, including containers and virtual machines, for software and service inventory changes Ex3: Maintain an inventory of the organization's systems	

To provide even more evidence of a secure cyber posture, a *penetration test* (*pentest*) may be performed. A pentest (discussed in Chapter 3) is *an active probing of the cybersecurity defenses of an organization for the purpose of improving security*. Pentests go beyond just confirming the presence of cybersecurity best practices by actively testing their effectiveness. *Vulnerability assessments* are another way to measure the cybersecurity of an organization. A vulnerability assessment is *a scan for known vulnerabilities on a computer system or network*. They can be automated to produce regular reports and alerts for cybersecurity personnel to review and address. Vulnerability scanners draw from a continually updated catalog of discovered vulnerabilities. While not as in-depth as pentests, they are much less expensive, can be performed frequently, and still provide valuable insights.

In addition to NIST, the *Cybersecurity and Infrastructure Agency* (*CISA*) is a United States government agency devoted to the cyber defense of our nation. Their cybersecurity mission is, "to defend and secure cyberspace by leading national efforts to drive and enable effective national cyber defense, resilience of national critical functions, and a robust technology ecosystem."[4] CISA was formally established in 2018 as a component of the Department of Homeland Security (DHS). They not only work with government agencies, but also partner with the private sector, sharing information about threats, providing risk management guidance, and promoting best practices. DHS also houses the *United States*

---

[3] Source: NIST Computer Security Resource Center Cybersecurity and Privacy Reference Tool

[4] cisa.gov website. *Cybersecurity Division - Mission.* Retrieved June 2025.

Computer Emergency Readiness Team (*US-CERT*). US-CERT was created in 2003 "to protect the Nation's Internet infrastructure by coordinating defense against and response to cyber attacks."[5] They host a website with resources for implementing cybersecurity, they have a national cyber emergency alert system, and they coordinate responses to major cyber incidents. All business leaders need to be aware of these cybersecurity resources.

Responsible organizations operate not only in accordance with the letter of the law but also according to best practices and an ethical code of conduct (more on this in Chapter 10). A leadership's ignorance of best practices, laws, rights, and ethics is not an acceptable defense in a court of law, and could be evidence of *gross negligence*. Gross negligence is *the willful disregard and failure to comply with best practices*. In the case of gross negligence, the organization carries a high risk of cyber incidents followed by litigation that could put them out of business or even land their leadership in jail. Leaders of an organization and others responsible for cybersecurity can defend themselves from civil and criminal lawsuits by demonstrating that they performed their *due diligence*. Due diligence is *a threshold based on what a "prudent man" would do to safeguard an organization*. This can be demonstrated by showing the results of an audit or pentest and by producing documentation proving that cybersecurity *policies* were in place and being followed. Policies are *written guidance that define how actions are to be performed*. These documents create a standard and define a repeatable process. They should be thoughtfully created with cybersecurity (and other concerns) in mind. Template policies addressing common business functions are widely available for free online and can be customized to suit the purposes of a particular organization.

To show they performed due diligence, leaders can also provide evidence of employee cybersecurity training and *raising awareness*. Raising awareness for cybersecurity is *small actions taken to regularly expose employees to cybersecurity threats and best practices*. An example of raising awareness is periodic announcements highlighting social engineering tactics. Additionally, organizations should regularly remind employees of expectations of ethical behavior. One way to do this is by posting ethical codes of conduct in conspicuous places around the organization.

5.2 Security Tradeoffs

"There's no such thing as a free lunch." - Popular saying

One of the most well-known axioms in cybersecurity is that there is no such thing as 100% security. This is literally true because there is no limit to the amount of protection resources that could be allocated to cybersecurity. But more importantly, the axiom highlights the difficulty of cybersecurity. The only way for organizations to achieve perfect cybersecurity would be by eliminating all of their dependencies on cyberspace. This is

---

[5] cisa.gov website. *US-CERT - Info Sheet*. Retrieved June 2025.

not a viable option in today's world—our economy depends on cyberspace and computers and networks are essential for business competitiveness. There is no going back to the "good old days" before cyberspace; getting rid of computers means going out of business.

Another axiom of cybersecurity is that security is not free—it always costs something. The costs of cybersecurity are more than just the monetary costs of purchasing cyber-related technology. Cybersecurity costs include personnel who are paid to research, implement, and maintain the technology. Costs include the time spent creating cybersecurity policies and training employees to follow them. Perhaps easiest to overlook, costs also include the inefficiencies introduced by cybersecurity technology and policies. These inefficiencies may be individually small, but they might impact every employee every day, so collectively they can add up quickly. The end result is less business productivity, and this is a real cybersecurity-related cost for organizations.

Implementing cybersecurity for any organization is a *cost center*. A cost center is *a part of a business that is not revenue producing*. Even for a business that provides cybersecurity services to other companies, managing their own cybersecurity is a cost center. In a perfect world, no money would need to be spent on cybersecurity, but as we saw in Chapters 3 and 4, that is definitely not the world we live in. Therefore, money and resources spent on cybersecurity are money and resources that could have been saved or spent elsewhere.

One way to illustrate the tradeoffs involved in cybersecurity is by charting cybersecurity against costs as Figure 5.1 illustrates. In general, the better cybersecurity, the higher costs. It is fairly easy to come up with solutions in the lower-left (low cost and weak cybersecurity) and upper-right (high cost and strong cybersecurity) quadrants. Take, for example, one core concern of cybersecurity: authenticating users (more on authentication in Chapter 8). Allowing employees to use simple passwords is in the lower-left quadrant of the figure. This solution has low costs of implementation and maintenance because simple passwords are easy for people to use. However, simple passwords are a weak form of authentication and put organizations at higher risk of cybersecurity incidents.

Forcing employees to use long passwords and to do an iris scan to login would improve cybersecurity. Multi-factor authentication (MFA) creates a strong security posture, but long passwords and the biometric iris scanners are expensive and a nuisance for employees (see Section 9.2.1.3). Plus, these measures can lead to frustrations and delays when things go wrong, like when the iris scanner has technical issues and users forget their passwords. Therefore, this option is in the upper-right quadrant.

The goal is to implement solutions in the upper-left quadrant when they can be found. Many would consider reasonable password requirements and MFA with a phone app to be in this quadrant. This solution provides strong security, moderate implementation and maintenance costs, and only a small amount of extra work for employees.

Figure 5.1

Cybersecurity tradeoffs showing different methods to authenticate users.

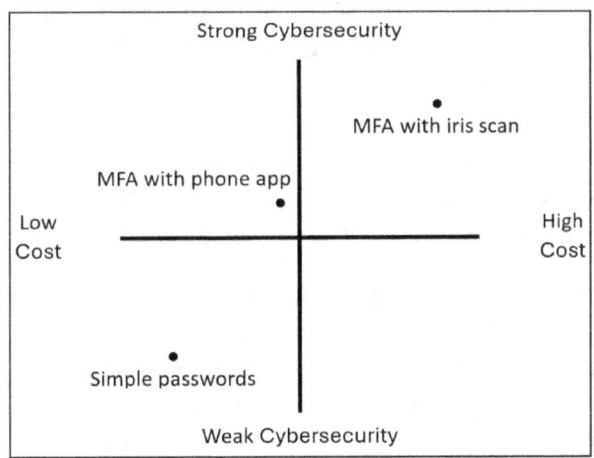

No solutions in the lower-right quadrant (high cost and weak cybersecurity) should be pursued. It is likely that a lower cost alternative exists that would provide the same level of security, or a same cost alternative exists that would provide better security, so it does not make sense to choose such a solution. Bad cybersecurity solutions are prevalent, however, because while costs are easy to compare, the relative security benefits of technology products are not. This makes the job of cybersecurity even more difficult. Some organizations adopt a *follow the leader* strategy. To follow the leader means *choosing the same cybersecurity solutions that a peer organization in the same sector uses*. One incentive to adopt this strategy is that if an organization experiences an incident due to a poor cybersecurity choice, they can argue that "everybody else is doing it" as a way to provide cover. Following the leader can be a good tactic for under-resourced organizations if they choose an appropriate leader to follow. However, this strategy can also be chosen out of laziness to save the work of investigating solutions, and it can cause poor cybersecurity practices to proliferate and create misconceptions of cybersecurity effectiveness.

Making these risk calculations (how much cybersecurity is enough?) and understanding these tradeoffs (are the extra costs worth it?) is central to cybersecurity risk management. Network and systems administrators who care most about cybersecurity are naturally pitted against their organization's management who care most about profits and employees who care most about getting their work done as quickly and easily as possible. Especially at the beginning of the cyberspace era, in most commercial businesses the trade-offs resulted in less costs and poorer cybersecurity. This is understandable because productivity is paramount because productivity leads to revenue. If businesses do not make a profit, they do not stay in business.

However, more recently businesses have started to realize the costs associated with poor cybersecurity. Businesses are a major target for hackers, and successful attacks costs businesses a significant amount of money. For example, if a business suffers a cyber attack, in addition to the direct damage done by the attack and the lost work and productivity, there will be costs to investigate the incident and remediate the damage and potential legal expenses as well. They may also have to pay fines if they were out of compliance with cybersecurity regulations, and in addition to these costs, they may suffer reputation harm that will damage future business prospects. When businesses face this reality, they recognize that when they spend money on cybersecurity, they are actually saving money in the long run. As more organizations adapt to this reality, costs across industries rise, and those costs are passed onto consumers in the form of higher prices. So in the end, society ends up bearing the costs of cyber-insecurity, and it is a drain on the national economy.

Military and intelligence organizations prioritize cybersecurity over productivity. They are not driven to make a profit, the data and systems they manage pose a national security threat, and they are a major target for elite hackers. Therefore, they are more likely to take significant cybersecurity measures that end up high and right in Figure 5.1. One unfortunately side-effect that can happen in this quadrant is if cybersecurity becomes too burdensome for employees, they may undercut security to make it easier for them to do their job, potentially making the organization less secure despite the high costs. For example, these types of organizations may have strict and cumbersome password requirements with short password expiration dates. This may lead to employees writing down their continually changing passwords and leaving them where others could potentially find them. Ironically, draconian password policies can increase other cybersecurity risks, like insider threats and evil maid attacks, and potentially reduce the organization's cybersecurity posture overall. In one real world example, a photo taken inside a Hawaiian government agency in 2017 and published in a national news story exposed the password "Warningpoint2." The password was written on a post-it note and stuck to a computer monitor in the background of the photo. Cybersecurity enthusiasts were quick to point out the poor cybersecurity practice.

Some cybersecurity measures are invisible to users and have no impact on employee productivity, like installing a high-quality versus a low-quality lock on the server room door. Of course, the better lock will cost more, so it will be relatively higher and to the right in the diagram in Figure 5.1. Therefore, this purchasing decision, too, involves a cost tradeoff.

5.3 The Cyber Risk Management Process

As we have seen, cybersecurity boils down to a risk calculation. There is no limit to the amount of resources that could be consumed by cybersecurity, but every penny spent and measure taken cuts into profits, so the big question is how should cybersecurity expenses be allocated and prioritized? To answer this question organizations engage in the *cyber risk management* process. Cyber risk management is *a detailed process of identifying cyber assets, enumerating how threats and vulnerabilities pose risks to assets, analyzing the severity of the risks, and then choosing how to handle the risks.* As discussed above, organizations like NIST provide detailed guidance on the risk management process. This section highlights a few of the major functions in the process.

5.3.1 Cyber Assets

The first consideration in cyber risk management is identifying and valuing *cyber assets*. Cyber assets are *computer systems and data of value*. Cyber assets must be identified and valued before they can be appropriately protected. Otherwise, organizations could end up spending more money protecting an asset than it would cost to replace it.

Computer systems are relatively straightforward to value because they are physical assets with a known market value. As part of the cyber risk management process, organizations should inventory all of their computer systems (see Table 5.3 above). Computer system inventories should include a record of every computer system owned by the organization, where it is located, and who is responsible for it. If this information is not compiled, computer systems could be lost or stolen without anyone noticing, and such losses pose a risk to the organization. Computer systems include employee workstations (e.g., laptops), servers, copiers, printers, and company-owned smartphones. These inventories must be kept up-to-date, so this is an ongoing process. When computers are replaced, the old ones need to be taken out of inventory and securely discarded.

Organizations also need to understand the value of their data. Data has value based on how the organization uses it as well how others value it, including customers and potential adversaries. One example of data that has value is *personally identifiable information (PII)*. PII is *data that can be used to identify a person and commit identity theft*. PII includes names, birthdays, addresses, phone numbers, and social security numbers. Most organizations store PII for their constituents so they can track and contact them. PII is not only valued by organizations but also by criminals because it can be used to commit *identity theft*. Identity theft *is fraudulent actions taken in someone else's name to obtain a financial benefit*.

When PII is divulged a data breach occurs. Cyber criminals may use the stolen PII directly, or they may sell it on the *black market* such as those found on the *dark web*. The black market is *a marketplace for stolen and illegal goods*—this includes data. The dark web is *a collection of websites accessible via specialized web browsers designed to protect the anonymity of the website hosts and clients*. PII is valued on a per record basis. The more records an organization has, the bigger target they are for hackers.

Data breaches have a negative impact on the people whose PII was stolen. Their privacy was violated and they are at heightened risk for identity theft. However, data breaches may not have as big of an impact on the breached organization itself. They may still have access to their constituent's data because attackers typically exfiltrate PII without deleting it. This dynamic is an example of an *externality*. An externality is *a cost borne by external parties that exceeds the cost borne by the party responsible for causing or preventing it*. This creates an asymmetry that makes incidents more likely, and ultimately, harms the public good. One way governments address externalities is by imposing fines and legal liability on companies to help correct the imbalance. Several such regulations have been attached to PII and other personal and private data, including educational, health, and financial records, driving up the value of this data to organizations and incentivizing them to invest more resources into protecting it. Understanding these regulations and laws and how they apply is an important part of security governance. Table 5.4 lists some of the major data privacy regulations in the United States.

Table 5.4

Major data privacy regulations in the United States.

Federal Regulation	Data Protected
FERPA - Family Educational and Privacy Rights Act (1974)	Student educational records
HIPAA - Health Insurance Portability and Accountability Act (1996)	Patient medical record
COPPA - Children's Online Privacy Protection Act (1998)	Data of children under thirteen
GLBA - Gramm-Leach-Bliley Act (1999)	Customer records of financial institutions

Intellectual property (IP) is another type of data that has value. IP is *proprietary information that provides a competitive advantage to an organization.* Organizations spend a considerable amount of resources to amass their IP and value it highly. For example, a pharmaceutical company may invest millions of dollars in the research and development of experimental drugs. The records and data they compile are an example of IP. Another example are blueprints for a jet or rocket. Organizations have other data of value besides IP, including their financial data, emails, and other business records. This data could be an attractive target to business competitors and also to criminals that might try to use it for blackmail or to embarrass an organization in a doxxing attack.

Some organizations, like the United States military and intelligence agencies, store data related to national security. The value of this data is based on the harm it would cause to the United States if it were divulged. This type of data is categorized and protected according to its classification (more on data classification in Chapter 8).

5.3.2 Cyber Threats

We examined cybersecurity threat actors and their attacks at length in Chapters 3 and 4. In this section we take a step back and look at threats from a wider angle before defining cyber threats in terms of cyber risk management. In general, a threat is anything that can cause harm, and they can be internal or external to an organization. For cybersecurity, this is the difference between insider threats and outside hackers. Both are serious cybersecurity threats.

Threats in general can also be natural or man-made. A natural threat is a threat due to nature, and a man-made threat is a threat caused by humans. Cybersecurity threats are limited to just man-made threats because cyber attacks are initiated by human beings. Therefore, while floods, fires, and earthquakes can be significant natural threats to an organization, and they can impact computer systems and data, strictly speaking, they are information technology (IT), not cybersecurity, threats.

A threat can also be intentional or accidental. Hackers deliberately attack organizations—cybersecurity threats are always intentional. Accidental threats certainly exist and can also cause harm to computer systems and data, but these are not cybersecurity threats. For example, an employee might drop his laptop, spill coffee on his keyboard, or even accidentally delete important files—all of these accidents are IT threats, not cybersecurity threats. Even in the cases where an employee inadvertently views another employee's private files or unwittingly publishes customer PII on the Internet, these are not strictly cybersecurity threats because they are not performed by a cyber threat actor with malicious intent.

While the distinction between intentional and accidental threats may seem pedantic, it reinforces the centrality of the human adversary to cybersecurity. And importantly, it does not impact cybersecurity practice. Cybersecurity measures put in place to protect against insider threats address accidental threats, too. If a well-meaning employee can accidentally cause a data breach, then an insider threat could easily do so and probably much worse.

Therefore, in terms of cybersecurity risk management, *a cyber threat is an action taken with malicious intent that discloses, alters, or denies access to a cyber asset.* Cyber threats are not limited to cyberspace only. A cyber threat can come through cyberspace or physical space and can cause undesirable consequences in either cyberspace or physical space or both. For example, installing malware remotely on an industrial control system (ICS) through a computer network that causes physical equipment to overheat and explode is an example of a threat through cyberspace that causes undesirable consequences in physical space. On the other hand, igniting a fire in the server room that causes an organization's website to go down is an example of a threat through physical space that causes undesirable consequences in cyberspace (and physical space, too, in this case).

A cyber threat actor is a person that poses a cyber threat. As part of the cyber risk management process, cyber threats should be enumerated. Chapter 3 described several different types of cyber threat actors, including criminal hackers, hacktivists, nation state hackers, and nuisance hackers. These categories should be examined from both an insider threat and external hacker perspective to understand who may be motivated to attack an organization. As a general rule, cyber threat actors are numerous in cyberspace and pose a serious threat to every organization.

5.3.3 Cyber Vulnerabilities

As we saw in Chapter 3, *a cyber vulnerability is a weakness.* Vulnerabilities are exploited by cyber threat actors. *An exploit is an action that takes advantage of a vulnerability to compromise security.*

Cyber threat modeling is a systematic approach to identifying cyber vulnerabilities. It uses brainstorming exercises and hypothetical scenarios and applies adversarial thinking to explore how cyber threat actors could find and exploit vulnerabilities within an organization. One way to do threat modeling is by examining each of the categories of *people, processes, technology, and facilities.* These four categories are the functional underpinnings of any organization. They are where vulnerabilities reside and also where cybersecurity is implemented.

5.3.3.1 People

An old adage warns that people are the weakest link in cybersecurity. Cyber threat actors often target employees as the easiest way to achieve their objectives. As we saw in Chapter 4, hackers use social engineering to deceive people into taking actions that undermine security. People are trusting, non-confrontational, and want to help, and these are vulnerabilities that hackers exploit by using deception and preying on a person's goodwill. Organizations need to consider how different categories of employees might be socially engineered to compromise cybersecurity. They also need to identify what other groups of people have access to cyber assets (e.g., the supply chain), and what harm could be done by insider threats.

5.3.3.2 Processes

Processes also have vulnerabilities. Hackers examine processes to find weaknesses. A weakness may be an exception to the process that was never considered, opening up a security loophole. As a non-cybersecurity example, airport security personnel follow a process for screening passengers before allowing them to board an airplane. One of their objectives is to prevent people from bringing explosive devices onboard a flight. A terrorist group in the early 2000s found a loophole in the security process: shoes were not screened. After an incident where a terrorist was caught mid-flight in the act of trying to ignite his "shoe bomb" (fortunately, he did not succeed), the screening process was modified so that passengers now have to remove their shoes (see Figure 5.2). For cyber threat modeling, organizations need to examine the different functions they perform and apply adversarial thinking to brainstorm how they could be subverted.

5.3.3.3 Technology

The history of cybersecurity has proven that cyberspace is rife with technology vulnerabilities as we saw in Section 4.1.2.2. There are vulnerabilities in software, hardware, and networks. These vulnerabilities are widespread because cyberspace is complex, and as we have seen, complexity is the enemy of security. Complexity creates darkness where vulnerabilities can hide. Most software vulnerabilities are unintentional bugs created by well-meaning programmers. The *Common Weakness Enumeration* (*CWE*) database is a publicly available catalog of software and hardware-related flaws that can lead to vulnerabilities. Table 5.5 lists the top five weaknesses for 2024. Figure 5.3 shows a portion of the CWE that describes the programming flaw that leads to cross-site scripting vulnerabilities.

Figure 5.2

A pair of shoes concealing explosives that were successfully snuck through airport security.

Table 5.5

The top five CWE's from 2024.

Rank	CWE ID	Description
1	CWE-79	Cross-Site Scripting (XSS)
2	CWE-787	Buffer Overflow (family of related weaknesses)
3	CWE-89	SQL Injection
4	CWE-352	Cross-Site Request Forgery (CSRF)
5	CWE-22	Path Traversal

Figure 5.3

The CWE describing the programming flaw that leads to cross-site scripting attacks.

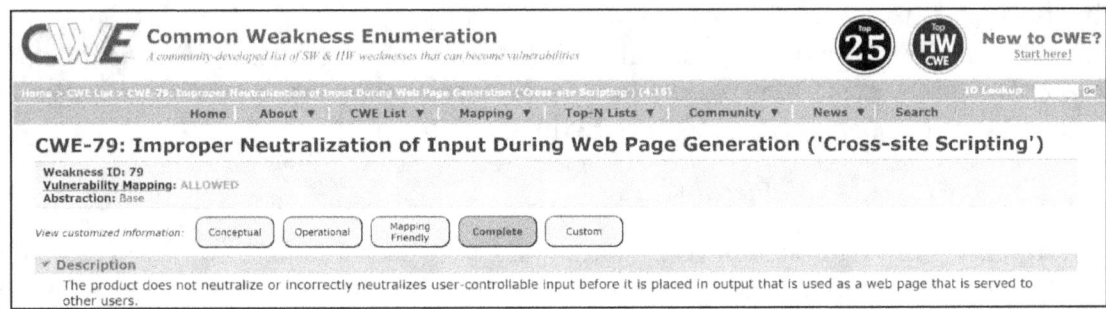

Vulnerabilities exist in hardware as well. *Side-channel attacks* can be used to compromise otherwise secure processes and algorithms. A side-channel attack is *an attack that exploits incidental information leakage.* An example is the electromagnetic emissions produced by processors when performing cryptographic calculations that can be measured with sensitive equipment and used to narrow down key value ranges. The *Rowhammer attack* exploits a vulnerability in memory chips that can be used to gain unauthorized modification access to data.

Vulnerabilities also exist in computer networks. Vulnerabilities in networking protocols, including authentication protocols, are frequently discovered and exploited by hackers to gain unauthorized access to computer systems and data. The *LAND (local area network denial) attack* is similar to the Ping of Death attack in that it causes a computer system to crash by sending it a malformed packet. In the LAND attack, a packet is sent with a source IP address that is spoofed to match the destination system's IP address. If the system is not programmed to catch this trick, it will enter into an infinite loop of sending replies to itself and will quickly consume all of its computing resources!

5.3.3.4 Facilities

Cyber vulnerabilities also exist in facilities. If hackers are able to physically access computer systems, they may be able to login and exfiltrate data or plant malware. An example of a facility-related vulnerability are Ethernet ports located in walls throughout office buildings. If the ports are active, a bad actor may be able to plug in his laptop in an inconspicuous area and gain access to the local area network. As part of cybersecurity, organizations need to examine their physical office space for vulnerabilities that could be exploited by cyber attackers that come onsite.

Cyber vulnerabilities need to be enumerated as part of the cyber risk management process. Chapter 4 covered multiple different ways hackers attack computer systems and data. Organizations need to be familiar with these techniques and the vulnerabilities that make them possible. Organizations should also perform pentests and vulnerability assessments to help identify vulnerabilities.

5.3.4 Risk Analysis

Cyber risks occur where threats and vulnerabilities intersect. *A cyber risk is the potential for a cyber threat actor to exploit a vulnerability that allows him to disclose, alter, or deny access to a cyber asset.* Risk is sometimes written as the formula:

Risk = Threats x Vulnerabilities

Both inputs on the right side of the equation are necessary because no risk exists when threats are present but not vulnerabilities, and vice versa. As a non-cyberspace example, some biological viruses pose no risk to certain people. This could be because the virus is present in their area but they are immune to it (i.e., the threat exists but not the vulnerability), or because they are susceptible to the virus, but it is not found in their area (i.e., the vulnerability exists but not the threat). Both the presence of the threat and the existence of the vulnerability are necessary for risk to occur.

After an organization has examined threats and vulnerabilities to identify risks, the next step is to determine their severity so they can be prioritized. Risk severity depends on two factors: the likelihood of the event occurring and its impact if it does occur. Risk severity can be written as the formula:

$$Risk\ Severity = Likelihood \times Impact$$

There are two approaches to analyzing risk severity and both are helpful. A *qualitative risk assessment* uses coarse-grained categories and a *quantitative risk assessment* uses fine-grained numerical values.

5.3.4.1 Qualitative Risk Assessment

A qualitative risk assessment creates scales for likelihood and impact. Scales can be custom defined to suit an organization's needs, but a typical scale ranges from 1 (low) to 5 (high). A risk's likelihood and impact is rated on the scale. The ratings are an estimate based on professional opinion and expertise. It is advisable to assign scores collaboratively in a group setting to help ensure that all the relevant factors are being considered. Once the two values are assigned, they are multiplied together to determine the risk severity. Groups are also defined for levels of risk severity, and may range from low (green), to medium (yellow), to high (red) (see Figure 5.4).

Figure 5.4

A qualitative risk assessment chart for grading risk severity.

Likelihood

Impact		1	2	3	4	5
	5	5	10	15	20	25
	4	4	8	12	16	20
	3	3	6	9	12	15
	2	2	4	6	8	10
	1	1	2	3	4	5

For example, a small business has an ecommerce website (a cyber asset) that is their main source of revenue. They have identified several risks to the website including distributed denial of service (DDoS) attacks that would make their website inaccessible. Based on their research of threat actors and their motivations, they believe a DDoS attack is unlikely and rate it a 2 on the likelihood scale. They also believe that DDoS attacks could take their site down for less than one week over the course of a year, and could disenfranchise customers and cost them some sales. They rate the severity of a DDoS attack as 3 (medium). Therefore, due to their qualitative risk assessment, they score the risk severity of DDoS attacks against their ecommerce website as:

DDoS Risk: 2 (likelihood) x 3 (impact) = 6 (risk severity)

Overall, this falls into the low risk category.

As another example, they also identified during a vulnerability assessment that their web server is unpatched and threat actors are actively exploiting a known vulnerability it has to gain root access to servers. Therefore, they have identified a risk of their web server being hijacked. If this were to happen, it could cause a data breach and also take their ecommerce site down for weeks and cost them numerous sales. For this risk, they rate the likelihood a 5 and the impact a 5 for an overall risk assessment score of 25:

Web Server Hijacking Risk: 5 (likelihood) x 5 (impact) = 25 (risk severity)

This risk scores in the high risk category.

Based on this type of analysis, applied across many different risks, organizations can intelligently prioritize risks.

A qualitative risk analysis sometimes can lead to counterintuitive conclusions. Taking a hypothetical non-cyberspace example, which city is at higher risk for low magnitude earthquakes, Boston or San Francisco? Assuming that earthquakes are three times more likely in San Francisco, but that the historic buildings in Boston are four times more likely to be damaged during a low magnitude earthquake, leads to the following risk severity scores:

San Francisco Earthquake Risk: 3 (likelihood) x 1 (impact) = 3 (risk severity)

Boston Earthquake Risk: 1 (likelihood) x 4 (impact) = 4 (risk severity)

Based on these assumptions, Boston is at higher risk for low magnitude earthquakes than San Francisco! This illustrates the fallacy of equating risks with threats. Threat does not equal risk. The risk severity calculation corrects this by incorporating impact so that risks can be properly assessed.

5.3.4.2 Quantitative Risk Assessment

Quantitative risk analysis is another way to assess risk severity. It is a fine-grained assessment that uses dollar amounts to quantify risks. With a quantitative risk analysis, a precise cost benefit analysis can be performed. There are several acronyms and equations involved in the quantitative risk analysis calculations, but only basic math is needed.

A cost benefit analysis is based on the *annualized loss expectancy (ALE)* calculation. The ALE is *the projected losses to a cyber asset due to a cyber risk over the course of a year*. A cost benefit analysis compares the ALE before controls are put in place (the *pre-* analysis), to the ALE assuming controls have been added (the *post-* analysis). If the cost of the controls exceed the projected savings, then the controls should not be pursued. ALE equals:

ALE = Single Loss Expectancy x Annualized Rate of Occurrence

The ALE calculation starts with the *single loss expectancy (SLE)* calculation. The SLE is *the projected losses to a cyber asset due to a cyber risk as a result of a single incident.* The SLE measures the impact, similar to the qualitative risk assessment above.

SLE = Asset Value x Exposure Factor

The asset value (AV) is the value of the cyber asset. The exposure factor (EF) is the percentage of the asset's value that will be compromised if the risk is realized. The EF is a percentage, and it is always between zero and one (inclusive). Zero means that none (0%) of the cyber asset's value will be lost, and one means that all (100%) of the cyber asset's value will be lost.

Lastly, the *annualized rate of occurrence (ARO)* needs to be determined. The ARO is *the expected annual frequency of the incident occurring.* The ARO measures the likelihood, similar to the qualitative risk assessment above.

Taking the example of the small business above that needs to determine the risk severity of a DDoS attack on their ecommerce web server, they first determine the SLE. The web server brings in $1,000,000 in revenue per year—this is the AV.

AV = $1,000,000

The next calculation is the EF. If the business suffers a DDoS attack, what percentage of its value will be compromised? Based on research, the business has determined that an average DDoS attack against a similar scale website takes the server offline for one week. For simplicity's sake, this analysis does not take into account other costs of a DDoS attack, such as reputational harm, loss of future sales, etc. One week is 1/52 of a year, so the exposure factor, with a little rounding, is 2%

EF = .02

Therefore, the SLE is $20,000.

SLE = $1,000,000 x .02 = $20,000

Next, the business needs to determine the number of times per year they can expect a DDoS attack to occur—this is the ARO. Based on market research and other available data, they determine that a DDoS attack happens to similar scale websites on average once every four years. This means the ARO is .25.

$$ARO = .25$$

With these inputs, based on their assumptions, they can calculate the ALE and quantify the risk in dollars of a DDoS attack against their web server.

$$ALE = \$20,000 \times .25 = \$5,000$$

In this scenario, the ALE is $5,000, meaning, they can expect to lose $5,000 per year due to the risk of DDoS attacks on their ecommerce web server. This is an annual cybersecurity cost they are absorbing.

Next, the organization evaluates potential controls, or safeguards, for DDoS attacks. Their ISP advertises DDoS protection services for $3,000 per year. Should they purchase this service? At first glance, it appears they should because the DDoS risk costs them an average of $5,000 per year but they can get the safeguard for $3,000 per year, so overall they would be saving $2,000 per year. This analysis is flawed, however, because it assumes that the safeguard lowers the risk to zero. This is not a valid assumption. The DDoS protection service does not completely eliminate the risk of the website going offline due to a DDoS attack—it only reduces the EF. Therefore, the organization needs to calculate the ALE with the safeguard in place.

With the DDoS protection service in place, it is expected that a DDoS attack will take their website offline for one day a year, reducing it from one week per year. This means the EF is now 1/365 which, with a little rounding, is .2%.

$$EF_{POST} = .002$$

None of the other factors have changed, so the SLE_{POST} and ALE_{POST} can be calculated.

$$SLE_{POST} = \$1,000,000 \times .002 = \$2,000$$

$$ALE_{POST} = \$2,000 \times .25 = \$500$$

With this information compiled, the organization can determine the *return on investment (ROI)* for the safeguard. The ROI is *the net savings that result from an investment*. If the ROI is positive, then it makes financial sense to adopt the safeguard.

$$ROI = ALE_{PRE} - ALE_{POST} - Annualized\ Safeguard\ Costs$$

The annualized safeguard costs (ASC) is the cost of a safeguard over the course of a year. Some safeguards have productivity and other costs in addition to out-of-pocket monetary costs. The ASC must include the cost to install and maintain the safeguard, plus other operational costs. For example, is it possible that the DDoS protection could occasionally block legitimate customers? If so, then that would need to be added to the

ACS. In this case, we can assume that the DDoS protection server adds no extra overhead cost. Therefore, the ROI can be calculated using only the annual fee.

$$ROI = \$5{,}000 - \$500 - \$3{,}000 = \$1{,}500$$

Because the ROI is positive ($1,500), this means that purchasing the safeguard will save the organization money over the course of a year and should be purchased. In this case, the organization has a new $3,000 per year expense for DDoS protection, but according to their calculations, they will end up saving $1,500 per year by making that purchase.

5.3.4.3 Risk Analysis Summary

Both qualitative and quantitative risk assessments examine risk severity based on likelihood and impact. The quantitative risk assessment is attractive because it uses precise figures. At the end of the day, organizations care about dollars and cents, and a quantitative risk assessment delivers that. However, the calculation is based on assumptions, and the assumptions are only as good as the available data. Therefore, the precision of the calculation can be deceptive.

A qualitative risk analysis is also based on assumptions and the available data, but it has a bigger "fudge factor" making it quicker and easier to perform. It does not attempt to deliver the exact ROI of an investment, but it sheds light on how risks should be prioritized. Both approaches are helpful and should lead to similar conclusions.

5.3.5 Handling Risk

Once risks have been identified and assessed, organizations need to determine how to handle the risks. There are four options for handling any given risk: *avoiding*, *transferring*, *mitigating*, and *accepting*.

5.3.5.1 Avoiding

Most cyber risks cannot be avoided. Avoiding a risk means *eliminating the risk as a possibility*. Since all risks have two components, threats and vulnerabilities, at least one of the two would need to be eliminated in order to avoid the risk. Most threats (i.e., human adversaries) are beyond an organization's ability to control, so most of the time there is not much that can be done with this half of the equation. In physical space a business could move locations to eliminate a non-cybersecurity threat like a hurricane, but in cyberspace, it is not possible to move out of the reach of threat actors.

However, there are some instances where an organization can eliminate a threat. One real-world example is the Sony Pictures hack of 2014. Sony Pictures came under a nation state threat from North Korea for a controversial film they made mocking the North Korean government. If Sony had decided not to release the film, the North Korean threat would have ceased, eliminating the cyber attack risk. However, Sony did release the film, and unfortunately for them, the risk was realized when they became the victim of a major doxxing attack.

Like threats, vulnerabilities are also largely inherent in conducting business in cyber-space. However, an organization can opt out of some technologies and business practices that introduce vulnerabilities. For example, an organization might determine a risk of IP theft due to vulnerabilities arising from their work from home policy. They could choose to avoid this particular risk by not allowing employees to work from home. Of course, the decision to cancel a work from home policy would have multiple ramifications beyond cy-bersecurity, making it a complex business decision. This helps to illustrate the tradeoffs involved in cybersecurity—cyber risk management decisions can have big impacts on the way companies do business.

Another way a company can avoid risks is by eliminating cyber assets. Data is an asset that is a big target for hackers. The more data a company stores, the more likely they are to be targeted. Companies typically prefer to save as much data as possible because data storage is cheap, and data has potential value. For example, data analyses could provide cost-saving insights. However, it is sometimes in a company's best interest to forego the potential value that data holds by permanently deleting it, making it inaccessible to them as well as to hackers. Companies need to evaluate the value of holding onto data versus the risk they accept by retaining it.

5.3.5.2 Transferring

Some cyber risks can be transferred. Transferring a risk means *passing the risk to an-other organization*. Insurance is the classic way to transfer risk. For example, when a person pays for car insurance, he is passing the financial liability of a car accident to the insurance company. The insurance company agrees to pay to repair the cars damaged in an accident and for any medical costs that result. The financial risk of a car accident is significant. If a person causes a car accident and does not have insurance, he could go bankrupt, and this is an unacceptable personal risk—this is why car insurance is such a vital industry.

Insurance companies write policies to *indemnify* their customers in case of a loss. In-demnify means *to compensate for a loss. The insurance policy is a contract stating what is covered by the insurance agency and under what circumstances.* Policies state any *exclusions* and the *premium* and the *deductible*. An exclusion is *a loss explicitly not cov-ered by the policy*. The premium is *the cost to purchase the insurance*, and the deductible is *the amount owed by the insured party for a covered incident*. Insurance premiums are based on the amount of *exposure*. Exposure is *the potential losses that could result from an incident*. The higher the exposure, the more expensive the insurance. To limit their own exposure, insurance companies might also include a *limit of liability* in their policies. The limit of liability is *the maximum amount the insurance company will pay in case of a loss*. Therefore, the insurer will pay for the damages minus the deductible and only up to the limit of liability.

The deductible is the insured party's *residual risk*. Residual risk is *the risk that remains after being mitigated or transferred*. The point of insurance is to reduce the risk to an acceptable amount, not necessarily to eliminate it. The higher the deductible the lower the premium. It is up to the insured party to determine how much financial risk they want to accept and for what types of losses, and to choose deductible and policy coverages appropriately.

Cyber insurance is a way to transfer cyber risks to an insurance company. Cyber insurance as an industry took decades to mature and for many experts to consider it a sound investment. Cyber insurers have to be experts at assessing cyber risks since they take on so much of it aggregately through their customers. They conduct research and detailed analyses of the available data, including the likelihood of incidents and their costs, performing calculations similar to the quantitative analysis assessment above. In the early days of cyberspace, making accurate assessments was difficult because of the ever-changing cybersecurity landscape and the dearth of reliable data. Also, the policies were not always clear about what types of incidents were covered and what exclusions applied. This made disputes over claims more likely, and disputes delay payments (if they come at all) and incur legal costs.

One typical exclusion in insurance policies, including cyber insurance, is losses due to acts of war. Insurance policies may state that any loss caused by a nation state in an act of war is not covered. In a highly publicized cyber insurance dispute, a giant pharmaceutical company named Merck suffered over a billion dollars in losses due to the 2017 NotPetya cyber attack. NotPetya was a Russian wiperware virus that targeted Ukraine, but many private companies, including Merck, were caught up in the collateral damage. When Merck filed their insurance claims, their insurance companies cited the war exclusion and refused to cover the losses, and this resulted in a long legal battle that was eventually settled out of court. This incident illustrates some of the complexities involved in cyber insurance policies. The risk of a company being directly harmed by an act of war between two distant countries is unlikely in any world other than cyberspace.

Cyber Insurance is now a major industry. Cyber insurers typically require their clients to comply with a checklist of cybersecurity best practices, whether from a well-known standards body like NIST or a custom checklist of their own. The assumption is that implementing cybersecurity best practices reduces the likelihood of cyber incidents, reducing the risk to the client and the insurance company. Failure to comply with the responsibilities may mean that a loss will not be covered.

Another way to transfer cyber risks is by contracting with a third party for some business functions. *Cloud computing* has revolutionized the way business is conducted. Cloud computing is *the practice of using third-party servers over the Internet for business purposes*. One of its benefits is that it allows businesses to transfer some cybersecurity risk to their cloud computing provider. For example, the cloud computing provider is the one that has to worry about securing access to their web servers. But just because a third party hosts an organization's data does not necessarily make cybersecurity incidents less likely, nor does it remove all liability from the customer. Exactly which party is responsible under what conditions is outlined in the cloud computing contract. The more risks the cloud computing provider accepts, the higher the costs for their service—again, there is no free lunch!

Many cloud providers store data for their clients in encrypted format and do not have the ability to decrypt it. Only the customers hold the necessary keys. This is a way that cloud providers can avoid risk. Encrypted data is worthless to cyber attackers. Cloud providers forego the ability to do data mining and other revenue generating activities based on customer data, but they also reduce their risk.

5.3.5.3 Mitigating

The most natural reaction to a cybersecurity risk is mitigating it. Mitigating a risk means *reducing the risk*. Cybersecurity risks are mitigated through the use of controls. There are many different categories of controls. In this chapter, we will cover only *preventative*, *detective*, *deterrent*, and *corrective* controls. Controls are implemented in the categories that are the functional underpinnings of an organization: people, processes, technology, and facilities.

5.3.5.3.1 Preventative

Preventative controls are *measures taken to prevent a risk from being realized*. Examples of preventative controls in physical space include locks, safes, and fences. These controls are designed to prevent unauthorized access to physical spaces. Physical security is also relevant for cybersecurity because of the risk of a threat actor gaining physical access to and tampering with computers. For such a risk to a company's servers (a valuable cyber asset), a preventative control would be an access control system for the server room. This is a facilities-focused control designed primarily to mitigate the risk of unauthorized access by preventing it.

Mitigating does not mean eliminating. Even with the access control system in place, it still may be possible for a person to gain unauthorized access to the server room, but the control reduces the likelihood of this occurring. As we saw above, there are cost tradeoffs for an organization to consider when evaluating controls. Access control solutions could be as simple as a bolt lock with a physical key or as complex as multi-factor authentication with a digital keypad and biometric scanner. The various options would have different costs of implementation, maintenance, and productivity, and some solutions would mitigate more risk than others. Should a company pay 50% more for a product that mitigates 10% more risk? Maybe. This is why either formally or informally, organizations need to conduct a risk assessment and a cost benefit analysis to determine the right options for them.

There is also the risk of gaining unauthorized access to an organization's computers through cyberspace. An example of an access control system for computer networks is a *firewall*. A firewall is *a software application or hardware appliance that allows or denies network traffic based on a set of rules* (more on firewalls in Chapter 9). Firewalls are like fences with a gate. They force entry through a single point where inbound and outbound network traffic can be inspected and potentially denied entry and exit.

There are numerous examples of technical preventative controls for various cybersecurity risks including encryption, vulnerability scans, pentests, and antivirus software. Raising awareness campaigns are people-focused preventative controls. Employee background checks and software patch management are process-focused preventative controls. Background checks can prevent bad actors from being hired and becoming insider threats.

5.3.5.3.2 Detective
Detective controls are *measures taken to detect incidents*. The prime example of a detective control in physical space is a security alarm. The alarm sounds when an intruder is detected, alerting security personnel to the threat. Some detective controls might catch a threat in the act, doubling as a preventative control, and others may detect an incident after it has occurred. A virus scan is a detective control in cyberspace. The goal of a *virus scan* is to identify whether a computer has been compromised with malware. Another example of a detective control in cyberspace is performing a *log analysis*. A log analysis for cybersecurity is *a review of system and network logs to identify malicious activity*. Other examples of detective controls include special-purpose ransomware scans that continually monitor files, and network scans that monitor systems sending and receiving Internet traffic.

5.3.5.3.3 Deterrent
Deterrent controls are *measures taken to discourage cyber threat actors from acting*. Examples of deterrent controls in physical space include armed security guards and security cameras. If a would-be criminal thinks he might be shot or at least identified and caught, he may be deterred from breaking into a business. Security guards and security cameras are also detective controls—many detective controls also act as deterrent controls since they make it more likely a bad actor will be caught. Perhaps the most effective deterrent for both physical security and cybersecurity are federal and state laws. Laws threaten prison time, fines, and other consequences for law breakers, and they discourage people from engaging in criminal activity.

For nation state cybersecurity the threat of a response can act as a deterrent: "if your country attacks us, we can and will attack you back." If the other nation finds the threat credible, they will be deterred from acting. Organizations can deter insider threats through the use of digital and physical notices reminding employees that all of their activity is being monitored.

Threat actors can also be deterred by making them believe that attacking is not worth their time. For example, if an organization is believed to have a strong cybersecurity posture, then a threat actor may avoid attacking them in favor of an easier target. Ransomware threats can be deterred by convincing would-be attackers that the organization will never pay a ransom under any circumstances.

5.3.5.3.4 Corrective

Corrective controls are *measures taken to recover after a cyber incident.* The next section reviews planning for failures in detail, but part of the process is to put measures in place ahead of time that will aid in the recovery of an incident. Data backups are an example of a corrective control. If data is lost due to a cybersecurity incident, it may be restorable through data backups. Contracting with an incident response company is another example of a corrective control. They can be hired ahead of time and put on *retainer* so that they can be activated quickly if they are needed. A retainer is *a fee paid in advance to secure future services if and when they are needed.*

5.3.5.3.5 Controls Summary

Controls are not mutually exclusive. They work together to mitigate risks. For a given risk, there may be multiple controls from the same and different categories implemented in different areas across the organization's infrastructure. While it is important to understand how and why a control can be effective, it is not really necessary to categorize them by type. Some controls act to mitigate risks in multiple different ways. Tables 5.6 and 5.7 provide examples of the four categories of controls and across the different areas of implementation.

Controls should be selected based on a cost benefit analysis. They should also be tested and monitored to help ensure they are functioning as designed.

Table 5.6

Example risk mitigations for each category of controls.

Risk	Preventative	Detective	Deterrent	Corrective
A nation state hacker installing a remote access trojan via a credential stealing attack.	Multi-factor authentication	Network scan	The threat of a military response	System snapshots
A nuisance hacker installing a wiper virus via an n-day exploit.	Software patching	Virus scan	Laws	Site backups
An insider threat stealing PII.	Employee background checks	Log analysis	Employee monitoring notices	Incident response planning
A criminal hacker installing ransomware via a phishing email attack.	Raising awareness campaign	Ransomware scan	Making a public vow to never pay ransoms	Data backups

Table 5.7

Example risk mitigations for each category of controls and place of implementation.

	People	Processes	Technology	Facilities
Preventative	Raising awareness	Vulnerability scanning	Software patching	Locks
Detective	Job sharing	Log analysis	Virus scanning	Alarms
Deterrent	Employee monitoring labels	Employee background checks	Logging	Lights
Corrective	Simulation exercises	Disaster recovery planning	Data backups	Backup sites

5.3.5.4 Accepting

An option always available to organizations is accepting a risk. Accepting a risk is *a deliberate decision to live with a risk*. In the qualitative risk analysis above, it is assumed that many risks with a low rating will be accepted. The organization believes that based on the low likelihood or low impact of the risk, the best decision is to accept it. This is a frequent occurrence in cybersecurity and is another way to prove the maxim that there is no such thing as 100% security. All organizations live with at least some cyber risk.

Even when a risk is mitigated, residual risk usually remains. In the quantitative risk analysis above, this was the role of the ALE_{POST} calculation—the annualized loss expectancy with the safeguard (mitigation) in place. If there is no residual risk after mitigation or transfer, then the risk has been avoided. Organizations typically transfer or mitigate a risk down to a level they are comfortable with, and then they accept the residual risk.

Even some risks that are substantial may have to be accepted. If there is no cost effective way to avoid, transfer, or mitigate a risk, or if the organization cannot afford to do so, then they have no choice but to accept it. Small businesses and startups may have to accept many such risks, and this makes them relatively more vulnerable to existential threats.

5.4 Planning for Failures

"An ounce of prevention is worth a pound of cure." - Popular saying

In some security contexts there is an expectation of 100% success. For example, the United States Secret Service is expected to have a perfect record for protecting the President. Maintaining a perfect record is *not* the expectation for cybersecurity. It is understood that there will be failures. Cybersecurity has been compared to a sporting match. It is assumed that sports will be competitive with each team scoring some points. This has always been the lived reality for cybersecurity. It is important to accept this reality because it is better to plan for reality than to live in a fantasy world. Expecting 100% success in cybersecurity is fantasy.

Therefore, organizational leadership should expect that a day will come when they are informed of a cyber incident. When that day comes, they need to have a plan in place to handle it. Even though it will be a bad day no matter what, it will be a much worse day if they are panicking in the moment of crisis and scrambling to respond. Panic and urgency are not conducive to good decision making.

Business continuity planning (BCP) is ensuring that a business can continue to operate in the wake of a disruption. BCP begins by performing a *business impact analysis (BIA)*. A BIA is *a method for determining how a cybersecurity incident will impact the organization.* The BIA identifies all critical assets and conducts a risk assessment. Part of a BIA is calculating the *maximum allowable downtime (MAD)* for specific cyber assets. The MAD is *the maximum amount of time an asset can be unavailable before the organization is severely impacted.* For example, the MAD for the payroll system may be two weeks. If employees miss a paycheck, they may stop coming to work. So it may be determined that payroll can be delayed two weeks at most. Based on this information, the organization can determine what corrective controls need to be in place so that the business can continue operating even if an incident takes their payroll system offline. An example of a control would be a back-up payroll system that can be activated in an emergency.

A major component of a BCP is a *disaster recovery plan (DRP)*. A DRP is *a formal document that details an organization's incident response process.* The DRP describes the major risks to an organization and their impact and the plans to respond and recover if the risks are realized. There are five phases to a DRP: *respond, activate, communicate, assess,* and *reconstitute* (see Table 5.8).

Table 5.8

Example phases of a disaster recovery plan.

Phase	Summary
Respond	Assess the situation and its severity.
Activate	Alert the necessary team members.
Communicate	Communicate to stakeholders as appropriate.
Assess	Perform a detailed assessment to understand the damage.
Reconstitute	Restore systems and assets.

As an example, organizational leadership might imagine the following scenario: an employee comes into work over a holiday and is the first person in the office. When he turns on his computer, he is greeted with a ransomware note on his desktop similar to the one in Figure 5.5. What happens next?

Figure 5.5

A screenshot of a ransomware note.

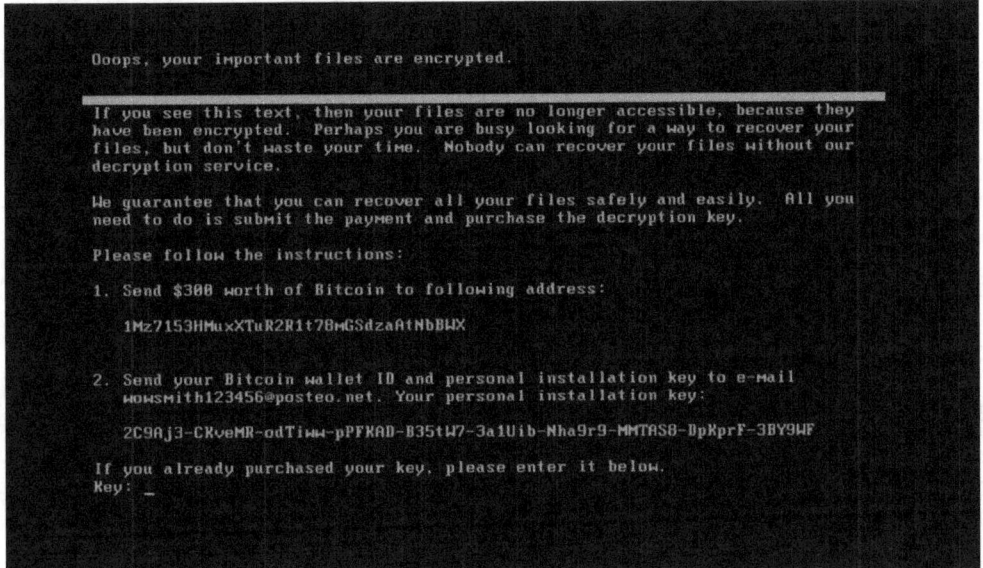

The first phase is to respond. Would an employee know who to contact in case of an emergency like this, especially on a holiday when nobody else is in the office? Time is of the essence, so the quicker he is able to notify leadership, the better. Imagining this scenario might help the organization understand the need to provide clear guidance to employees of what to do when an incident happens. Covering what to do could be part of the new employee orientation process, and employees could be periodically reminded in emails or as part of annual security trainings—these are examples of controls. The company could even establish an emergency phone that is manned 24 / 7 / 365 (24 hours a day, 7 days a week, 365 days a year) in case an incident occurs and make sure all employees know the number to call.

Once a person in leadership is notified, he needs to quickly assess the situation to determine if it constitutes an emergency. In this case, he might ask the employee to login to a different computer to see if the ransomware note appears there as well. The leader might try to login remotely to see if he can access key systems.

Once it is determined that this is a real ransomware incident, the next phase begins: the necessary team members must be alerted. The first to be called would be cybersecurity and IT staff members who could respond quickly and perhaps reduce further damage. It may be prudent to take the office network offline and to shut down computer systems. During this process personnel need to be careful not to destroy any evidence that could help determine what happened and who is responsible. Next, key leadership needs to be informed so they can start carrying out the plan. Part of the leadership could include members of the board of directors and legal counsel. There are people that specialize in cybersecurity *incident response*—they show up on site immediately after an incident and help the organization investigate and recover. This is a field of cybersecurity adjacent to ethical hacking and cyber defense—workers in this field know the tricks of the hacking trade and are experts in cyber forensics. If the organization has a contract with such an individual or company, they would need to be contacted right away.

It is important in this step that everybody can be reached. Email could be down due to the incident, and even if not, people may not be checking their email, so phone calls, text messages, and other methods of communication need to be identified. Minimally, peoples' personal phone numbers need to be readily available. If numerous people need to be contacted, a call chain can be established that gives different people responsibility for contacting others.

The next phase is to communicate. Communication will continue through the remainder of the response and recovery as progress is made and updates become available. Leadership must decide who else needs to be informed, when, and what level of detail needs to be shared. At some point, all the employees of the organization might need to be made aware that an incident has occurred. They may be asked to not come into the office until further notice. It is also possible that customers need to be notified. This could be through a mass email, a social media post, or through a notice posted on the company website. The company also needs to determine how they should respond to media requests. What should an employee say if a reporter contacts him? If he is not coached ahead of time, he may divulge sensitive information or make inappropriate statements that make the situation worse.

After communicating, the next phase is a detailed assessment. How did the incident happen? Is it contained? Is the original vulnerability fixed? This could be facilitated by a company that specializes in incident response and by law enforcement.

Lastly, the organization can begin reconstituting. As they do this exercise and imagine what it would take to become fully operational again in the wake of an incident like this, they can determine what corrective controls would need to be in place. The final part of the process is performing a *post-mortem*. A post-mortem in a business context *is a review of an incident after the fact to improve the process going forward.* While a post-mortem after an incident will always identify things that could have gone better, too many "I wish we had thought of that" moments indicate poor preparation and a deficient DRP.

This overview covers only the basics of a DRP. An actual DRP would need to be more detailed and cover more aspects of the incident response and recovery process, including when and how to notify law enforcement. The point of a DRP is that these plans are made before an incident occurs, when the right people are in the room and there is no urgency or panic. Once a DRP is created, it needs to be tested, personnel must be trained, and everybody in the organization must be made aware of the high-level plan. A DRP can be tested by doing tabletop exercises, simulations, and possibly even unannounced drills. Lastly, the DRP needs to be revisited often. As the organization changes and risks change, a DRP can become outdated. If that is the case, it may not provide much help when it is called upon in the time of crisis.

5.5 Conclusion

Organizational leaders in the C-suite focus much of their attention on risk management. For the long term sustainability and growth of their company, they need to study the risks they face and address them appropriately. Many of the risks are not cyber related. For example, their office location may be hit with a natural disaster. They may lose a large customer's business. A key supplier may go bankrupt. While they may be unpleasantly surprised when something goes wrong, good leadership should not be unprepared. Their goal is to identify and handle the risks they face by putting appropriate controls in place.

Because of the central role technology plays in organizations, cyber risk management has become an increasingly important focus for organizational leaders. This chapter has covered the importance of cyber risk management and the basics of how it should be performed. By following a process such as the one outlined in this chapter, an organization can thoughtfully determine how much cybersecurity they need and how much they should spend on it.

Chapter 6

6. The Skill of Cybersecurity: Adversarial Thinking

"[Detectives] consider only their own ideas of ingenuity; and, in searching for anything hidden, [think only about how] they would have hidden it…but when the cunning of the individual felon is diverse in character from their own, the felon foils them."
- The Purloined Letter *by Edgar Allan Poe*

Cybersecurity is made up of really only two things: computers and adversaries.[1] Take away the adversaries, and we have a world with computers, and those computers need programmed, installed, and maintained. Problems will occur due to programming mistakes (i.e., bugs) and hardware failures. Floods, fires, and storms will happen. Users will accidentally delete important files. Power outages will occur. This world will require plenty of IT support, and backups and other safeguards will be necessary. However, this world would not need cybersecurity.

On the other hand, take away computers, and we have a world with adversaries in it. (This was the actual world until the latter part of the 20th century!) Attackers will steal, kill, and destroy. They will take advantage of others, violate their rights, and harm them. This world will need security guards, police forces, militaries, and a criminal justice system with laws and regulations. However, this world would not need cybersecurity.

Only in a world with both computers and adversaries do we need cybersecurity (see Figure 6.1). This is the world we live in. IT support is still needed in our world and can be efficiently managed based on probabilities. How often does equipment fail on average?

---

[1] This chapter draws on material from two of the author's journal articles:
S. Hamman, K. Hopkinson, R. Markham, A. Chaplik, and G. Metzler, "Teaching game theory to improve strategic reasoning in cybersecurity students," IEEE Transactions on Education, vol. 60, no. 3, pp. 205-211, 2017.
S. Hamman and K. Hopkinson, "Teaching adversarial thinking for cybersecurity," Journal of The Colloquium for Information System Security Education, vol. 4, no. 1, pp. 93-110, 2016.

How likely are power outages? What acts of nature might occur and how frequently? But IT support is not enough. With intelligent adversaries, the probabilities of random events and acts of nature are immaterial. Adversaries will manufacture the perfect storm so that they can attack when their target is at its weakest. This world will still need the normal IT functions of data backups and emergency power supplies, but it will need a lot more (see Table 6.1).

Figure 6.1

The essence of cybersecurity: computers and adversaries.

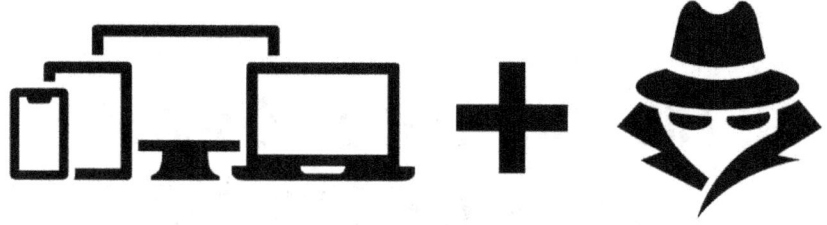

Table 6.1

A comparison of IT and cybersecurity.

IT in a World WITHOUT Hackers	Cybersecurity in a World WITH Hackers
Random mistakes and failures result in data loss and the need for restoring from backups	Deliberate attacks destroy production data *and* backups
Operating systems crash unexpectedly	Malware takes over operating systems
An IT help desk is needed to troubleshoot computer problems	A security operations team is needed to try to prevent catastrophes
Simple log audits can be used to determine the cause of incidents	In-depth cyber forensics are need to determine the cause of incidents

Cyber adversaries are what differentiates cybersecurity from other academic disciplines such as computer science, IT management, and engineering. At its core, cybersecurity is an adversarial conflict between attackers (i.e., hackers) and defenders. In order to practice cybersecurity effectively, one must pay careful attention to what hackers are thinking. What are their objectives? How might they go about achieving them? This mindset is known as *adversarial thinking*.

Perhaps because cybersecurity arose out of the technical discipline of computer science, there tends to be less emphasis placed on the human aspects of cybersecurity. The point of this chapter is to help correct that imbalance. This chapter hammers home the simple fact that without cyber adversaries, there would be no such thing as cybersecurity. In order to defend cyberspace, we need to think like our adversaries.

6.1 Adversarial Thinking Defined

"And here Alice began to get rather sleepy, and went on saying to herself, in a dreamy sort of way, 'Do cats eat bats? Do cats eat bats?' and sometimes, 'Do bats eat cats?' for, you see, as she couldn't answer either question, it didn't much matter which way she put it." - Alice's Adventures in Wonderland *by Lewis Carroll*

Certain academic disciplines are synonymous with a particular way of thinking (see Table 6.2). In order to excel at them, you must possess a particular mindset. Take, for example, the discipline of mathematics. Mathematicians start with axioms and then make logical inferences to prove theorems. They use proof techniques like proof by induction, proof by contradiction, and direct proofs, all of which require rigorous logical thinking. Therefore, logical thinking and the discipline of mathematics go hand-in-hand.

Another example is the discipline of computer science. Computer scientists design algorithms. Algorithms are a sequence of detailed instructions for solving abstract problems. Algorithms accept inputs and produce correct outputs. The best computer scientists excel at algorithmic thinking, and this is the distinctive characteristic of the discipline.

The discipline of cybersecurity also requires logical and algorithmic thinking, but these are not its primary emphasis. Because of the centrality of hackers to cybersecurity, adversarial thinking is the hallmark of cybersecurity and what distinguishes it from other disciplines. The skill of adversarial thinking is essential for cybersecurity, and those that excel at it will do well in the field. It is the fundamental skill of cybersecurity.

Table 6.2

Hallmarks of academic disciplines.

Discipline	Approach	Fundamental Mindset
Mathematics	Constructing proofs	Logical Thinking
Computer Science	Writing programs	Algorithmic Thinking
Cybersecurity	Security practices	Adversarial Thinking

But what exactly is adversarial thinking? When the term is used, in many cases it is not defined at all, taking it for granted that adversarial thinking merely means thinking like a cyber adversary (i.e., a hacker). However, this raises the obvious question, what is unique about the way hackers think? If we cannot answer this question accurately, we cannot hope to impart it and assess it.

The discipline of cognitive psychology studies the mind and what it means to think, and it can help unpack what it means to think like a hacker. Robert Sternberg is a well-known cognitive psychologist who proposed a theory of intelligence called *the triarchic theory*. The triarchic theory identifies three distinct aspects of the intellect and is similar to a tripart theory of the mind developed by Aristotle millenia before him. Sternberg's three areas are: the analytical, the creative, and the practical. The analytical area captures the popular notion of intelligence and corresponds with IQ. It includes mathematical ability and logical reasoning. It is book smarts. The creative area of the intellect includes the ability to make unique connections and to see the world in original ways. Artists, authors, and musicians excel in this aspect of the intellect. It is imaginative and creative thinking. Practical intelligence includes the ability to plan, strategize, and accomplish goals. CEOs and military leaders have high degrees of practical intelligence. It is street smarts (see Table 6.3).

Table 6.3

Sternberg's triarchic theory.

Area	Description	Popular Conception	Exemplar
Analytical	Mathematical ability and logical reasoning	Book smarts	Einstein
Creative	The ability to make unique connections and original insights	Creative ability	Van Gogh
Practical	The ability to plan, strategize, and overcome obstacles to accomplish goals	Street smarts	Napoleon

To help illustrate how these areas of the intellect are distinct, here is a quick exercise to isolate each way of thinking in your own mind. First, for analytical thinking, consider the question: *how many months old are you?* If you take this question seriously and try to work it out, and then step back and review your thought process, you will isolate the analytical portion of your intellect. First, you likely multiplied your age by twelve, the number of months in a year. The mental math of multiplying the numbers in your head may have involved several substeps and adding. Having calculated that number and set it aside, you may have then counted the number of months since your most recent birthday. Lastly, you recalled the first number and added it to the second to determine the answer. This is analytical thinking in practice, and it comes more naturally to some than to others!

Next, for an exercise in creative thinking, consider the question: *what do Santa Claus and Abraham Lincoln have in common?* Again, if you work it out, and then step back and analyze how your thoughts proceeded, you will get a sense for how creative thinking works. You likely created two compartments, first thinking about Santa Claus and his attributes, then Abraham Lincoln and his, and then you started trying to link the two. Maybe you quickly found a general commonality like they are both men, but not satisfied with such a boring connection, you dove back in to try and find something more interesting. After several false steps maybe you realized they both have iconic facial hair or something even more creative. The aha! moment when you know you have identified a unique link is a fun payoff to creative thinking. Creative thinking is the process of making unique connections.

Lastly, to isolate practical thinking, consider the question: *how could you pull off an epic surprise birthday party for your best friend?* If you take the time to really think this through, you will feel your mind engaging in practical thinking. After figuring out how much time you have before your friend's birthday, you work backwards to determine a feasible timeline. Then you start thinking about who to invite and how to invite them without your friend finding out. You also realize you need a venue and that you need to plan entertainment, food, and refreshments. You might consider delegating some of these tasks to others, trying to determine who would be a good fit for which jobs and how you can enlist their help. You also start thinking about under what pretense you can get your friend to the venue without spoiling the surprise. Along the way you play out scenarios and imagine obstacles, risks, and constraints, and start planning how you can work around them. This is practical thinking.

These three ways of thinking are different and feel different. They are meant to capture types of intelligence that all people possess to one degree or another. To illustrate the triarchic model's explanatory power, Sternberg provides an example he has seen many times as a university professor. Some of his undergraduate students excel at college and appear to be a natural fit for graduate school, but when they get there, they fail. What explains how they could be so successful in one arena yet fail in another so closely related? He believes it could be that these students have high analytical intelligence which works well in the structured world of undergraduate education, but they lack either the creative or practical intelligence necessary to excel in graduate school. Creativity is needed to perform original research, and practical intelligence is needed to navigate and achieve the many necessary milestones.

Now that we understand Sterberg's triarchic theory, we can apply it to hackers to develop a definition of adversarial thinking. How do the three areas map to attributes of hackers? What contributions do book smarts, creativity, and street smarts make to a hacker's success? (see Figure 6.2)

Figure 6.2
The triarchic theory applied to hackers.

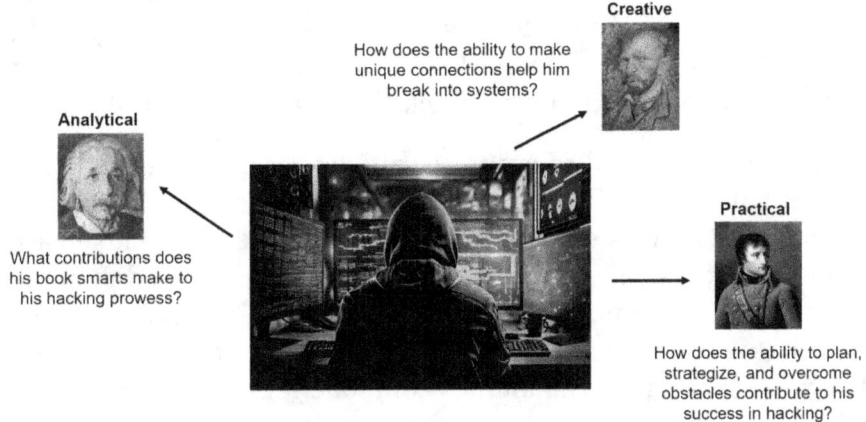

For book smarts, hacking involves detailed knowledge of many technical aspects of computing, including computer networking, programming, and operating systems. In Sternberg's paradigm, this technical knack exhibited by hackers ties into the analytical component of their intellect. Their analytical gifts translate into an unusual facility with computers and technology. Having strong technical abilities is vital to hackers because this is the tool of their trade.

Creativity is at the core of *the hacker mindset*—a phrase coined to highlight how hackers approach situations in unconventional ways. While fiction writers excel at creating original stories that capture the imagination, hackers excel at creating original exploits that manipulate technology in unexpected ways. Both are manifestations of the same root—they involve seeing the world in a unique way, and the ability to put disparate information together in a new form. While most technologists are concerned with making systems work, hackers are obsessed with pushing the limits of systems and exploring possibilities that most people would never consider. This aspect of hacking is the main connection between the pejorative way the term hacker is used today, and the original, complimentary term from a previous era which connoted a person's uncanny ability to bend computers to his will.

The practical component of Sternberg's triarchic theory is the aspect of the intellect that involves planning, strategizing, maneuvering, and overcoming obstacles to accomplish goals. As we saw in Chapter 4, while script kiddies are known to indiscriminately fire point-and-click exploits at random in hopes of finding unpatched systems, more highly skilled hackers select targets, conduct reconnaissance, carefully plan their attacks, and meticulously cover their tracks. In general, hackers try to use their time and resources wisely, and they strive to outwit their targets. It is their practical intelligence that helps them anticipate defenses and obstacles so that they can accomplish their objectives and not get caught.

Putting these insights together, we can derive a precise definition of adversarial thinking. Adversarial thinking is *the ability to embody the technological capabilities, the unconventional perspectives, and the strategic reasoning of hackers.* To the extent a person can do this, he will be able to compete with hackers on a level playing field (analytical), find and fix vulnerabilities before hackers have an opportunity to exploit them (creative), and anticipate future attacks and thwart attacks in progress (practical).

If adversarial thinking is central to cybersecurity practice, then it must also play a key role in cybersecurity education. How is it being addressed in the classroom? Most cybersecurity curriculum guidelines stress the importance of teaching adversarial thinking, but they do not provide specific learning outcomes. This makes it difficult to know how to actually address it. Our definition provides the necessary precision to derive practical learning outcomes and assessments (see Table 6.4). Examining these learning outcomes, it is clear that the technological capabilities component of adversarial thinking is heavily emphasized in cyber curriculums through classes like operating systems, computer networks, and programming courses. The unconventional perspectives component is typically taught by example—in fact, it is probably more likely to be caught than taught. Students learn about creative attack vectors such as SQL injection, network packet manipulation, and cross-site scripting. The practical component of adversarial thinking is addressed through threat modeling exercises, but little emphasis has been placed on helping to improve students' ability to think strategically. This is a blind spot because strategic thinking is well-understood and can be taught and acquired relatively easily.

Table 6.4

Adversarial thinking learning outcomes.

Dimension	Learning Outcome	Application
Technological capabilities	Understand computer technology at a deep level (e.g., networking protocols, programming languages, and operating systems)	Leveling the playing field
Unconventional perspectives	Identify unconventional uses of software and protocols that could be exploited as attack vectors by hackers	Employing the hacker mindset
Strategic reasoning	Anticipate the strategic actions of hackers, including where, when, and how they might attack, and their tactics for evading detection	Anticipating and thwarting attacks

This book addresses the technological capabilities aspect of adversarial thinking in several chapters, and it touches on the unconventional perspectives aspect in places. The remainder of this chapter, however, focuses exclusively on strategic reasoning. The goal is to help develop the reader's ability to anticipate the strategic actions of hackers by teaching basic game theory concepts. Learning game theory is a proven way to improve one's strategic reasoning abilities.

6.2 Game Theory

"Strategic thinking seems to be more like learning to windsurf, ski, or fly an airplane, activities that require people to learn skills which are unnatural but teachable, and less like weight-lifting or dunking a basketball, where performance is constrained by physical limits." - Colin Camerer

There are games of skill, games of chance, and games of strategy. *Game theory* deals with games of strategy. Game theory is not primarily concerned with sports, board games, and video games, although these activities may involve strategy. Game theory was developed to analyze economic scenarios, and is still a staple of the discipline of economics today. It helps answer questions like: how many widgets should we produce next year and how much should we charge per widget? The answers to questions like these involve strategy because companies compete with one another to sell products. Each company makes plans without knowing what the other companies in the market are going to do and try to anticipate how they can outdo one another. Game theory is also taught in other academic disciplines on college campuses, including military studies programs. Unfortunately, game theory is not typically addressed in cybersecurity education, although the academic literature contains many examples of game theory being applied to cybersecurity. Because cybersecurity is adversarial in its essence, game theory is a natural fit for cybersecurity students.

Game theory was formally established as an academic discipline with the 1944 publication of *Theory of Games and Economic Behavior* by John von Neumann (yes, the same von Neumann from Chapter 2) and Oskar Morgenstern. *John Nash* went on to make major contributions to the field in the 1950s. Game theory is a mathematically rigorous approach to analyzing strategic contests (i.e., *games*). It can be defined as *the study of interdependent decision making between multiple players where each player strives to maximize his own utility*. This definition highlights the three primary ingredients of a game: *players, interdependent choices*, and *utility preferences*. The players are *the actors in the game*. Interdependent choices means that *the outcome for each player depends in part on the choices made by the other players*. Utility preferences are *an ordering of the outcomes from least to most desirable*. There are many types of games, including: two player and multi-player, cooperative and non-cooperative, zero-sum and non-zero sum, one shot and repeated play, etc. This chapter covers only basic terminology and general principles.

The Hacker's Dilemma

Trudy and Eve, two hackers, are nabbed by law enforcement and accused of a serious cybercrime. On the way to the police station, they both agree to say nothing to the cops. They know that the cops may have evidence of their cyber intrusion into Acme Bank, but probably not enough to convict them of their salami attack, *where Trudy and Eve diverted small amounts of money in "transactions fees" from millions of bank transfers over several months, eventually accumulating a huge sum of money in an off-shore bank account.*

Arriving at the police station, they are placed in separate interrogation rooms, and Eve is presented with indisputable evidence of her having gained unauthorized access to the Acme network of computers. The interrogator then offered Eve a deal, "Look, we have enough here to lock you up for a year under the Computer Fraud and Abuse Act (CFAA), but I am willing to let you off on probation if you just tell me about Trudy's involvement in the salami attack."

What should Eve do? She had already agreed with Trudy to keep her mouth shut, but now she faces a hacker's dilemma: should she keep her agreement with Trudy or betray her?

Assume that Eve knows that Trudy is being offered the same deal in the other room. Therefore, she realizes that they each have the same two choices: sticking to their agreement (*cooperating*), or ratting one another out (*defecting*). If Trudy and Eve cooperate with one another, the cops will be unable to convict them of the salami attack, and they will both get tried and convicted under the CFAA, resulting in one year of prison time. If they both defect from their agreement and rat one another out, they will both be found guilty of the bigger cybercrime and share a five year prison sentence. If one cooperates while the other defects, the defector will get just probation (no prison time) while the one who keeps the agreement will be convicted as the sole perpetrator of the salami attack and get ten years in prison.

Of those scenarios, Eve's utility preferences are clear: the more prison time she serves, the less she likes the outcome! In order for Eve to make the best possible choice, she should consider what Trudy is going to do and how she should respond. What should Eve do if Trudy defects? If Eve also defects she gets five years in prison, but if she cooperates she gets ten years. Therefore, if Trudy defects, Eve should also defect. But what if Trudy cooperates? If Eve defects, she just gets probation, but if she cooperates, she gets one year in prison. Clearly, if Trudy cooperates she should defect—no prison time is much better than a year in prison! This analysis shows that, no matter what Trudy does, whether she cooperates or defects, Eve should defect to maximize her utility preferences. Since everything is the same from Trudy's perspective, she should also defect from their agreement. This is the solution to the game: they should both defect, rat one another out, and end up with five years in prison. However, this is ironic because if they just stick to their original agreement, they would both end up with just one year in prison instead of five—a much better outcome.

The hacker's dilemma is a cyber-themed retelling of *the prisoner's dilemma*—one of the most famous games in game theory. It has been dramatically portrayed in literature, films, television shows, and even in a popular British game show. What makes it fascinating is the unfortunate result for the players. Is this bad outcome for both players really the solution to the game? Why is that?

Figure 6.3

The hacker's dilemma in normal form.

Figure 6.3 shows the hacker's dilemma represented in *normal form* game theory syntax. The normal form grid captures all of the essential elements of the game: the players, their interdependent choices, and their utility preferences. Each of the four squares represent combinations of choices. In Figure 6.3, the top-left square is where Eve and Trudy both choose to cooperate. The other squares represent different combinations of cooperate and defect. In each square, the numbers represent utility preferences. The higher the number the more they prefer it. Utility is like money—the more the better! Since her name is listed on the left, Eve is the *row player*, and since she is listed at the top, Trudy is the *column player* in this game. Their choices are listed by their names. In this game, they each have the same two choices: cooperate (C) or defect (D). (In other games, players may have more than two choices, and they may have different choices from each other.) Each square contains two numbers. The row player's utility is the first number, and the column player's utility is the second number. What combination of choices does the bottom-left square represent? (Eve: Defect, Trudy: Cooperate.) What are Eve and Trudy's utility preferences for this outcome? (Eve: four, Trudy: one.) Are there any other outcomes that Eve prefers more than this one? (No. Four is the highest row number in the grid, so it is her top choice—this is the result where she gets no prison time.) Does Trudy prefer any outcomes above this one? (Yes. One is the lowest column number in the grid, meaning this is Trudy's worst outcome—this combination is where Trudy has to spend ten years in prison.)

It would seem natural for the players to end up with the [C, C] outcome. This is what they agreed to do ahead of time, and it is the best of the two "fair" outcomes. They both get their second best choice. However, from both players' perspectives, if they think they might land at [C, C], there is a big incentive to defect! No prison time is much preferable to an entire year in prison. Therefore, they are both being pulled towards defecting like a magnet. Ultimately, this results in a worse outcome for both of them. Cooperating is an unstable position that is difficult to hold, especially if one player has even the slightest suspicion that the other might defect.

[D, D] is the only stable outcome in this game, meaning, from this square, neither player can unilaterally change his or her choice and end up with a better outcome. For example, if either Eve or Trudy were to change to cooperate while the other remained at defect, the one who changed would receive ten years in prison rather than five—a worse outcome. This equilibrium dynamic is called the *Nash equilibrium* after John Nash who proved that all finite games have at least one stable point like this (although in some games, the equilibrium is in mixed strategies, meaning that the players must assign probabilities to choices). In this game, this is the only Nash equilibrium. In all the other squares, at least one of the players could unilaterally change his or her choice and end up with a better outcome.

The prisoner's dilemma story is obviously contrived and appears to be irrelevant to real life. Most of us do not anticipate ever ending up in the backseat of a cop car making an agreement with our partner in crime on the way to the police station! But the tension illustrated in the prisoner's dilemma actually occurs frequently in the real world. For example, doping in sports (e.g., cheating by taking performance enhancing drugs), is a real-life prisoner's dilemma. If no athletes dope, everybody would be better off because they would not have to deal with the health consequences of taking performance enhancing drugs, and all of the athletes would still be competing on a level playing field. This is the [C, C] outcome. However, this leaves the door open for cheaters (defectors) to gain an upper-hand and outperform better athletes! In sports, not doping is an unstable outcome—there is an incentive pulling athletes towards cheating, especially if they suspect other athletes may be cheating. This results in many defectors, making everybody worse off. Former Major League Baseball player Lenny Dykstra captured this tension well in a surprisingly candid interview. When he was asked if he ever used a relatively mild performance enhancing drug called human growth hormone (HGH) when he played baseball, he replied, "I put that in my cereal man. It was in my cereal...Come on, HGH? Nah, we're talking about the good stuff, you know? Deca durabolin and testosterone and anadrol [strong steroids]. We're talking about the difference of making thirty million or getting a real job and working and making sixty thousand...*Do you want the guy next to you taking them and you're not going to take them*?"[2] (emphasis added)

The prisoner's dilemma tension is also felt by college students. They face a prisoner's dilemma when it comes to academic integrity. It is tempting for college students to compromise their integrity if they suspect that other students may be doing less work but earning better grades due to cheating. To be a cooperator while others defect feels unfair, and it provides a convenient rationalization for cheating. The prisoner's dilemma perfectly illustrates why character is prized by society and those that exhibit it are highly esteemed—it highlights temptation and the real costs that sometimes come from choosing character.

6.2.1 Real Life Through the Lens of Game Theory

"All models are wrong, but some are useful" - George E.P. Box

The previous section started with a theoretical game and then showed how it applies to real life. In this section, we begin with a real-life situation that on the surface appears like it has nothing to do with game theory, but then we analyze it in terms of players, interdependent choices, and utility preferences to try and predict the outcome.

The Hebrew Bible records the story of King Solomon, an Israelite king who was renowned throughout the ancient world for his wisdom. His wise insight is famously illustrated in an account known as *Solomon's Wise Ruling*.

---

[2] From an interview on The Herd with Colin Cowherd, May 17, 2016.

Solomon's Wise Ruling

"Now two prostitutes came to King Solomon and stood before him. One of them said, 'Pardon me, my lord. This woman and I live in the same house, and I had a baby while she was there with me. The third day after my child was born, this woman also had a baby. We were alone; there was no one in the house but the two of us.

'During the night this woman's son died because she lay on him. So she got up in the middle of the night and took my son from my side while I your servant was asleep. She put him by her breast and put her dead son by my breast. The next morning, I got up to nurse my son—and he was dead! But when I looked at him closely in the morning light, I saw that it wasn't the son I had borne.'

The other woman said, 'No! The living one is my son; the dead one is yours.'

But the first one insisted, 'No! The dead one is yours; the living one is mine.' And so they argued before the king.

The king said, 'This one says, "My son is alive and your son is dead," while that one says, "No! Your son is dead and mine is alive."'

Then the king said, 'Bring me a sword.' So they brought a sword for the king. He then gave an order: 'Cut the living child in two and give half to one and half to the other.'" (1 Kings 3:16-25)

Note the predicament for the wise judge and king, Solomon. He is dealing with a case of she-said, she-said, with no forensic evidence or eye witness accounts to help determine the truth. Game theory, which would not be invented until thousands of years later, can help us appreciate King Solomon's intuitive understanding of human rationality. We can analyze this story from a game theoretical perspective to predict how it is going to end.[3] To do this, like any game, we need to define the players, their choices, and their utility preferences.

The players in this game are the real mom (*Mom*) and the imposter (*Imposter*). Once Solomon issues his verdict, each woman has a decision to make. They can accept the judge's ruling or defy it. We can assume that both women would have been familiar with Deuteronomy 17:8-12 on Israelite judicial practices. It states that a judge's verdict is final, it must be followed to the letter, and the penalty for failing to do so is death. Therefore, defying a judge's order was not something to be done lightly! We will refer to this option as the *Reject* choice and to accepting the verdict as the *Accept* choice. Their choices are interdependent because the combinations of their choices result in different outcomes (e.g., who lives, who gets to keep the baby, etc.). Some outcomes are more preferable than others, and they can be ranked to determine utility values.

---

[3] This illustration is adapted from the book:
Brams, S.J. (2003). *Biblical Games: Game Theory and the Hebrew Bible*. MIT Press.

Obviously, if they both accept the ruling and Solomon carries out his decision, the baby would die—neither woman would receive the baby in that case. If one woman rejects the ruling she is giving the baby to the other woman and risking the death penalty for herself. Even though the consequences for both women are symmetrical, what makes this game interesting is that they do not prefer the outcomes the same. This is because one of the two women is the living baby's real mother. We can assume that the real mom loves her newborn baby and values her baby's life above her own life—after all, she is fighting for her baby in court. This would not be true for the imposter who has already lost her baby and stolen this other baby. She values her own life above the life of the baby. Figure 6.4 shows what this game looks like in normal form, and it illustrates the asymmetry in the utility preferences. What combination of choices does the top-left square represent? (Mom: Accept, Imposter: Accept.) What is the outcome of this combination of choices? (Solomon carries out his ruling and the baby dies.) How do the women rank this outcome? (Even though she does not get to keep the baby, this is Imposter's second best outcome because she loses nothing. It is Mom's worst outcome because her baby dies.)

Figure 6.4

Solomon's wise ruling in normal form.

Given these assumptions the game can be analyzed to identify the Nash equilibrium. To solve the game, start from Mom's perspective and consider what she should do. In order for Mom to decide what she should do, she should first consider what Imposter might do and how she should respond. This is called *best response analysis* and is a helpful approach for solving games. If Imposter Rejects, what should Mom do? (Accept.) If Imposter Accepts, what should Mom do? (Reject.) Unlike in the prisoner's dilemma, this approach has not clarified the situation—depending on Imposter's choice, Mom should either Accept or Reject. However, Mom can go a level deeper. Putting herself in Imposter's shoes, Mom can see what Imposter should do based on what Mom might do. If Mom Accepts, what should Imposter do? (Accept.) What if Mom Rejects? (Accept.) Now we are getting somewhere! No matter what Mom does, she can see that Imposter should Accept. This means that Reject is a *dominated strategy* for Imposter. A dominated strategy is *a strategy that will never be chosen*. With this insight, Mom can then determine what to do. Knowing that Imposter will choose Accept, what should Mom do? (Reject.)

Therefore, the solution to the game is Imposter: Accept, Mom: Reject. This is the Nash equilibrium, because from this square, neither player can unilaterally change her choice and receive a better outcome.

Where we left off in the story above, the king had just announced his verdict, "Cut the living child in two and give half to one and half to the other." Now let's see how the story ends:

"The woman whose son was alive [Mom] was deeply moved out of love for her son and said to the king, 'Please, my lord, give her the living baby! Don't kill him!' [Reject]

But the other [Imposter] said, 'Neither I nor you shall have him. Cut him in two!' [Accept]

Then the king gave his ruling: 'Give the living baby to the first woman. Do not kill him; she is his mother.'

When all Israel heard the verdict the king had given, they held the king in awe, because they saw that he had wisdom from God to administer justice." (1 Kings 3:26-28)

Game theory accurately predicted the women's choices. It turns out King Solomon had a trick up his sleeve; so, happily, the ending was not what anybody could have expected—the real mother received her baby back alive. The story does not say what happened to Imposter, but we can assume the ending was not so happy for her.

This is an illustration of how game theory can be applied to real-life situations. It takes some careful thought, and may involve some assumptions, but game theory can be a helpful tool for methodically analyzing complex real-world scenarios.

6.2.2 Game Theory Summary

In both the hacker's dilemma and Solomon's wise ruling, the games were solved by viewing the situation through the choices available to the other player. Eve determined what she should do by thinking about what Trudy might do, just like Mom figured out what she should do based on the Imposter's choices. Best response analysis is very important for strategic thinking—it is putting yourself in the shoes of your adversary.

The most important takeaway for us is to see that when analyzing a strategic scenario, it pays to consider the world from the perspective of your opponents. It is helpful to take the time to enumerate their choices and to consider how they would rank the different outcomes.

Game theory is the study of interdependent decision making between multiple players where each player strives to maximize his own utility. Real life strategic situations can be analyzed through the lens of game theory by identifying the players, the interdependent choices available to them, and their utility preferences. Game theoretical analysis proceeds by viewing the situation through the choices available to the other player. In the next section we will dig deeper into what it looks like to explore strategic situations from our adversary's perspective by learning about behavioral game theory.

6.3 Behavioral Game Theory

"I'll tell you in a minute. First, let's drink. Me from my glass, and you from yours."
- Vizzini in The Princess Bride *film*

One of the underlying assumptions of game theory is *player perfect rationality*. This means that players behave perfectly rationally when making strategic choices. This assumption makes it possible to analyze and solve games mathematically. It assumes that players will work through all the options logically. In this section, we are going to learn about behavioral game theory which challenges this assumption.

The Traveling Hacker's Dilemma

Two black hats, Veryl and Ruth Ann, are traveling back from a hacking conference in Las Vegas. At the conference they both purchased a special purpose code cracking device from a hacking tools dealer. Fittingly, the dealer only accepted cash and did not provide receipts. At the airport they reluctantly agreed to check their devices because they were not allowed to carry them on the airplane—the ticket counter agent deemed them suspicious-looking. Unfortunately, when they arrived at their home airport, they found themselves in the luggage claim office because their devices never made it to the baggage claim carousel.

After determining the devices disappeared without a trace and there was no hope of recovery, the luggage claim agent offered to compensate them for their losses. Unfortunately, Veryl and Ruth Ann did not have receipts, and the agent had no way of verifying how much the devices were really worth. So he came up with a plan. He gave Veryl and Ruth Ann each a piece of paper and asked them to separately write down the value of the code cracking device. He knew the devices were worth no more than $500 and no less than $100, but his goal was to determine their actual value. If they both put down the same number, he would accept that as the value and pay it out to both of them. However, if they put down different numbers, he would consider the true value the lesser amount, and he would reward the person who put down the lower number for being honest with a $50 bonus and penalize the other for being deceitful by reducing his or her payout by $50.

Rather than putting down the amount he actually paid, Veryl saw an opportunity to make a little extra cash! His first thought was to put down $500 thinking that Ruth Ann would probably do the same—then they could both take home $500. But, on the other hand, being a hacker and thinking deviously, if he knew Ruth Ann would go with $500, he should actually write down $499. This would result in him taking home the $50 bonus and $549 overall. But what if Ruth Ann was even more devious and submitted $498? That would mean if Veryl wrote down $499, he would get only $448 because he would have to absorb the $50 penalty. Maybe Veryl should write down $497 just to be on the safe side, but again, what if Ruth Ann was thinking the same thing!

The traveling hacker's dilemma is a cyber-themed retelling of a game theoretical game called *the traveler's dilemma*. There are two players, Veryl and Ruth Ann, and they each have numerous choices—they can put down any dollar amount between $100 and $500. Their utility preferences are tied to their payout—they both want to take home as much money as possible. Figure 6.5 shows a small portion of the normal form version of this game. Interestingly, no matter which of these squares you start in, at least one of the players has an incentive to unilaterally change his or her answer. For example, in the square where both players submit $499, both of them have an incentive to change their answer to $498 because this would result in the $50 bonus. Like with the prisoner's dilemma, there is a magnet pulling the players downward and to the right in the grid. None of these squares represent stable choices, and none of them are the solution to this game. The Nash equilibrium is actually the very bottom right square in the overall grid of the game. This is the square where both Veryl and Ruth Ann submit $100 and receive a $100 payout. From here, unlike all the other squares, neither player has an incentive to change his or her answer. If Veryl knows that Ruth Ann is going to submit $100, then any choice between $101 and $500 would result in only a $50 payout because he would have to absorb the penalty. So If Ruth Ann submits $100, then Veryl should also submit $100 so he can take home $100. The solution to the game for both players is to submit $100.

Figure 6.5

The traveling hacker's dilemma in normal form (partial game).

Ruth Ann

		500	499	498
	500	500, 500	449, 549	448, 548
Veryl	499	549, 449	499, 499	448, 548
	498	548, 448	548, 448	498, 498

It takes many iterations to start at the top-left square [500, 500] and find the Nash equilibrium of this game. With each iteration, strategies are eliminated. This is an example of the successive elimination of dominated strategies. Eventually, there is only one strategy remaining [100, 100]. This is the game theoretical solution to the game. However, when game theorists conduct studies with actual people and have them play this game, they never choose the Nash equilibrium strategy, and their end result is better than if they had chosen it. In other words, they deviate from what game theory says they should do and are rewarded for it. So is game theory wrong?

There have been critics of analytical game theory from the beginning—those that maintained it was not accurate at predicting what real people would do in many types of strategic situations. The traveling hacker's dilemma is a prime example of a type of game where analytical game theory fails to provide a helpful analysis. For decades researchers have done experiments to observe what real people actually do when faced with strategic scenarios. This approach was initially called *experimental game theory*, but it eventually became known as *behavioral game theory*. It does not supplant traditional, analytical game theory, but it proves that there is an alternative that is more helpful in some contexts. Table 6.5 provides a summary comparison between analytical game theory and behavioral game theory.

Table 6.5

A comparison of analytical and behavioral game theory.

	Analytical	Behavioral
Method	Deductive	Inductive
Approach	Theoretical	Empirical
History	Formalized in the 1940s by von Neumann and Morganstern	Coined in the 2000s but built on experimental game theory
Accurate Predictions	Many repeated-play games	Many one-shot games
Paradigmatic Game	The prisoner's dilemma	The traveler's dilemma
Key Contribution	Nash equilibrium	Level-*k* reasoning

One of behavioral game theory's major contributions is the concept of *level-*k *reasoning*. Level-*k* reasoning is *the process of iteratively thinking about what the other player might do and how to best respond*. Through many experiments it is well established that most people engage in some degree of level-*k* reasoning when faced with making a strategic decision. In level-*k* reasoning, the most obvious, instinctive strategy is the level-0 choice. Expecting your opponent to choose the most obvious strategy is the level-1 choice. Expecting your opponent *to expect you* to choose the most obvious strategy is the level-2 choice, and so on—the levels go on *ad infinitum*. The level-*k* type implicitly assumes his opponent is a level-*(k* - 1) type.

In the traveling hacker's dilemma, Veryl was using level-*k* reasoning when he thought that Ruth Ann may submit $500 (level-0) so he should submit $499 (level-1). But then it occurred to him Ruth Ann may anticipate him doing that and submit $498 (level-2), so Veryl should submit $497 (level-3), etc. A more familiar game that illustrates level-k reasoning is the rock paper scissors hand game. In this game two players simultaneously choose either rock, paper, or scissors. Rock is the level-0 choice because it is the first word in the name of the game and the players' hands are in the rock position as they count down to the draw. Paper is the level-1 choice because paper beats rock, and scissors is the level-2 choice because scissors beats paper. Since there are only three choices, the level-3 choice circles back to rock and so on. A humorous illustration of level-*k* reasoning is portrayed in the film *The Princess Bride*. In the battle of wits scene, the man in black faces off against the evil genius, Vizzini. Vizzini must decide which of two cups is poisoned, and he uses level-k reasoning to an absurd degree as he goes back and forth eliminating each cup as a possibility.

In order to perform level-k reasoning, the level-0 strategy must be clearly defined. In some games, the level-0 strategy is not obvious, so not all games lend themselves to this type of analysis. Additionally, in repeated play games, the level-k strategies change after each play. This is because players learn from one another and their expectations for the level-0 strategy (i.e., the most obvious strategy) are reset. For example, in rock paper scissors, for the first play, scissors is the level-2 strategy, but in subsequent plays, the winning strategy from the previous play arguably becomes the level-0 strategy for the next play.

Experimental results show that most people choose level-0 or level-1 strategies and rarely if ever go beyond the level-3 strategy. They may stop descending because going deeper becomes too confusing or because they assume others will not keep going. Studies show that level-2 and level-3 strategies are the sweet spot because this anticipates what most people are going to do and one-ups them. As is hopefully clear, level-k reasoning is helpful in performing a strategic analysis, but it is not an exact science. It would suggest that in the traveling hacker's dilemma, Veryl should submit either $498 (level-2) or $497 (level-3), but of course, this does not guarantee an optimal outcome. However, it is highly likely that applying level-k reasoning in such a game would result in a much better outcome than the Nash equilibrium solution.

6.3.1 Level-k Reasoning in Security Games

The *hide and seek game* is helpful for thinking about various ways that level-k reasoning can apply to strategic situations.[4] In this game, one player (the *hider*) has hidden money in one of four boxes arranged in a row (see Figure 6.6). The other player (the *seeker*) has one chance to guess where the money is hidden. If the seeker guesses correctly, he wins the money, otherwise the hider gets to keep the money.

Game theory predicts that the seekers will win 25% of the time since there are four choices, and technically, all four choices are equally likely. But experiments show that seekers win 33% of the time. Clearly, the second box changes the dynamics of this game. It illustrates focal point biases—people tend to focus on unusual features. In this game, players immediately focus on the B box because it is different from the other three.

Therefore, it creates a kind of level-0 strategy—it is the obvious place to hide or seek something. So even in a game like this, level-k reasoning plays a role. Game theory researchers have also found that when people are presented with choices laid out in a row they predictably avoid end points and focus on the middle. This paired with the fact that the B box stands out, is what makes the third box the most likely to be chosen next—it is

---

[4] This section draws on findings from the following journal article:
A. Rubinstein, "Experience from a course in game theory: pre- and post-class problem sets as a didactic device," Games and Economic Behavior, vol. 28, iss. 1, pp. 155-170, 1999.

the level-1 strategy. 54% of seekers and 45% of hiders choose the third box. Interestingly, this game also illustrates asymmetry between hiders and seekers—hiders select the right-most box twice as often as seekers (22% to 11%). This box is the level-2 choice in the game.

Figure 6.6

The hide-and-seek game.

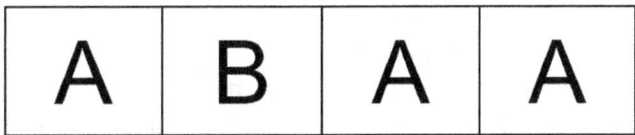

Asymmetry like this naturally exists in security games. *A security game is a game the-oretical game involving an attacker and a defender.* In security games defenders tend to start at level-0 whereas attackers tend to start at level-1. In other words, the attacker's most obvious strategy is actually a level-1 strategy. They start at a more strategic level be-cause they instinctively think about the other player's perspective. Meanwhile, defenders naturally focus on what they are trying to protect instead of the adversary's perspective, and they often make the predictable choice (i.e., the level-0 strategy). Predictable choices are not good for cybersecurity because intelligent adversaries can anticipate them.

The Colonel Blotto Game

At dusk and on opposite sides of a valley, Colonels Alto and Blotto survey the terrain. They know at dawn battles will commence over three distinct strategic positions. They each have nine companies of soldiers, and in the cover of darkness, they must allocate their soldiers to battlefields so that fighting can ensue over each position at first light. Since their forces are equally matched, the battlefields will go to the side that allocates more soldiers. The colonels have no way of knowing what the other side is planning. How many companies of soldiers should Colonel Blotto allocate to each of the three battlefields?

In 1921 French mathematician Émile Borel outlined a strategic contest that later became known as *the Colonel Blotto game*. In the Colonel Blotto game, players allocate soldiers to battlefields, and the player who allocates the most soldiers to a battlefield wins that bat-tlefield. Whether the player allocates one more or one hundred more soldiers, the result is the same. Therefore, it pays to win battlefields by small margins because this frees up more resources to allocate to other battlefields.

The Colonel Blotto game is a fundamental model of scarce resource allocation. Econo-mists have applied it to the analysis of electoral competitions in which the candidates (the colonels) compete over battleground states (battlefields) and must decide how much campaign money (soldiers) to spend on each state. The quantity of each state's electoral votes determines its utility. The candidates' budgets are limited and they do not know ahead of time how much campaign money their opponent has allocated to each state, but the idea is that whichever side allocates the most money to a particular state will win that state. This is clearly a simplifying assumption, but this type of analysis can still provide helpful insight into making strategic choices. There are many variations of the basic Colonel Blotto game, including the numbers of battlefields and soldiers and the values of the battlefields, and there are also different ways to resolve ties, including not awarding the utility to either player or splitting it between the players.

The Colonel Blotto game is important because scarce resource allocation is an everyday phenomenon. It is especially critical in security contexts because, as we know, there is no such thing as 100% security. Protective resources are always limited. Defenders need to allocate their man hours and dollars as efficiently as possible to get the "biggest bang for their buck."

In the specific Colonel Blotto game outlined above (see Figure 6.7), the three battlefields labeled X, Y, and Z, are all equally valuable and each colonel wants to win as many battlefields as possible. Because companies of soldiers cannot be broken up, and each colonel has the same number of companies, it is impossible to win all three battlefields. The best each colonel can hope for is to win two which would result in an overall victory.

Figure 6.7

The Colonel Blotto game.

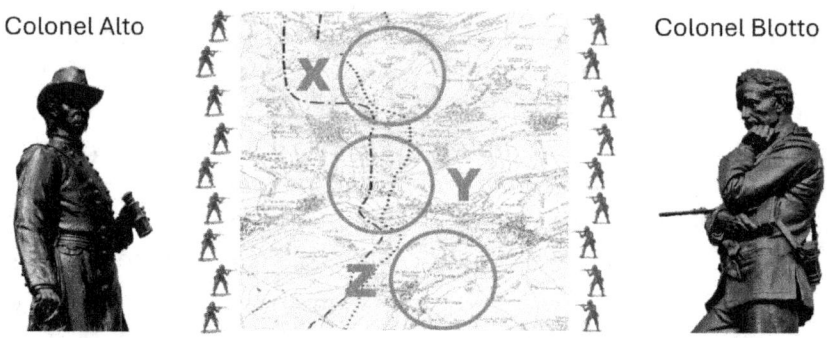

Since there are nine companies and three battlefields, the instinctual strategy is to allocate three companies per battlefield [3, 3, 3]. This is known as the *proportional allocation strategy*. While it is indeed mathematically efficient, it is not very strategic—it is the level-0 strategy in this game. Many different strategies could qualify as level-1 responses to the level-0 strategy, but the most straightforward might be [1, 4, 4]. This strategy loses the first battlefield but wins the second and third for an overall victory. Anticipating this strategy would lead to a level-2 strategy and so on. The Colonel Blotto game involves level-k reasoning in multiple dimensions. How many battlefields should be prioritized? Which battlefields should be prioritized (this involves focal point biases)? How many soldiers should be allocated to "abandoned" battlefields? The Colonel Blotto game sheds light on the complexity of the scarce resource allocation problem. It makes it clear there are no foolproof or simple solutions. It also provides valuable insights because it helps in the rigorous analysis of security scenarios.

6.3.2 Behavioral Game Theory Summary

Behavioral game theory is a better predictor of strategic choices than analytical game theory for many situations because there is a limit to the degree of rationality people apply. Behavior game theory's concept of level-k reasoning is a helpful way to approach strategic contests. Not all strategic contests lend themselves to level-k reasoning because there may not be an obvious, level-0 strategy. In some situations there are multiple dimensions of level-k reasoning in play. Studies show that two or three levels of reasoning performs well in most games because it anticipates the natural strategic choices of others.

6.4 Conclusion

Cybersecurity, at its essence, is an adversarial conflict—without adversaries, there is no such thing as cybersecurity. Therefore, adversarial thinking is the hallmark of the discipline. Furthermore, it is the fundamental skill of cybersecurity—those who excel at it will be prized cyber defenders.

This chapter has shown that adversarial thinking has three distinct components that map to Sternberg's triarchic theory of intelligence. Most of cybersecurity education focuses on the first component: technological capabilities. In order to practice adversarial thinking for cybersecurity, technological capabilities are indeed vital—this levels the playing field between the attackers and the defenders. The second component, unconventional perspectives, AKA the hacker mindset, is also widely acknowledged as important for cybersecurity education. Cyber students are taught creative attack vectors through case studies and labs and are encouraged to practice outside-the-box approaches in capture-the-flag competitions. The third component, strategic reasoning, does not receive as much attention, but it is no less important. Cybersecurity practitioners need to be able to think like a hacker when it comes to planning and strategizing. In an effort to improve the reader's strategic reasoning abilities, this chapter presented some basic game theory concepts from both analytical and behavioral game theory. The biggest takeaway is that one must consider strategic situations from the perspective of the adversary, not primarily from one's own perspective. In cybersecurity, if we get so focused on what we are trying to defend or any one best practice, technology, or tool, that we forget about the human adversary, we do so at our own peril. We must always remember the reason it all exists: hackers! They are intelligent and study their targets, trying to anticipate what the defenders are thinking and how to outwit them.

This chapter has also made it clear that when it comes to complex, real-world cybersecurity scenarios, the point is not to apply game theory to find the solution, as if only one solution exists. Rather strategic reasoning should be the lens that informs all of the day-to-day and practical decisions that must be made. Cybersecurity practitioners need to pause often and consider the perspective of hackers. What do hackers think the defenders are thinking? Strategic reasoning improves our ability to anticipate and thwart the activities of our adversaries. The goal of this chapter in associating cybersecurity with game theory is to produce enduring strategic mindedness, improving cybersecurity practice over the long haul.

 **Chapter 7**

7. The Bedrock of Cybersecurity: Cryptography

"For our scenarios we suppose that A and B (also known as Alice and Bob) are two users of a public-key cryptosystem." - Alice and Bob's introduction to the world in "A Method for Obtaining Digital Signatures and Public-Key Cryptosystems" by Rivest, Shamir, and Adleman

In the real world, valuables like cash, gems, precious metals, and important papers are kept behind lock and key to create a barrier for would-be thieves. Even if the bad guys are able to get physically close to the bounty, they still have to breach the final barrier in order to get their hands on it. Banks purchase huge and expensive steel vaults to protect their cash reserves and safety deposit boxes. Businesses invest in secure rooms with sophisticated access controls. Homeowners buy safes and conceal them in the walls of their houses. In physical space, steel, safes, locks, and keys are synonymous with security.

The analog to locks and keys in cyberspace is cryptography. Unlike physical space valuables, data cannot be discreetly stored and tucked away out of sight. In order for its value to be realized, it must be readily accessible. Physically securing a computer in a locked room is a good idea, but if the computer is online, the data on that computer is still vulnerable. Because of the connectedness of cyberspace, valuable data is tantalizingly close to being compromised constantly. Data of incalculable value swirls around the Internet all the time, passing through untrusted "hands" on the way from one endpoint to another. Cyberspace data cannot be secured by locking it up. Instead it is kept secure by temporarily transforming it into something inscrutable and useless. This is what cryptography does.

The word *cryptography* comes from two Greek words meaning "hidden writing." It is *the art and science of scrambling and unscrambling information using a secret to keep it private*. The goal is that only people possessing some secret knowledge would be able to undo the scrambling and recover the original information. In computer cryptography, the scrambling process is called *encryption* and the unscrambling process is called *decryption*. Encryption turns the original, readable information, *plaintext*, into unreadable *ciphertext*. Encryption and decryption algorithms are parameterized with a *key*. The key is the secret knowledge that is needed to recover the original data (see Figure 7.1).

Figure 7.1

Alice encrypts a message to Bob using their shared key.

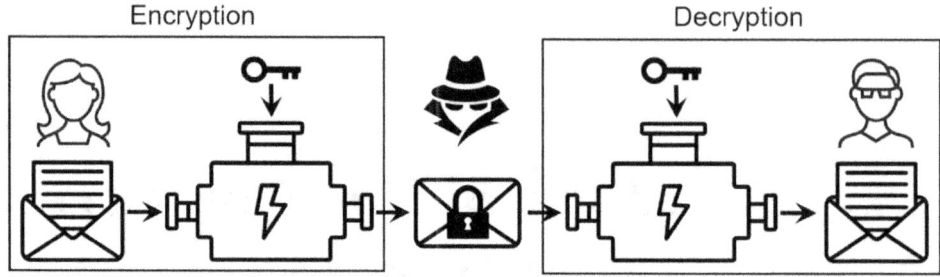

Cybersecurity and cryptography are inextricable, just like locks and keys are synonymous with security in physical space. Cryptography is the bedrock of cybersecurity because it is the main way that confidentiality and integrity are preserved in cyberspace—two of three main CIA triad pillars covered in Chapter 4. This chapter provides an overview of cryptography starting with its history and then exploring how cryptography is implemented in cyberspace.

7.1 Classic Cryptography

> *"Few persons can be made to believe that it is not quite an easy thing to invent a method of secret writing which shall baffle investigation. Yet it may be roundly asserted that human ingenuity cannot concoct a cipher which human ingenuity cannot resolve." - "A Few Words on Secret Writing" by Edgar Allan Poe*

Many cryptographic schemes have been invented throughout human history. This section covers only a few of the major developments. The techniques learned in classical cryptography are directly applicable to computer cryptography in the modern era. In this section we will examine some implementation details of classic cryptography to drive home the fundamental techniques that are used.

7.1.1 Letter Substitution: Ciphers

> *"E however predominates so remarkably that an individual sentence of any length is rarely seen, in which it is not the prevailing character." - "The Gold Bug" by Edgar Allan Poe*

One of the most famous early examples of cryptography is from Julius Caesar. He encoded his military orders using a shift of the Greek alphabet—substituting the original letter with the letter three places forward in the alphabet. In the English language, the *Caesar cipher*, as it became known, would transform the plaintext message *attack at dawn* into the ciphertext *DWWDFNDWGDZQ*.[1] A *cipher* is an encryption scheme that operates on the level of letters, creating mappings from a plaintext alphabet to one or more ciphertext alphabets. Table 7.1 shows the full alphabet mapping for the Caesar cipher. Obviously, Caesar was concerned that his correspondence might be intercepted, and if that were to occur, he did not want the enemy to be able to read his military orders. The axiom "knowledge is power" is never more true than in wartime, so there is a long tradition of militaries employing cryptography in an effort to conceal as much information as possible from the enemy. One of the advantages of the Caesar cipher is that it is easy and inexpensive both in time and resources to use. The scheme is easy to remember and messages can be both encrypted and decrypted quickly. On the other hand, its effectiveness is dubious. One wonders if Caesar's adversaries would have been able to reverse engineer his scheme to recover the plaintext. This is an example of *cryptanalysis—the art and science of decrypting ciphertext when not in possession of the secret key.* It is also called *code cracking*—an attack against ciphertext to reveal the encrypted message.

Table 7.1

Caesar cipher plaintext-to-ciphertext alphabet mapping.

Plaintext Alphabet	a	b	c	d	e	f	g	h	i	j	k	l	m	n	o	p	q	r	s	t	u	v	w	x	y	z
Ciphertext Alphabet	D	E	F	G	H	I	J	K	L	M	N	O	P	Q	R	S	T	U	V	W	X	Y	Z	A	B	C

The Caesar cipher is an example of an *alphabetic shift cipher*. The Caesar cipher uses a shift of three, but in the English alphabet, shifts from [0-25] all produce valid mappings. Because the letters wrap around the end (see how *x* is mapped to *A* in Table 7.1), shifts beyond twenty-five are equivalent to ones from zero to twenty-five (e.g., twenty-six is the same as zero, twenty-seven is the same as one, etc.). In these ciphers, the shift is the key—the secret knowledge that the corresponding parties share. The convention is that encryption shifts forward in the alphabet and decrypting shifts backwards to recover the original text. The cipher known as ROT-13 uses a shift of thirteen. It has the advantage of not having to remember whether encryption shifts forwards or backwards because the result is the same. ROT-13 is sometimes used in Internet forums to conceal spoiler alerts. Readers will not accidentally be exposed to spoilers but those that want to know the information can easily decode the ciphertext.

---

[1] This text follows the convention of using lowercase letters for plaintext and uppercase letters for ciphertext. Also, spaces are removed from ciphertext to conceal word patterns—this makes cryptanalysis more difficult for the adversary, but poses only a minor inconvenience for the decrypting correspondent. These are conventions only and not constraints for real-world cryptography.

In ancient times the Hebrews employed a similar scheme known as the *Atbash cipher*. The Atbash cipher maps letters to the opposite end of the alphabet (see Table 7.2). Twice in the Hebrew Bible, the word *Babylon* is encrypted as the word *Sheshak* (see Jeremiah 25:26 and 51:41). Babylon in Hebrew is *BBL* (*bet bet lamed*), and the letters at the opposite end of the Hebrew alphabet are *SHSHK* (*shin shin kaph*), or *Sheshak*. Reflecting its Hebrew origin, the Atbash cipher got its name from the first two pairs of letter mappings in Hebrew—A (*aleph*) to T (*taw*) and B (*bet*) to SH (*shin*). Using the English alphabet, it would be called the "Azby cipher." In the Atbash cipher, *attack at dawn* maps to *ZGGZX-PZGWZDM*. To decrypt the message, the correspondent finds the letter in the ciphertext alphabet and maps it back to the plaintext letter. Going in the reverse direction undoes the encryption and returns the original plaintext message.

Table 7.2

Atbash cipher plaintext-to-ciphertext alphabet mapping.

Plaintext Alphabet	a	b	c	d	e	f	g	h	i	j	k	l	m	n	o	p	q	r	s	t	u	v	w	x	y	z
Ciphertext Alphabet	Z	Y	X	W	V	U	T	S	R	Q	P	O	N	M	L	K	J	I	H	G	F	E	D	C	B	A

Ciphers are not limited to formulaic mappings like alphabetic shifts or patterns. Any mapping from a plaintext alphabet to a ciphertext alphabet is valid, and there are a mind-boggling number of such mappings. The total number of mappings is the number of permutations of the English alphabet. There are 26! permutations, or over 4×10^{26} unique mappings. Table 7.3 shows one such random permutation. In this cipher, *attack at dawn* would turn into the ciphertext *WOOWJUWOMWDX*. Because it is not based on a simple pattern, the key for this cipher is more difficult to memorize. It is the twenty-six ciphertext alphabet letters in order: *WRJMPVBCTKUFNXAZEQSOHYDLIG*.

Table 7.3

A randomly generated plaintext-to-ciphertext alphabet mapping.

Plaintext Alphabet	a	b	c	d	e	f	g	h	i	j	k	l	m	n	o	p	q	r	s	t	u	v	w	x	y	z
Ciphertext Alphabet	W	R	J	M	P	V	B	C	T	K	U	F	N	X	A	Z	E	Q	S	O	H	Y	D	L	I	G

All of these ciphers are examples of *monoalphabetic substitution ciphers*. A monoalphabetic substitution cipher is *a cipher that uses one plaintext-to-ciphertext alphabet mapping* (*monos* means *single* in Greek). The ciphertext characters are not limited to English letters like in the above examples. Numerical digits or any other symbol, including groups of symbols, can be used. The *pigpen cipher* is an ancient cipher that uses geometric shapes for ciphertext symbols.[2] The key for the pigpen cipher is easy to recreate from memory because it maps to logically laid out grids that resemble pigpens (see Figure 7.2). Sir Arthur Conan Doyle used stick figures for his ciphertext alphabet in his short Sherlock Holmes story *The Adventure of the Dancing Men* (see Figure 7.3).

Figure 7.2

The key for the pigpen cipher.

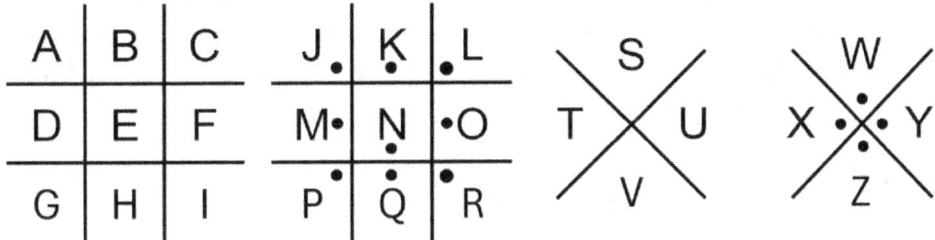

Figure 7.3

Dancing men (top) and pigpen ciphertext (bottom) for "attack at dawn."

An important security property of a cipher is its *keyspace*. The keyspace is *the number of possible keys*. If an adversary has ciphertext and knows or can guess the type of cipher being employed, he can try to decrypt it by guessing the key. This is called a *brute-force key search attack*. A measure of a cipher's strength, therefore, is the amount of time it would take an attacker to guess the key. This is the advertised strength of the cryptographic technique. When designing a cryptographic algorithm, the math for brute-force key search attacks should always be considered since attackers may attempt it. When adversaries use cryptanalysis to find a quicker way to reveal ciphertext messages than a brute-force keyspace attack would take, the cryptographic technique is said to be *broken*, and it should no longer be used.

This formula provides the worst case time for an attacker to perform a brute-force key search attack:

keyspace / keys per second

---

[2] See Figure 3.3 for an example of the pigpen cipher being used (hint: the first word is *defending*).

For example, for the alphabetic shift cipher, which has a keyspace of twenty-six, if we estimate that each key takes a person one minute to check by hand, then the worst case time to crack the ciphertext by hand is:

26 keys / 1/60 keys per second = 1560 seconds

This is twenty-six minutes. This math assumes that the attacker will find the key on his last guess. However, it is just as likely that he will find the key on his first guess. For this reason, to provide a more realistic estimate, we assume that the attacker will succeed after trying half the keys—this is the average case. Therefore, this is the formula used to determine a cipher's key strength:

keyspace / keys per second / 2

For the alphabetic shift cipher given our assumptions, this is:

26 keys / 1/60 keys per second / 2 = 780 seconds

As we saw above, the keyspace for the general monoalphabetic substitution cipher that uses the English alphabet for ciphertext characters is approximately 4×10^{26}. If this cipher were attacked with a supercomputer, we can estimate that it might be possible to test one trillion (10^{12}) keys per second. Therefore, it would take:

4×10^{26} keys / 10^{12} keys per second / 2 = 2×10^{14} seconds

This many seconds is bigger than it looks—it exceeds six million years! Needless to say, the ancient monoalphabetic substitution cipher is secure against a brute-force key search attack even in the modern computer era.

However, the Arabs figured out a way in the 10th century AD to successfully attack the monoalphabetic substitution without resorting to a brute-force key search. For example, all of the ciphertexts for *attack at dawn* above share a pattern regardless of the key. In Figure 7.4, the digits represent unique characters. In these twelve characters of ciphertext, seven unique characters are used, and two characters appear multiple times. What emerges is a pattern, and this pattern can be analyzed to narrow down and possibly even reveal the plaintext message. The ciphertext symbols disguise the original letters, but the distinctive characteristics of the letters still shine through their disguises.

Figure 7.4

Monoalphabetic ciphertext pattern for "attack at dawn."

122134125167

The twenty-six letters of the English alphabet all have a unique personality—some are more "shy" than others and some groups of letters tend to "hang out" together! For example, the letters *E T A O I N S H R D L U* are the most popular letters, in order, in English text (approximately). The letter *E* is by far the most frequently occurring, and for texts a paragraph or more of length, it will easily stand out, even if it is "wearing the disguise" of a different letter or a dancing stick man. Also, some pairs (*bigrams*) and triplets (*trigrams*) of letters appear much more frequently than others. For example, the trigram *THE* is the most frequently occurring three letter combination in English text. The more ciphertext characters, the more the pattern emerges. In our short *attack at dawn* example, the second and third ciphertext letters are the same, and this narrows down the plaintext letter possibilities for this character because some English letters do not occur back-to-back. Taking this paragraph (not counting this sentence) as a representative sample of typical letter frequencies, *E* appears the most (115 occurrences) followed by *T* (93 occurrences).

Claude Shannon, one of the founding fathers of computing, studied the theoretical underpinnings of cryptography in the 1940s as part of the United States World War II effort, and he published a paper that was declassified in 2013. In this paper, Shannon showed that for the English language, it takes about thirty letters of a monoalphabetic substitution ciphertext to uniquely determine a message. The longer the message, the more fodder for the cryptanalysts, and the easier it is to crack the ciphertext. Because it does not sufficiently conceal the pattern of the underlying language, the monoalphabetic substitution cipher is not secure and should never be used to protect valuable information.

The next evolution in classical cryptography was the *polyalphabetic substitution cipher*. In this scheme, multiple ciphertext alphabets are mapped to the plaintext alphabet (*poly* means *many* in Greek). The *Vigenère cipher* is a polyalphabetic substitution cipher that uses a cleverly-designed table to aid in encrypting and decrypting (see Table 7.4). The key in this cipher is the number of ciphertext alphabets used and their order, and it can be remembered with a simple codeword or phrase. For example, the codeword *SECRET*. This key uses six ciphertext alphabets. To encrypt a plaintext message, the *S* ciphertext alphabet encrypts the first letter, *E* the second letter, *C* the third letter, and so on. When the key runs out, it circles back to the beginning, so the seventh letter of ciphertext is also encrypted with the *S* ciphertext alphabet, the eighth with *E*, and so on. In the Vigenère table, the key is used to index the column and the plaintext letter to index the row, and the letter at the intersection is the ciphertext letter. Using the Vigenère cipher to encrypt the message *attack at dawn* with the key *SECRET* produces the ciphertext *SXVRGDSXFRAG*. To check this, note that the second ciphertext letter, *X*, is at the intersection of the *E* column (the second letter in the key) and the *t* row (the second letter in the plaintext) in the table. Also note that the third letter of the plaintext is also a *t*, but the ciphertext for it is the letter *V*, not the letter *X*. This demonstrates that polyalphabetic substitution ciphers do a better job of concealing patterns in the plaintext

Table 7.4

Vigenère table for encoding all possible polyalphabetic substitutions.

	A	B	C	D	E	F	G	H	I	J	K	L	M	N	O	P	Q	R	S	T	U	V	W	X	Y	Z
a	A	B	C	D	E	F	G	H	I	J	K	L	M	N	O	P	Q	R	S	T	U	V	W	X	Y	Z
b	B	C	D	E	F	G	H	I	J	K	L	M	N	O	P	Q	R	S	T	U	V	W	X	Y	Z	A
c	C	D	E	F	G	H	I	J	K	L	M	N	O	P	Q	R	S	T	U	V	W	X	Y	Z	A	B
d	D	E	F	G	H	I	J	K	L	M	N	O	P	Q	R	S	T	U	V	W	X	Y	Z	A	B	C
e	E	F	G	H	I	J	K	L	M	N	O	P	Q	R	S	T	U	V	W	X	Y	Z	A	B	C	D
f	F	G	H	I	J	K	L	M	N	O	P	Q	R	S	T	U	V	W	X	Y	Z	A	B	C	D	E
g	G	H	I	J	K	L	M	N	O	P	Q	R	S	T	U	V	W	X	Y	Z	A	B	C	D	E	F
h	H	I	J	K	L	M	N	O	P	Q	R	S	T	U	V	W	X	Y	Z	A	B	C	D	E	F	G
i	I	J	K	L	M	N	O	P	Q	R	S	T	U	V	W	X	Y	Z	A	B	C	D	E	F	G	H
j	J	K	L	M	N	O	P	Q	R	S	T	U	V	W	X	Y	Z	A	B	C	D	E	F	G	H	I
k	K	L	M	N	O	P	Q	R	S	T	U	V	W	X	Y	Z	A	B	C	D	E	F	G	H	I	J
l	L	M	N	O	P	Q	R	S	T	U	V	W	X	Y	Z	A	B	C	D	E	F	G	H	I	J	K
m	M	N	O	P	Q	R	S	T	U	V	W	X	Y	Z	A	B	C	D	E	F	G	H	I	J	K	L
n	N	O	P	Q	R	S	T	U	V	W	X	Y	Z	A	B	C	D	E	F	G	H	I	J	K	L	M
o	O	P	Q	R	S	T	U	V	W	X	Y	Z	A	B	C	D	E	F	G	H	I	J	K	L	M	N
p	P	Q	R	S	T	U	V	W	X	Y	Z	A	B	C	D	E	F	G	H	I	J	K	L	M	N	O
q	Q	R	S	T	U	V	W	X	Y	Z	A	B	C	D	E	F	G	H	I	J	K	L	M	N	O	P
r	R	S	T	U	V	W	X	Y	Z	A	B	C	D	E	F	G	H	I	J	K	L	M	N	O	P	Q
s	S	T	U	V	W	X	Y	Z	A	B	C	D	E	F	G	H	I	J	K	L	M	N	O	P	Q	R
t	T	U	V	W	X	Y	Z	A	B	C	D	E	F	G	H	I	J	K	L	M	N	O	P	Q	R	S
u	U	V	W	X	Y	Z	A	B	C	D	E	F	G	H	I	J	K	L	M	N	O	P	Q	R	S	T
v	V	W	X	Y	Z	A	B	C	D	E	F	G	H	I	J	K	L	M	N	O	P	Q	R	S	T	U
w	W	X	Y	Z	A	B	C	D	E	F	G	H	I	J	K	L	M	N	O	P	Q	R	S	T	U	V
x	X	Y	Z	A	B	C	D	E	F	G	H	I	J	K	L	M	N	O	P	Q	R	S	T	U	V	W
y	Y	Z	A	B	C	D	E	F	G	H	I	J	K	L	M	N	O	P	Q	R	S	T	U	V	W	X
z	Z	A	B	C	D	E	F	G	H	I	J	K	L	M	N	O	P	Q	R	S	T	U	V	W	X	Y

While this is a much stronger scheme than the monoalphabetic substitution cipher, it is still vulnerable to cryptanalysis because the key is recycled over and over again. Therefore, the underlying plaintext language pattern emerges in the ciphertext, except now it emerges at the periodicity of the key length—it just requires more ciphertext to reveal it. Charles Babbage, the first person to theorize a general purpose computer, wrote about how the Vigenère cipher could be broken through this type of cryptanalysis in 1854. The Vigenère cipher is a stronger cipher than the monoalphabetic substitution cipher, and more ciphertext is needed to recover the plaintext message using cryptanalysis—the longer the key, the more ciphertext is needed.

The next evolution beyond Vigenère-type ciphers was a truly revolutionary breakthrough in cryptography: the *one-time pad*. First described in 1882 but operationalized by *Gilbert Vernam* in 1917, the one-time pad is a polyalphabetic substitution cipher with two special properties: the key is random (i.e., it is not tied to a word like *SECRET)*, and the key is the same length as the plaintext message. Because it encapsulates all possible letter substitutions, the Vigenère table can be used with the one-time pad. Using this scheme, the random key *JHECNISMAXAE* encrypts *attack at dawn* to *JAXCPSSFDXWR*. Unlike the previous examples, *and unlike all other cryptographic schemes ever invented*, this ciphertext contains no pattern whatsoever. Claude Shannon proved that the one-time pad is *information-theoretic secure*. This means that it is impervious to cryptanalysis and is not even susceptible to a brute-force key search attack!

In other words, even if an adversary were to intercept the ciphertext *JAXCPSSFDXWR*, he would learn *no information* about the plaintext message other than the length. From the adversary's perspective, every twelve character plaintext message is equally probable. To see this, assume he tries every key. For a message of this length, there are $26^{12} \approx 9 \times 10^{16}$ possible keys. Given the assumptions we made for the supercomputer above, every key could in fact be attempted in just over a day—that is not the problem. The problem is that the keys will produce every possible twelve character message, not just the actual plaintext message. The output will include the actual plaintext *attack at dawn,* but it will also include countless other valid plaintexts such as *depart at dusk*, and *send supplies*. From this collection of possible plaintexts, the attacker has no way of narrowing them down to the actual plaintext message—there is no basis for eliminating any of the guesses. Therefore, the attacker learns nothing and has wasted his time performing the brute-force key search.

This is not the case for any other cipher. All other ciphertext-key pairings contain a pattern of the underlying message, and for messages of any length, the correct key is the only key that "unlocks" the pattern. All the wrong keys produce nonsensical plaintexts and can be ruled out one-by-one until the correct key is identified.

The great American writer Edgar Allan Poe was also an expert in cryptography. In the quote at the beginning of this chapter, Poe stated that there will never be a cipher that "human ingenuity cannot resolve." While he was a genius of extraordinary talent, the later invention of the one-time pad proved Poe wrong! It is unbreakable. Even in a hypothetical world where an alien race possesses advanced mathematical techniques and technology, and are capable of breaking all of our strongest modern cryptographic systems, the good old one-time pad will still be beyond their grasp!

Since that is the case, it would appear that cryptography has been solved—all people everywhere, including in cyberspace, should always just use one-time pads for encrypting their secret messages. But this is not what has happened. The reason is because cryptographic schemes that look great on paper have to be implemented in the real world, and there are practical issues with implementing the two special properties of the one-time pad scheme. The one-time pad's name comes from a notepad of paper filled with random characters. This is the *keystream* used to encrypt and decrypt messages. A keystream is *a sequence of characters used to encrypt and decrypt messages on a per character basis.* A source of randomness must be used to produce the keystream so that no pattern of any kind can emerge—producing the high number of random characters needed is not trivial. Two identical copies of the notepad must be made—one for the encrypting party and one for the decrypting party—and both must be kept secure at all times. The keystream is consumed quickly because every plaintext letter requires a keystream character, and once a keystream has been used, it must be discarded—this is why it is called a *one-time* pad. So bottomline, one-time pads are difficult to implement.

As a real-world example, the one-time pad encryption scheme was used by the Soviet Union for nuclear espionage purposes in the 1940s. Unfortunately for them, when they ran out of keystream, the Soviet spies reused portions of their one-time pads. This is called using a one-time pad *in-depth*, and it forfeits the provably secure property. It turns the one-time pad into a type of Vigenère cipher, albeit one with a really long key. The *Venona project* was an United States signals counterintelligence operation that collected and analyzed the Soviet's ciphertext correspondence. Due to the misuse of the one-time pad scheme, the team was able to decode many of the Soviet's messages, exposing important details of the spying operation. For even such a vital, sophisticated, and well-financed operation, fully implementing the one-time pad scheme proved too onerous.

7.1.2 Word Substitution: Codebooks

> *"The United States…[made] use of a resource that virtually no other combatant had: pools of tongues so recondite that almost no one else in the world understood them. These were the American Indian languages, which are isolated both geographically and linguistically." - on the United States use of code talkers during both World Wars in* The Codebreakers *by David Kahn*

All of the above cryptographic schemes are ciphers that employ substitution at the level of letters. Another cryptographic scheme called *codebooks* also uses substitution, but it does so at the level of words. In these schemes, two codebooks are created containing plaintext-to-codeword mappings. One book is ordered by the plaintext words and is used for encrypting, and the other is ordered by codewords and is used for decrypting. Unlike with ciphers, for codebook schemes the vocabulary is limited to the words for which mappings exist, so forethought is needed to know which plaintext words to include. Table 7.5 shows a partial codebook that uses random three digit numbers as codewords. In this scheme, *attack at dawn* becomes *915301717*.

Table 7.5

A partial codebook.

Plaintext Codebook

Plaintext	Codeword
at	301
attack	915
dawn	717

Codeword Codebook

Codeword	Plaintext
301	at
717	dawn
915	attack

Codebooks essentially create a new language. A famous example of a codebook actually used a real language: the Navajo Code Talkers. During World War II, English bilingual Navajos used their native language to communicate military messages over the radio waves. Because of the open nature of the radio spectrum their messages were intercepted by enemy forces, but the enemy was never able to successfully decode them. The Navajos substituted some of their native words for modern military terms, such as "buzzard" for "bomber" and "iron fish" for "submarine." Because the structure of their language is unusual and is spoken by so few people, it was an ideal spoken codebook—enemy forces had trouble isolating words and could not identify the language nor translate it. One enormous advantage of this technique was its practicality and efficiency. Code Talkers could be trained easily and messages could be encrypted and decrypted quickly as Navajos translated in their heads in real time.

7.1.3 Letter and Word Transposition

> *"A few weeks earlier, Manton Marble, one of Tilden's closest political advisors, had written an open letter to the* New York Sun *contrasting dark Republican practices with Tilden's station in 'the keen bright sunlight of publicity.' Whitelaw Reid, the* Tribune's *brilliant editor...inserted the cipher telegrams in editorials as subtle commentaries on Marble's letter."* - The Codebreakers *by David Kahn*

Substitution at the level of letters and words is not the only mechanism for creating cryptographic schemes. In addition to substitution, cryptography can also be employed with *transposition*. Transposition is the rearranging of plaintext letters or words to transform them into ciphertext. Table 7.6 shows a *double transposition cipher*. The plaintext message is written out in a template table, filling it from left-to-right and top-to-bottom. Then the columns and rows of the table are rearranged and the result is written out to produce ciphertext. The size of the template table and the permutation of the columns and rows is the key. In the example, the key is *BDCA321* and *attack at dawn* is transformed into *ANWDKTACTATA*. In this scheme, the letters are not disguised, but the underlying language pattern is.

Table 7.6

Example double transposition cipher plaintext-to-ciphertext mapping.

	Plaintext			
	A	B	C	D
1	a	t	t	a
2	c	k	a	t
3	d	a	w	n

	Ciphertext			
	B	D	C	A
3	A	N	W	D
2	K	T	A	C
1	T	A	T	A

In the 1876 United States Presidential election between Republican candidate Rutherford B. Hayes and Democratic candidate Samuel J. Tilden, political operatives used cryptography to protect the confidentiality of messages they sent over telegraph wires. The election turned out to be extraordinarily close and the results were highly contested. In heated rhetoric both the Republican and Democratic parties accused the other side of election interference. Unfortunately for the Democrats, some of their ciphertext telegrams were preserved and acquired by Whitelaw Reid, the Republican editor of the *New York Tribune*, who published them in their encrypted form. Soon, Reid, with the help of his readers, was able to crack the Democrats' ciphers. He published the plaintext under the headline, "The Captured Cipher Telegrams," exposing a major political scandal. One of the cryptographic schemes used by the Democratic operatives was a partial codebook with word transposition. Table 7.7 shows the ciphertext and decrypted plaintext of one telegram. The damning telegram is about bribing an election official for $200,000 in order to get Tilden elected. *Moses* was the codeword for Manton Marble, the Democrat operative that sent the telegram.

Table 7.7

The ciphertext telegram sent by a Democrat operative in the aftermath of the 1876 United States presidential election and its plaintext decryption.

Ciphertext telegram	CERTIFICATE REQUIRED TO MOSES DECISION HAVE LONDON HOUR FOR BOLIVIA OF JUST AND EDINBURGH AT MOSELLE HAND A ANY OVER GLASGOW FRANCE RECEIVED RUSSIA OF
Plaintext decryption	Have just received a proposition to hand over at any hour required Tilden decision of canvassing board and certificate of Governor Stearns for two hundred thousand. Manton Marble.

7.1.4 Combinations

Substitution and transposition are the two basic primitives of every cryptographic scheme, ancient and modern, including computer cryptography. The two techniques can be used separately or in combination, either in an *additive* fashion or as a *product*. Additive means that the two techniques are used together in some proportion in a single scheme. Product means multiple cryptographic schemes are used in sequence. For example, in a product, one cryptographic technique is applied first to transform the plaintext into intermediary ciphertext, and then another is applied to the intermediary ciphertext to transform it into the final ciphertext. Monoalphabetic substitution (MS) can be used with double transposition (DT) to create a stronger cryptographic scheme: MS × DT = MSDT. MSDT would not conceal the single letter frequencies (e.g., the letter E would still be easily identifiable), but it would obscure bigram and trigram frequencies, and it would force cryptanalysts to do more work. MSDT is a new cryptographic scheme that is stronger than either of the individual schemes alone. It also forces the communicating parties to do more work to encrypt and decrypt messages. Cryptographic schemes can always do more substituting and transposing, but it comes at a cost.

The same cryptographic scheme can also be applied multiple times, but the result is not necessarily stronger cryptography. For example, applying the monoalphabetic substitution cipher twice using two different keys produces ciphertext that could have just been directly produced with a third key, i.e., $MS_{KEY-A} \times MS_{KEY-B} = MS_{KEY-C}$. The third key is just a different key in the monoalphabetic substitution keyspace—it is not a "better key." Double MS requires twice as much work encrypting and decrypting messages, but it yields *no extra cryptographic strength*! A fun example of a product of a cryptographic scheme is the dreaded *double ROT-13* cipher. Double ROT-13 = *ROT-13 × ROT-13*. Applying ROT-13 twice yields plaintext! "They must have been using double ROT-13 encryption" is a facetious way of saying that cryptography should have been employed to conceal sensitive information, but to the embarrassment of the communicating parties, it was not and their plaintext messages were exposed.

7.2 Computer Cryptography

"The development of computer controlled communication networks promises effortless and inexpensive contact between people or computers on opposite sides of the world...these contacts must be made secure against both eavesdropping and the injection of illegitimate messages...Contemporary cryptography is unable to meet the requirements." - "New Directions in Cryptography" by Diffie and Hellman

The primary way that confidentiality (preventing the unauthorized reading of messages) is preserved in cyberspace is through computer cryptography.

In the classic cryptography examples above, English letters and words were encrypted to create ciphertext. But computers do not "understand" English. As we learned in Chapter 2, they are restricted to operating on only two types of signals: on and off. We commonly refer to these signals as 1s and 0s, or bits. So, to a computer, *attack at dawn* looks like this:

011000010111010001110100011000010110001101101011001000000110000101110100000100000110010001100001011101110110111

Note that this string of 1s and 0s is not ciphertext—it is a well-known standard for encoding the English alphabet into the binary number system for use by digital computers. The standard used in this case is ASCII. ASCII uses eight bits to represent the twenty-six letters of the English alphabet in addition to numerical digits, punctuation marks, and other symbols. In a similar way, all types of information (not just the English alphabet) can be encoded as bits. This is what makes it possible for computers to process and store photos, videos, programs, music, etc.

The goal of encryption is to make the original information completely unintelligible so that confidentiality can be preserved. This means that an encrypted audio file should be indistinguishable from an encrypted video file, text file, or any other type of file. No matter what type of data is encrypted, the ciphertext is a random-looking string of 1s and 0s. If encrypted data is accessed by an unauthorized person, nothing should be learned about the underlying data. Only people with the correct key should be able to unlock (technically, decrypt) the ciphertext and recover the original data.

Computer cryptography in one sense is extremely simple because it boils down to encrypting only 1s and 0s. It does not matter what the underlying data is or how it is encoded. Just like in classical cryptography, encrypting bits means creating a reversible mapping from one set of bits (the plaintext) to a new set of bits (the ciphertext). And also like classical cryptography, there are only two fundamental techniques available for doing this: substituting 1s and 0s for other 1s and 0s and transposing 1s and 0s.

Computer cryptography is implemented in software (i.e., computer programs) and hardware. There are several components involved in computer cryptography, including algorithms, data encoding, and protocols. For this reason, cryptography is implemented in computers as a *cryptosystem*. A cryptosystem encapsulates the basic cryptographic technique along with all the supporting components needed to implement it. The attack surface of a cryptosystem is much larger than just the basic cryptographic algorithm, and the added complexity of the software and protocols creates additional vulnerabilities. Even if the fundamental cryptography is secure, the implemented cryptosystem may turn out to be unsecure.

7.2.1 Symmetric Key Cryptography

> "'And what is the key for?' the boy would ask. 'What is it the key of? What will it open?' 'That nobody knows,' his aunt would reply. 'He has to find that out.'" - The Golden Key *by George MacDonald*

Prior to 1976, all cryptography involved a secret key shared among the communicating parties. But that year a paper was published by *Whitfield Diffie* and *Martin Hellman* titled "New Directions in Cryptography" that forever bifurcated cryptography. The paper begins with a bold proclamation: "We stand today on the brink of a revolution in cryptography." Indeed, the paper did revolutionize cryptography by introducing a new paradigm to the world: the concept that different keys could be used for encrypting and decrypting messages. The new paradigm became known as *public key cryptography* in order to differentiate it from the traditional approach. The traditional approach of using a shared key also now needed a new name and became known as *symmetric key cryptography*.

All of classic cryptography is symmetric key cryptography because the communicating parties share the same key. In the computer era, public key and symmetric key cryptography coexist. This section examines symmetric key cryptography and the next section examines public key cryptography.

Symmetric key cryptography is based on the foundational techniques covered in the previous section. It is broken down into two main approaches: *stream ciphers* and *block ciphers*.

7.2.1.1 Stream Ciphers

"So the boat was left to drift down the stream as it would, till it glided gently in among the waving rushes." - Through the Looking-Glass *by Lewis Carroll*

Stream ciphers are an implementation in computers of the classic polyalphabetic substitution cipher. Each bit of the plaintext is either flipped (i.e., changed from a 1 to a 0 or vice versa) or kept the same based on a series of ciphertext alphabets. Because there are only two characters in the plaintext and ciphertext alphabets, the Vigenère-like table for stream ciphers is small and simple. Computers accomplish this reversible mapping with the Boolean logic operator, *Exclusive OR* (XOR). XOR takes two bits as input and produces one bit as output. It outputs a 1 when exactly one of the input bits is a 1 and 0 otherwise. Table 7.8 shows the XOR truth table.

Table 7.8

The Exclusive OR truth table used in stream cipher encryption and decryption.

	0	1
0	0	1
1	1	0

Stream ciphers approximate a one-time pad encryption scheme. Computers face the same practical problems of producing and distributing one-time pads that were highlighted above. Instead of a real one-time pad they use a keystream. Importantly, the keystream is not the key. A key is used as input into the keystream generator algorithm to construct the keystream for the communicating parties. The keystream acts as the one-time pad, and it is generated from the key and a keystream algorithm on the fly (see Figure 7.5). This means the communicating parties do not actually have to store a one-time pad like in classical cryptography. They can programmatically generate the one-time pad as needed.

Figure 7.5

A keystream generator takes a short key as input and produces a long stream of pseudo-random bits as output.

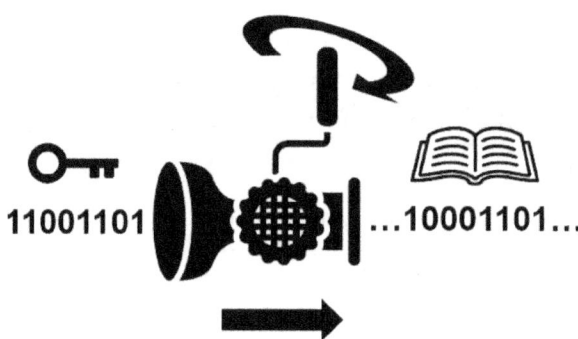

The keystream algorithm is *deterministic*. Deterministic means that the same input produces the same output every time. When the same key is used with the same keystream generator, the first n bits of the keysteam is always the exact same string—there is no source of randomness. Therefore, keystreams do not adhere to the one-time pad requirement of being random. Because stream ciphers trade the provable security of one-time pads for deterministic keystreams, they are sometimes known as *pseudo one-time pads*.

While not technically random, the keystream must have the same properties of a random binary string. This means any given bit has an equal chance of being a 1 or a 0 and for long strings of bits, half will be 1s and half will be 0s. The keystream must also be long since it is consumed quickly, so the algorithm needs to output a practically endless stream of bits. Because the keystream algorithm is a closed system, it is finite, and this means it must eventually repeat. Therefore, another property of a keystream algorithm is that it needs to have a really long cycle. Otherwise, it could be used in-depth, violating a core principle of one-time pads.

Another way that a keystream could be used in depth is to encrypt multiple messages with the same key starting in the same place in the keystream. These messages would re-use the same portion of the keystream. To prevent this from occurring, stream ciphers should use a *nonce* in addition to a key. A nonce is *a randomly generated string of bits also known as a "number used once."* The nonce is mixed with the key in a programmatic way, and this generates a new key to input into the keystream generator. Since the key is new, the result is a new, never-before-seen keystream. When using nonces with stream ciphers, every message is encrypted with a unique "one-time pad," and there is no danger of using it in-depth.

In order to produce the same keystream to decrypt the message, the receiver needs to know the nonce. Conveniently, the nonce does not need to be kept secret! It can be sent in plaintext along with the message. Knowing the nonce does not help the adversary because the nonce must be mixed with the key in order to produce the keystream. Since the adversary does not know the key, he still cannot produce the correct keystream (see Figure 7.6).

Figure 7.6

A cryptosystem using a nonce.

Nonces are used frequently in cryptosystems. They are easy to use and provide enhanced security even though they are not a secret. The "number used once" property means that nonces should never be reused. They are generated randomly, used once, and then thrown out. Keeping a list of used nonces to make sure that new nonces do not match previously generated ones would be prohibitive, and fortunately, it is not necessary. Nonces are chosen at random from a sufficiently large pool of possibilities, that probabilistically, the same one will never be generated more than once. For example, if nonces are sixteen bytes long (128 bits), the total pool of nonces is $2^{128} \approx 10^{38}$. You would need to produce approximately $2^{64} \approx 10^{19}$ nonces in order for a repeat to become probable. Even if you needed one million nonces per second every second, it would take over 300,000 years before you would need to start worrying about a repeat!

When using a stream cipher, the communicating parties input their shared key into a keystream algorithm to produce the keystream. For encryption, the keystream is XORd with the plaintext string to produce ciphertext. For decryption, the keystream is XORd with the ciphertext to produce plaintext (see Figure 7.7).

Figure 7.7

Example stream cipher encryption and decryption.

	Encryption							
	1	1	0	0	0	0	1	[Plaintext]
$\oplus$	0	0	0	1	0	1	0	[Keystream]
	1	1	0	1	0	1	1	[Ciphertext]

	Decryption							
	1	1	0	1	0	1	1	[Ciphertext]
$\oplus$	0	0	0	1	0	1	0	[Keystream]
	1	1	0	0	0	0	1	[Plaintext]

But does XOR encryption really work to preserve the confidentiality of messages? Taking the example in Figure 7.7, assume an adversary observes the ciphertext 1101011. From the adversary's perspective, since he does not know the secret keystream, every possible combination of seven bits of keystream (128 different combinations) is equally likely. By trying all 128 combinations, he will produce 128 different putative plaintexts, including the correct one. However, there is no reliable means by which he can rule out any of the possible plaintexts because no one is more likely than any other—this is the same situation as with the one-time pad. Therefore, confidentiality is preserved for as long as the keystream remains secret.

7.2.1.2 Block Ciphers

"You see, we are like blocks of stone out of which the Sculptor carves the forms of men. The blows of his chisel, which hurt us so much are what make us perfect." - The Problem of Pain *by C.S. Lewis*

Whereas stream ciphers are an implementation of a polyalphabetic substitution cipher, block ciphers are like codebooks. Block ciphers operate on a fixed size set, or *block*, of plaintext bits at a time known as a *plaintext block*. Via substitution and transposition, they produce a new set of encrypted bits of the same size known as a *ciphertext block*. Plaintext blocks can be thought of as plaintext words and ciphertext blocks as codewords.

Unlike classic crypto, however, there is no need to actually compile and keep codebooks to map plaintext words to codewords and vice versa. In computer cryptography, the "codewords" are produced algorithmically on the fly, just like stream ciphers produce "one-time pads" on the fly. Unique codewords are produced based on the cryptosystem and the key shared by the communicating parties. The encryption algorithm produces the plaintext word-to-codeword mapping, and the decryption algorithm produces the codeword-to-plaintext word mapping. No lookups are necessary—the process is entirely algorithmic.

Block cipher algorithms transpose and substitute the plaintext bits into a completely different and unrecognizable set of ciphertext bits. If two plaintext blocks are encrypted with the same algorithm and the same key and differ by only a single bit, the resulting ciphertext blocks will be completely different. The relationship between two ciphertext blocks should reveal no information about the relationship between the corresponding plaintext blocks—such is the degree of scrambling that must be performed.

Because the decryption process must recover the original plaintext bits, the encryption process is *lossless*. This means that none of the original information is lost—it is all there but in an unrecognizable form. This is a crucial property of a cryptographic algorithm because the plaintext must be 100% recoverable by the decryption algorithm.

The standard cipher recommended in the United States since the early 2000s is called *AES (Advanced Encryption Standard)*. AES is a block cipher that uses either a 128, 192, or 256 bit key. Figure 7.8 shows the string *attack at dawn* being encrypted and decrypted with AES 256 using the password *SECRET*.

Figure 7.8

Using OpenSSL to encrypt and decrypt a message.

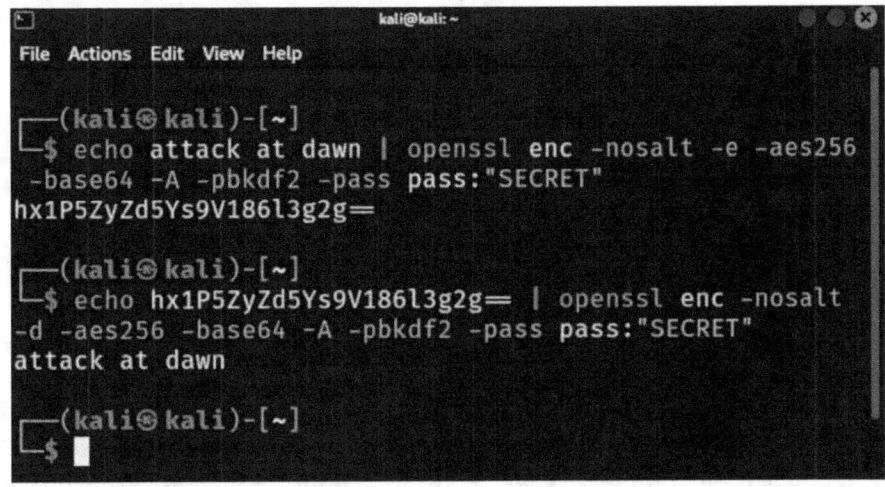

Block sizes can technically be any length, but AES uses a 128 bit block size. To encrypt, the input is chunked into 128 bit blocks. The first block of the plaintext (P_1) is fed into the encryption algorithm along with the key to produce the first ciphertext block (C_1). Then the next block (P_2) is transformed into C_2, and so on until all of the plaintext blocks are consumed. As many ciphertext blocks as are needed to encrypt the plaintext will be produced by this process. The last block of plaintext is almost always a partial block. This is because plaintexts can be any length, and the chance of them being a multiple of the blocksize is low. A *padding* protocol must be defined to deal with partial blocks. Padding is filler bits that the cryptosystem adds to the final plaintext block to make it exactly the block size. The padding protocol defines the scheme for adding padding in such a way that the padded bits can be positively identified and stripped away by the decryption process. Adding and stripping away padding is performed by block cipher cryptosystems under the hood—users never see the padding.

Like keystream algorithms, block cipher encryption is deterministic. Given the same encryption algorithm and the same key, the same plaintext block will always produce the same ciphertext block. This is a weakness because it reveals information about the underlying plaintext. If an adversary observes the same ciphertext blocks, he knows the corresponding plaintext blocks are also the same. This is a serious issue if a plaintext file has long sections of repeated bits. Figure 7.9 shows a plaintext image file of a black and white treasure map and the resulting ciphertext.[3] Although it is not quite as clear, the path to the X that marks the spot of the hidden treasure can be easily identified in the encrypted file! The problem is not with the encryption algorithm or the key—both are industry strength—but with the fact that the image has many repeated sections of white

---

[3] In Figures 7.9 and 7.10, the plaintext image header was copied to the encrypted file so the encrypted image could be rendered—no other data was modified.

and black bits and they all produce the same shade of ciphertext bits that preserve the underlying image.

Figure 7.9

A treasure map encrypted with a block cipher with no block chaining.

To fix this weakness block ciphers use a technique called *block chaining*. Block chaining incorporates the previous block into the encryption of each subsequent block. Because there is no previous block for P_1, block ciphers employ an *initialization vector (IV)*. IVs are like nonces. The IV is randomly generated by the sending party and sent in the clear along with the ciphertext blocks—it is not a secret value. When using block chaining, even if the same key is used to encrypt the same plaintext, as long as the IV's are different, all the resulting ciphertext blocks will be different. This is because the IV is mixed into the encryption of P_1, and P_1 is mixed into the encryption of P_2, and P_2 is mixed into the encryption of P_3, etc. In this way, the IV permeates all the way through and impacts all of the subsequent blocks from P_1 to P_n. Figure 7.10 shows the same Figure 7.9 image encrypted with the same algorithm and key, but this time block chaining is used with an IV, and now the treasure is clearly safe from prying eyes!

Figure 7.10

A treasure map encrypted with a block cipher using block chaining.

7.2.2 Public Key Cryptography

"I walked downstairs to get a Coke, and almost forgot about the idea. I remembered I had been thinking about something interesting, but couldn't quite recall what it was. Then it came back in a real adrenaline rush of excitement." - Whitfield Diffie recalling the moment he conceptualized public key cryptography in The Code Book *by Simon Singh*

Symmetric key cryptography relies on the communicating parties sharing a secret, the key. Obtaining the shared key is a significant hurdle because it means that for people to communicate securely, they must first have communicated securely to exchange the key. This is known as the *key distribution problem*. Keys must be generated and securely distributed and this is not easy to do. Ironically, cryptography is only needed when people cannot communicate securely, and that is why key distribution is a serious issue.

Key management is also difficult. A unique key is needed for every communicating pair. The formula to calculate the number of keys needed for n people to communicate with one another on a one-on-one basis is:

$$(n \times (n - 1)) / 2$$

This means five people would need ten keys, fifty people would need 1,225 keys, and 500 people would need 124,750 keys. Clearly, this does not scale well.

For communicating over the Internet, key distribution and key management are deal breakers—there is no good way to distribute and manage the needed shared keys. Fortunately, around the same time Internet technology was maturing in the 1970s, Diffie and Hellman discovered public key cryptography. In their seminal paper, they explained a new way to perform cryptography but were unable to provide an implementation. They introduced the concept of two keys that were inverses of one another. What one encrypted, the other could decrypt. They stated that it should be easy to create a key pair but infeasible given one of the keys to calculate its inverse. Thus, the concept of public key cryptography was born. Within a year, three researchers from the Massachusetts Institute of Technology, *Ron Rivest*, *Adi Shamir*, and *Leonard Adleman*, inspired by Diffie and Hellman's paper, devised a practical scheme to actually implement public key cryptography. This scheme became known as *RSA* after the last initials of the three researchers. RSA is based on number theory and utilizes mathematical equations that are easy to calculate in one direction, but are extremely time consuming to calculate in the reverse direction.

RSA is entirely mathematical. In physical space it would seem weird to perform math on messages—what is "attack at dawn" squared? But in cyberspace, this is completely natural because all data is a string of 1s and 0s, and any message or file can be treated as a sequence of base two numbers and used in mathematical operations.

RSA is elegant and simple to implement in software. Anybody can quickly generate a key pair on the privacy of their own computer (see Figure 7.11). Once they have a key pair, they can widely advertise one of the keys, the *public key*, and keep the other key, the *private key*, secret (see Figure 7.12). Files encrypted with the public key can only be decrypted with the private key. One physical space picture of this would be a lock with two keyholes for two different keys—one for locking and the other for unlocking. In cyber-space, plaintext messages are transformed into ciphertext by a mathematical operation parameterized by the public key. The private key is applied to the ciphertext in the same way, and since it is the public key's inverse, it cancels out the public key transformation and reveals the message. It is like multiplying x by seven and then one seventh—multiplying by one seventh reverses the change caused by the seven, leaving x. But unlike this example, in RSA, the math is more complex and the public key does not immediately give away the private key. It can be calculated, but not before the sun dies out.

Figure 7.11

Using OpenSSL to generate an RSA key pair.

Figure 7.12

Bob generates a key pair, keeps the private key secret, and advertises the public key to the world.

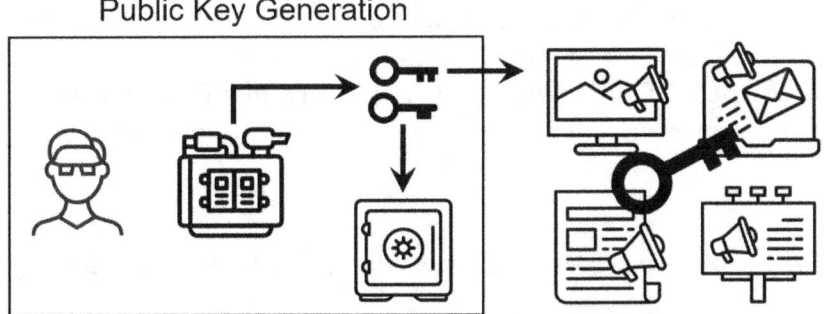

Public Key Generation

If Alice has a message that she would like to send securely to Bob using public key cryptography, she would encrypt her message with Bob's public key. We assume that she has or can obtain Bob's public key even if they have never met because public keys are publicly available. After she encrypts her message with Bob's public key, the only way to decrypt it is with Bob's private key. Public key cryptography assumes that private keys are secure, so this means nobody other than Bob can decrypt the message, not even Alice. After Alice's message is "locked" with Bob's public key, only Bob's private key can "unlock" it (see Figure 7.13). An adversary that observes the encrypted message and has the key it was encrypted with (i.e., Bob's public key), would still be unable to decrypt it.

Figure 7.13

Alice encrypts a message to Bob using Bob's public key and Bob decrypts it with his private key.

Encryption   Decryption

7.2.2.1 Message Signing

> *"'If you didn't sign it,' said the King, 'that only makes the matter worse. You must have meant some mischief, or else you'd have signed your name like an honest man.'" - Alice's Adventures in Wonderland by Lewis Carroll*

Because public and private keys are inverses, it is also possible for a person to apply their private key to a message. This results in message *signing*. Signing a message does not keep the message confidential, since the public key needed to "unlock" the message is not a secret, but it does mean that if the message is successfully "unlocked" with a person's public key, then it must have been signed by that person's private key. This provides authentication in the same way written signatures do in physical space. If Bob has a message that he wants to prove he wrote, he can sign the message with his private key and then make the signed message public. At this point, anybody can verify that Bob produced the message by applying Bob's public key to it (see Figure 7.14). Signing also provides *non-repudiation*. Since we assume that the only person with access to Bob's private key is Bob, then Bob *must have* signed it and cannot deny doing so. Of course, in a court of law, this may or may not pass muster, since Bob's private key could have been compromised.

Figure 7.14

Bob signs a message with his private key that anyone can verify with his public key.

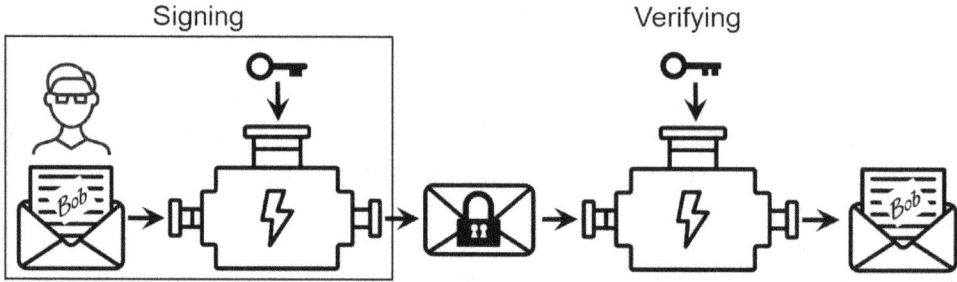

With public key cryptography, the communicating parties do not need to meet ahead of time in a secure environment in order to exchange a shared secret. People can communicate securely even if they have never met since public keys are publicly available. This solves the key distribution problem that plagues symmetric key cryptography, and it makes the modern Internet possible. On the Internet, public key cryptography is used constantly to secure network communications. Email messages, personal records, passwords, and financial transactions propagate between endpoints over untrusted channels but still remain confidential due to public key cryptography.

7.2.2.2 Key Management and Efficiency

> *"A public-key cryptosystem needs no private couriers; the keys can be distributed over the insecure communications channel." - "A Method for Obtaining Digital Signatures and Public-Key Cryptosystems" by Rivest, Shamir, and Adleman*

Public key cryptography key management also scales well. The formula to calculate the number of keys needed for n people to communicate with one another on a one-on-one basis is:

$$n \times 2$$

This is because each person needs their own key pair and that is it! Five people would need ten keys, fifty people would need 100 keys, and 500 people would need 1,000 keys. Symmetric key cryptography is quadratic (n^2) in the number of keys required and public key cryptography is linear ($2n$).

One downside of public key cryptography is that it is not nearly as efficient as symmetric key cryptography—it takes much longer to encrypt and decrypt messages. For example, symmetric keys are typically either 128, 192, and 256 bits long, but public keys need to be 1,024, 2,048, and 4,096 bits long to achieve similar levels of cryptographic strength, respectively. This is a humongous difference and translates into considerably more computer processing. Therefore, public key cryptography is typically limited to exchanging short symmetric keys. If Alice wants to send a secret message to Bob, she first generates a random symmetric key, encrypts it with Bob's public key, and sends it to him. Then, she encrypts the actual message using symmetric key cryptography with the new secret key she just generated. Bob first decrypts the secret symmetric key with his private key, and then uses the secret key to decrypt Alice's message. This creates a best of both worlds scenario: public key cryptography is used to distribute the symmetric key and the more efficient symmetric key cryptography is used for encrypting and decrypting the message.

7.2.2.3 The Public Key Infrastructure

Public key cryptography solves the key distribution problem that plagues symmetric key cryptography, but it has its own significant weakness. Since public key cryptography does not assume that the communicating parties have ever met, how can they be sure they are communicating with the correct party? This is an authentication problem. For example, in Figure 7.12, it is assumed that Alice has Bob's public key. On the one hand, this should not be an issue because Bob's public key is public and can be posted online or sent to Alice. But, on the other hand, how does Alice know for sure that she actually has Bob's public key and not someone else's? What if Mallory sent Alice her public key and claimed that it was Bob's public key? How would Alice know the difference between Mallory's public key and Bob's? This is called the *key binding problem*. The key binding problem involves tying an identity to a public key. If it is not addressed, public key cryptography is vulnerable to impersonation and *man-in-the-middle attacks*. In a man-in-the-middle attack, a malicious actor undetectably intercepts all the messages sent in both directions between the communicating parties. As an example, when Bob sends Alice his public key over an untrusted channel, Mallory intercepts Bob's public key and forwards her own public key to Alice instead, claiming it is Bob's. Alice accepts Mallory's key as Bob's, and encrypts her message to Bob using Mallory's public key and sends her message to Bob. Mallory intercepts the encrypted return message, and since she has the corresponding private key, decrypts it, reads it, and potentially modifies it before re-encrypting it using Bob's public key and forwarding it to him. Bob decrypts the message with his private key not realizing it was actually sent to him by Mallory, not Alice.

The key binding problem is solved on the Internet with a system called the *public key infrastructure (PKI)*. The PKI is *a trust hierarchy that uses signed certificates to bind keys to real-life identities*. In order to bind a public key to a real-life identity, public keys are presented to a *certificate authority (CA)*. A CA is *a third-party that is trusted to verify real-life identities*. Once the CA verifies the identity, it uses its private key to sign a *public key certificate* tying the identity to the public key. As long as the CA is trusted, then the public key in the certificate can be trusted as genuine. The man-in-the-middle attack outlined above worked because Alice did not demand a public key certificate from Bob. If she had, Mallory would not have been able to impersonate Bob because she would not have been able to obtain a valid certificate binding her public key (Mallory's) to Bob's identity. Mallory could get a certificate binding her public key to her own identity, but if she sent that certificate to Alice, then Alice could easily verify that it was not Bob's public key. On the other hand, if Mallory sent Bob's actual public key certificate to Alice, then Alice would accept it as valid, and use the public key on the certificate to encrypt her message. But this would foil Mallory's attack because she would not be able to decrypt the return message since she does not have access to Bob's private key.

7.2.2.4 RSA and Quantum Computing

As alluded to in Chapter 2, quantum computing algorithms have been devised that on paper will break RSA. In other words, a person in possession of a quantum computer would be able to read any text encrypted with RSA and would be able to fraudulently sign messages. This poses a serious risk to the Internet as we know it. Because no quantum computers yet exist, it is not certain that these algorithms will work. It is also not certain that a quantum computer big enough to perform these calculations will ever be built, but because the potential exists, at the time of this writing, work is ongoing to transition to post-quantum cryptography. The hope is that new algorithms will be implemented that have the same security properties as RSA, but that will not be vulnerable to the threat of quantum computers.

7.2.3 Cryptographic Hashing

Symmetric and public key cryptography are concerned with preserving confidentiality in cyberspace—keeping secret information secret. Data integrity—preventing, or at least detecting, the unauthorized writing of data—is achieved through a closely related technique called *cryptographic hashing*. Cryptographic hashing is a form of cryptography that creates digital fingerprints for data objects. The data object could be anything including a simple text string, a document, an executable, a hard drive—any digital information whatsoever. *Hash functions* take data as input and output a *hash* of the data. A hash is *a short, fixed-length binary string that uniquely represents a data object*. Hash functions are deterministic: they always produce the same hash given the same input data. Table 7.9 shows multiple different hash function outputs for the data input string *attack at dawn*. Since the early 2000s, *SHA-256 (Secure Hashing Algorithm 2 - 256 bit output)* has been the standard recommended hash function in the United States.

Table 7.9

Hashes in hexadecimal format.

Hash Function	String	Hash in Hexadecimal	Bits
MD5	attack at dawn	`fa961e42a869e6e045cae5f9fd20cadc`	128
SHA-1	attack at dawn	`e91fbe6fe58c9c0a57b26f1b82577769` `afdaf960`	160
SHA-256	attack at dawn	`d502810c71aeb17e5ea1cbf930b46b87` `bb645a75df45f500230d061992aeb90a`	256

7.2.3.1 Hash Properties

In physical space a fingerprint "stands in" for a person—even though it is just a small fragment, it uniquely identifies a person. In the same way, a hash uniquely identifies a data object. Like a fingerprint, hashes are *lossy*. As they crunch data, they lose information. Because hash functions are lossy, there is no way to go from a hash back to the original data object. Symmetric and public key encryption are both *lossless* because the original data must be recoverable, but this is not true of hash functions. This helps in making them much more efficient than the encryption process. Efficiency is important for hash functions since they need to be able to quickly process large data inputs. Unlike ciphertext, there is no way to reverse engineer or cryptanalyze hashes to recover the original data—this is a feature of hashes called the *one-way property*.

Hashes also must be unique. There cannot be two different data objects that have the same hash. This property is called *collision resistance*. It is mind-boggling that no *hash collisions* could ever be found (i.e., different data objects having the same hash). After all, hash strings are relatively short and there are an infinite number of data objects. But as we saw above with key lengths, powers of two grow quickly, and the number of "hash buckets" for a modern day hash function is practically infinite. Below we explore the math for the collision resistance property of hashes.

7.2.3.2 Data and Message Integrity

These properties of hash functions, one-way and collision resistance, have surprisingly many uses. One use is data integrity. For example, the hash for an executable can be stored in a secure environment, and the program can be periodically hashed to see if it produces the same hash string as the one on file. If it does, the program has not been tampered with, but if it does not, one can know definitively that the executable has been modified in some way—this is due to the collision resistance property of hash functions. The hash would not reveal the exact modification, but it would *detect* that a modification of some sort has occurred. In some instances, this would mean that the program can no longer be trusted. Table 7.10 shows hash results for slightly different strings. The first and second string are 112 bits long and differ by only one bit—the third bit is a 1 in the first string and a 0 in the second. The other 111 bits are identical. The hashes of these strings are 256 bits, and they differ by 124 bits—a difference of almost 50% of the bits.[4] A one bit difference in inputs produced nearly a 50% difference in outputs! On average, any two hashes will be approximately 50% different, and therefore, they provide no information about the data they represent. When examining two hashes, one cannot conclude *anything* about the relationship between the two data objects they represent—not the relative sizes of the data, the types of data, nor their degree of similarity.

Table 7.10

Hashes for similar strings showing the number of bits in the hash that are different.

Hash Function	String	Hash in Hexadecimal (256 bits)	Hamming Distance
SHA-256	attack at dawn	d502810c71aeb17e5ea1cbf930b46b87 bb645a75df45f500230d061992aeb90a	–
SHA-256	Attack at dawn	fa43ee5373a7f36d491be9e57617eae1 8ad62f3f85d5ceef78ee39fe00ce5eeb	124 bits
SHA-256	attack at dawn.	9156781e12f9522c7e8c5aef869a43a4 95be33b3f42c8c8fe9206dd7e2458b39	120 bits
SHA-256	attackatdawn	2503a8d58acc095633918ce909e24309 255972539d2d93319c13c29f3024d960	145 bits

Hashes can also be used for message integrity assuming that the communicating parties have a shared secret. For example, if Alice wants to send Bob a message and does not care about keeping the message confidential, but does care about message integrity (i.e., ensuring that any modifications to her message will be detected), she can send her message to Bob along with a *HMAC (hashed message authentication code)* of the message. An HMAC is *the hash of a message that has been combined with a shared secret.* When Bob receives the message, he can use the shared secret to generate an HMAC in the same way Alice did and compare his to the one that he received from Alice. In other words, he redoes the same calculation as Alice and verifies his results are the same. If the two HMACs match, then Bob knows the message has not been tampered with. If an adversary intercepts Alice's message and changes it, then the HMAC that Bob will generate will no longer match the HMAC he received. The shared secret is vital in this scheme because the adversary also has the ability to modify the HMAC. However, since the assumption is that the adversary does not know the shared secret, then he will not be able to combine it with the modified message to create a valid HMAC, and this will expose his tampering.

7.2.3.3 Committing to a Secret Value

> *"hash: ZghOT0eRm4U9s password: p/q2-q4! hash: gfVwhuAMF0Trw password: dmac" - the hashes and cracked passwords for Ken Thompson and Dennis Ritchie, respectively, on an early UNIX system, cracked decades later*

---

[4] The difference between two equal length bit strings is called the Hamming distance.

Hashes are also useful for committing to a secret value without revealing the value. This is why hashes are used in password-based authentication systems. It is a bad practice to store users' passwords, but this creates a dilemma because in order for a computer to authenticate a user with a password, the computer has to verify that the user knows the password. But how can a computer verify that a user knows his password if the computer itself does not know the password? This can be accomplished with hashes. Instead of storing passwords, computers store the hashes of passwords. Because hashes are one-way, they do not reveal the passwords they represent. When a user is authenticated, he inputs his password and the computer hashes the password and compares this hash to the hash stored on file. If the hashes match, then due to the collision resistance property of hash functions, the computer verifies that the user knows the password. When the computer stores a user's password hash, this commits the user to a specific secret value—no other value has the same hash. Only the user's password will produce this hash. Committing to secret values has many other uses as well, including in auctions and negotiations where no party wants to reveal their offer first, fearing they will be one-upped by the other parties. In this environment, the hashes of offers can be revealed with no danger of exposing the actual offers, yet the hashes still commit the parties to their offers.

Hashes cannot be reversed due to the one-way property. This means that there is no way to determine what data produced a hash just given a hash. However, hashes *are* vulnerable to *forward search attacks*. In a forward search attack, hashes are calculated for putative strings to find the string that produced the hash—in other words, instead of working backwards from the hash to the string (which is not possible), the attack works *forward* from putative strings to the hash. This is how password cracking attacks work. If a user's password hash is known, and it is suspected the person used a word from the dictionary as his password, then an adversary can hash every word in the dictionary and compare the hashes to the password hash. If he finds a match, then he knows the user's password.

To thwart forward search attacks, when hashes are used to commit to a secret value, the secret value should be mixed with random data. Linux operating systems do this by adding *salt* to a user's password before hashing it. Salt is *a short random string that is combined with a user's password before it is hashed.* Unlike the shared secret used in HMACs, salt values are stored in plaintext. They are generated when users create their passwords. When a user inputs a password to be authenticated (more on this in Chapter 8), the user's salt value is retrieved and mixed with the inputted string before it is hashed. This does not prevent forward search attacks on Linux password hashes, but it forces attackers to commit to a unique salt for every cracking attempt. This means all of the putative passwords an attacker hashes are only valid for one password, and all their work would need to be redone for every subsequent hash the attacker wants to crack. Windows operating systems do not use salt, so a single forward search attack is valid for *every* Windows password hash—this is a huge reward for the effort. Needless to say, massive files are available online containing password-hash combinations for Windows passwords—this makes "cracking" them easy because the work is already done!

Technically, forward searches are always a possible attack vector, just like brute-force key search attacks with cryptosystems, but they can be mathematically eliminated as a serious threat—more on password cracking in Chapter 9.

7.2.3.4 Blockchains and Hash Chains

"What is needed is an electronic payment system based on cryptographic proof instead of trust, allowing any two willing parties to transact directly with each other without the need for a trusted third party." - "Bitcoin: A Peer-to-Peer Electronic Cash System" by Satoshi Nakamoto

Another significant use of hashes is with *blockchains*. A blockchain is *a technology used to create a trusted public record in a low-trust environment. Bitcoin is a cryptocurrency that is built on a blockchain.* Each block in the blockchain contains transactions that are recorded on a shared ledger. Blocks are added regularly—in the case of Bitcoin, a new block is added approximately every ten minutes. Every block is hashed, and its hash value is incorporated into the next block, which is also hashed, thereby chaining the blocks together. In this way every block is tied to the previous block going all the way back to the *genesis block* (the very first block in a blockchain).

If a transaction were changed in any block, it would change the hash of that block, and since the original hash value was incorporated into the next block, it would change the hash of *every* subsequent block. To prevent an untrusted party from changing a transaction in an old block, generating a new hash for that block, and then updating the hash of every subsequent block to maintain the integrity of the chain, Bitcoin incorporates a *proof of work*—another useful application of hash functions. Because hash functions produce unpredictable hash values, there is a 50% chance that the first bit in a hash will be a zero. This is just like when flipping a fair coin, there is a 50% chance it will come up heads. The rule for Bitcoin is that the hash for every block must begin with a string of consecutive zeros of a certain length. A random value, similar to a nonce, is added to blocks so this property can be achieved. Random values are generated and tried until the hash of the block begins with the requisite number of zeros. Probabilistically, for a string of zeros of length n, it would take on the order of 2^n guesses to produce a conforming hash—it is exponential in the number of zero bits. This is the same math for repeatedly flipping a coin until it comes up heads n times in a row—it will eventually happen but will take approximately 2^n tries (i.e., to get ten heads in a row would take around 1,024 tries). Therefore, when a hash is produced that has this property, it demonstrates that the party that generated the hash spent time doing so. This is the proof of work, and it prevents parties from quickly changing blocks. In fact, the chain grows at the same rate that conforming hashes are produced, so changing a transaction in an old block and then working to catch back up is impossible.[5]

This is similar to a technique used to defend against password cracking called *key stretching*. In key stretching, a string is hashed, and the resulting hash is hashed, and so on, creating a *hash chain*. When key stretching is used for password hashes, a hash chain may be 10,000 links long or more, and the last hash in the chain is the one stored in the authentication database. The consequence of this is that when a user inputs a password to be authenticated, the input string must be hashed the chain length number of times before it can be determined if the user entered the correct password. For a hash chain of length 10,000, this requires 10,000 times the effort to authenticate each login attempt! However, hash functions are efficient, and as we know, computers are blazingly fast, so the slowdown is not even noticed by the user—maybe it is the difference between one microsecond and one hundredth of a second—in human time, authentication still occurs "instantaneously." But why is key stretching used? Key stretching slows down password cracking attempts by a factor of the length of the hash chain. Attackers generate a multitude of putative passwords (i.e., possible passwords) to crack a password hash (see Section 9.2.1.1). Suppose attackers can generate one trillion hashes in one day, and the average password hash takes one trillion hash attempts to crack. Assuming this math, attackers can crack a password in a day. However, if this hash were key stretched with length 10,000, then it would require the attackers to generate 10,000 times one trillion hashes to crack the password, and this would take them 10,000 days (more than twenty-seven years)! The extra work required for key stretching imposes an asymmetrical penalty on attackers because of the volume of hashes they need to generate.

7.2.3.5 Keystream Generator
Lastly, hashes can be used as a keystream generator. If two parties have a shared secret, they can use the hash of the secret as a keystream and XOR their plaintext message with it to produce ciphertext. Since hashes are relatively short and keystreams need to be long, an index number can be appended to their shared secret to produce as many keystream bits as necessary. A simple scheme is to increment the secret, e.g., SHA-256("SECRET"), SHA-256("SECRET1"), SHA-256("SECRET2"), etc. Table 7.11 shows how a hash function could be used to encrypt the message *attack at dawn* using the word *SECRET* as the shared secret.

7.2.3.6 Attacks on Hash Functions
There are two main attacks that can be performed on hash functions to break their collision resistance property. The first is a brute-force attack. A hash collision is guaranteed if "hash buckets plus one" hashes are generated due to the pigeonhole principle. This is the worst case scenario because it assumes that every new hash generated will fall into an empty bucket, and then once all the buckets are filled with exactly one hash, the next

---

[5] Technically, if somebody commandeered more than 50% of the computational power of the Bitcoin network, it would be possible to revise a block and catch back up.

hash generated will result in the first collision. But probabilistically, a collision will be found much sooner than this, and that is the case that needs to be used to calculate brute-force attack resistance. This is the same idea as the math used for cracking cryptographic keys—we want to focus on the average case.

Table 7.11

Using a hash function as a keystream generator for a stream cipher.

	Input	Binary	Notes
Plaintext	attack at dawn	0110000101110100011101000110 0001011000110110101100100000 0110000101110100001000000110 0100011000010111011101101110	8-bit ASCII, 112 bits
Keystream	SHA-256("SECRET")	0000100100010111101100010011 1010100100001001000110010001 0101110101010100101101100011 0011011011110100010110010000	first 112 bits of the hash
Ciphertext	Plaintext ⊕ Keystream	0110100001100011110001010101 1011111100111111101010110001 0011110000100000100101100101 0111000011100011001011111110	112 bits

A typical hash length for a modern hash function is 256 bits. 256 bits create 2^{256} different hash buckets. How many hashes would need to be generated to make finding a collision likely? We are looking for any two values that have the same hash. To make this exercise more accessible, we can ask the same type of question in a more familiar way: how many people need to be in a room to make it likely that any two have the same birthday? (same day and month, not year) There are 366 possible "birthday buckets" (including leap year babies), so the worst case would be 367 people. But on average, we would only need twenty-three people in order to have a better than half chance of finding a birthday collision! This number is much smaller than most people assume, and that is why this is called the *birthday paradox*. The reason it works is because twenty-three people yield 253 birthday comparisons:

$$23 \times 22 / 2 = 253$$

Every time a person is added, his birthday gets to be compared to *all* the previous persons' birthdays. So the second person's birthday is compared with the first person's birthday, the third person's birthday is compared with the first and second persons' birthdays, and so on, until the twenty-third person's birthday is compared with all twenty-two previous persons' birthdays. In this way, the number of comparisons grows quickly. 253 comparisons is well over half the number of possible birthdays, so it is not surprising that a collision is likely to happen after that many comparisons.

The birthday paradox math is the same math for finding a hash collision. It can be quickly approximated by taking the square root of the number of buckets. This make sense because the number of comparisons are squared:

$$(n \times (n - 1)) / 2 \simeq n^2$$

For the birthday problem, taking the square root of the number of birthdays yields:

$$\sqrt{366} = 19.13$$

This is a close approximation to the actual answer of twenty-three.

Now back to the original question: how many hashes would need to be generated to make finding a collision likely for a 256 bit hash function?

$$\sqrt{2^{256}} = 2^{128}$$

2^{128} may seem small, but it is actually an incomprehensibly huge number. It is not possible to create enough hashes to make a hash collision probable even if given all the computing power in the world and all the time remaining before the sun runs out of energy to do it.

Besides the brute-force attack, the other type of attack on hash functions is a cryptanalytic attack. Nothing about hash functions is secret. Their internal operations can be analyzed to dissect how they manipulate bits to produce hashes. If there is a flaw in the algorithm, then it may be possible to cleverly manufacture a collision. In other words, it is not necessary to produce countless hashes hoping to find a collision like in a brute-force attack. Input values can be chosen carefully to make collisions more likely. Two famous hash functions have fallen to cryptanalytic attacks. *MD5* (*Message Digest 5*) is a 128 bit hash function created in 1991. By 1996 a weakness was discovered in its design, so its continued use was discouraged. However, it is still a popular hash function in contexts where security is not critical (e.g., in capture-the-flag contests). *SHA-1* (*Secure Hash Algorithm 1*) is a 160 bit hash function created in 1993 and flaws were also soon discovered in its design. In the 2000s experts recommended that its use be discontinued. Google famously produced the first actual SHA-1 collision in 2017 in what they called the *SHAttered attack*.

7.3 Steganography

"A boat beneath a sunny sky / Lingering onward dreamily / In an evening of July — / Children three that nestle near, / Eager eye and willing ear…" - "A Boat Beneath a Sunny Sky" by Lewis Carroll

Cryptography is closely related to *steganography*. Steganography (sometimes abbreviated *stego*) comes from two Greek words meaning "covered writing." It is *the art and science of hiding information in plain sight*. It can be a message within a message or an unexpected method of communication. The goal with steganography is for communicating parties to transmit information in an open channel without onlookers noticing. The only necessary and sufficient condition is that the communicating parties have access to a shared resource that at least one of them can modify—this allows for one-way communication. If multiple parties can modify the shared resource, then multi-way communication can take place.

In steganography, the information is not scrambled, like in cryptography, instead it is "covered." A simple example of steganography is an acrostic—a message where the first letter of every word or line spells out a word or phrase. This is common in poetry, like in the poem, "A Boat Beneath a Sunny Sky" where the first letter of every line spells out the full name of the real-life inspiration for the character, Alice. As another example, even if the following battlefield correspondence was intercepted by enemy forces, they would find it unworthy of any serious attention. However, it contains a top-secret hidden message:

> ATTN: THE TROOPS ARE COOKING KABOBS AT THE DOCKS AT WEDNESDAY NOON

A classic example of steganography from real life was carried out by Vietnam prisoner of war, and later United States Senator, Jeremiah Denton. He was held by the North Vietnamese, and they forced him to participate in a propaganda interview praising the North Vietnamese's treatment of prisoners so they could televise it to the world and "prove" that they were abiding by international humanitarian law. Denton did say the words they gave him for the video camera, but while he was doing so he was also blinking his eyes to communicate in Morse code the word *TORTURE* to describe what was actually going on in the prisons (see Figure 7.15). The North Vietnamese did not detect the covered message, and they released the video. For those with eyes to see it, like the United States Naval Intelligence, the real message got through.

Figure 7.15

Prisoner of war Jeremiah Denton saying all the right words while blinking the word TORTURE in Morse code.

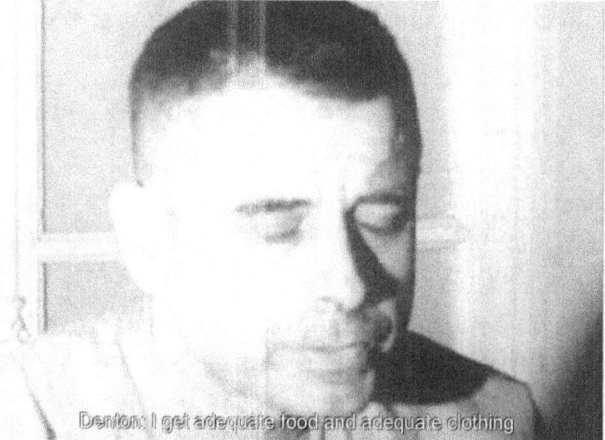

Denton: I get adequate food and adequate clothing

There are an infinitude of ways to perform steganography in both classic and computer cryptography, and that is its strength. People can always come up with new and ingenious ways to communicate secretly in plain sight. One big advantage of steganography over cryptography is that it does not reveal the presence of secret communication. When cryptography is used, the assumption is that eavesdroppers could observe the ciphertext messages and identify the sender and the receiver of the messages. This arouses suspicion and invites further scrutiny because the assumption is that the communicating parties have something to hide. If steganography is successful, the onlookers are oblivious to the secret communication and no suspicions are raised.

In cyberspace, one of the canonical ways to hide a message in plain sight is to encode the message in an image. As we saw in Chapter 2, the RGB color model is a 24-bit system. Because eight bits are used per color, this means each pixel has up to 255 parts red, green, and blue. Therefore, the RGB format is capable of encoding more than sixteen million colors ranging from RGB[0, 0, 0] (white) to RGB[255, 255, 255] (black). In this scheme there are many slightly different shades of every color, and these differences are indistinguishable to the naked eye—this is the avenue for steganography.

Figure 7.16

Two different colors by three bits that are indistinguishable to the human eye.

In Figure 7.16 the two purple blocks look the same but are actually different colors:

> Left: RGB[100, 25, 200]: 011001000001100111001000
> Right: RGB[101, 24, 201]: 011001010001100011001001

To a computer these colors are different because three out of the twenty-four bits are different. But to the human eye, they appear to be the same color purple because the RGB contributions are different by only one part each. This makes it possible to hide meaningful bits in an image without actually changing the appearance of the image. This is done by selecting some of the pixels and changing the 1s and 0s as needed to encode the secret message. The recipient can then examine the selected pixels to reconstruct the secret message. To an eavesdropper, nothing seems out of the ordinary—the image looks normal—so the secret message goes by unnoticed, which is the entire point of steganography.

Figure 7.17

Two different colors by three bits that are clearly not the same to the human eye.

Using this technique, the bits chosen for hiding the message matters. In Figure 7.17 the two blocks are clearly different colors:

> Left: RGB[100, 25, 200]: 011001000001100111001000
> Right: RGB[228, 153, 72]: 111001001001100101001000

To a computer, these colors again differ by only three bits, but this time, the most significant bit of each byte has been changed. This creates a difference of 128 parts each of red, green, and blue contributions. Modifying an image in this way would give away the presence of a secret message. Therefore, this image stego technique is known as *LSB RGB steganography*. LSB stands for *least significant bit*—the bit representing the ones place (2^0) in each byte.

LSB RGB techniques can be used to encode up to one secret message bit per image byte. It is not uncommon for images to be several megabytes in size, and this allows a significant number of secret message bits to be encoded, making this technique very efficient. It is even possible to encode a secret image within an image!

Besides steganography for message passing, another major use is in creating covert channels. As discussed in Section 4.1.4.1, hackers may use a covert channel to exfiltrate data off a network. This makes exfiltration slower and more complex, but it lowers the risk of detection. Steganography is also used locally on computers to conceal the presence of sensitive data. For example, a hacker may store incriminating evidence on a hard drive using steganographic techniques instead of just relying on cryptography. Using cryptography can reveal that he has something to hide and risks that the encryption could be cracked or the key could be discovered or coerced from him. On the other hand, if investigators only find unencrypted data and do not detect the use of steganography, they may be more likely to believe the suspect's pleas of innocence. In Poe's short story, "The Purloined Letter," the police suspect a villain is in possession of a blackmail letter and turn his apartment upside down looking for it, but it turns out it is hiding in plain sight right out on his desk—he was depending on steganography to protect him.

There are an infinite number of other ways to use computers to encode secret messages in all types of files including images, audio, video, and as we saw in Section 4.1.4.1, in network traffic. Algorithms have been developed to detect when steganography is being used, so these techniques are not guaranteed to go unnoticed, but they can be highly effective for smuggling information out of networks and for communicating in secret over open channels. One caveat, however, should be noted: if a stego technique involves out-of-the-ordinary communication patterns, it could tip observers off that something "fishy" might be going on. For example, if Alice and Bob start emailing innocent-looking images to one another out of the blue for no apparent reason, it could arouse suspicion and invite further investigation. Ideally, steganography should be incorporated into existing and routine methods and patterns of communication.

Because of its complete dependence on security by obscurity, using steganography can be risky. If the steganographic technique is exposed, then the secret message is divulged. For extra insurance, steganography and cryptography can be used together. In another of Poe's short stories, "The Gold Bug," a treasure map is written in invisible ink (steganography), and the text on it is encrypted with a monoalphabetic cipher (cryptography). Once the plaintext is finally revealed, it is a riddle that must be solved to find the hidden treasure—yet another layer of insurance!

7.4 Principles of Cryptography

> *"Few false ideas have more firmly gripped the minds of so many intelligent men that the one that, if they just tried, they could invent a cipher that no one could break."* - The Codebreakers *by David Kahn*

This section outlines a few guiding principles for cryptography based on lessons learned over its long history.

7.4.1 Kerckhoffs's Principle

> *"The enemy knows the system."* - *Claude Shannon's summary of Kerckhoffs's principle*

Auguste Kerckhoffs studied military cryptography in Europe in the 1800s. He outlined a counter-intuitive principle that became known as *Kerckoffs's principle* that still guides cryptography today. The principle states that it should be assumed that the enemy knows and has access to the cryptographic system being used. For this reason, cryptosystems should be parameterized with a secret value, a key, and the security should reside entirely in the secrecy of the key. Steganography depends entirely on security by obscurity in direct violation of Kerckhoffs's principle. The difference is that steganography counts on concealing the presence of the secret communication, while cryptography assumes that eavesdroppers will observe and intercept the ciphertext and attempt to cryptanalyze it to reveal the message. This is called a *ciphertext-only attack*.

Kerckhoffs was primarily concerned with how cryptography is actually implemented and used in the real world. He realized that cryptosystems inevitably fall into enemy hands. This can happen due to mistakes, espionage, capture, or to reverse engineering. When this happens, the cryptosystem should not have to be abandoned. Rather, it should be assumed up front that this is going to happen, and if and when it does, since the security of the cryptosystem does not depend on the secrecy of the cryptosystem, communications can proceed as usual.

It is much more difficult to create a secure cryptosystem that the enemy has access to, but this is exactly what Kerckhoffs's principle requires. In the examples above it is assumed the adversary not only has access to the encrypted messages, but also to the cryptosystem being used. The adversary can examine the cryptosystem to try and find weaknesses that could aid in the cryptanalysis of messages. The only thing the adversary does not have is the secret key. But because he has access to the decryption algorithm, he can attempt a brute-force key search attack to find the key. This is the reason why the keyspace must be large enough to make success improbable.

In the case of public key cryptography, the adversary also has the key that was used to encrypt the message and can generate ciphertexts of his own with the key. This is known as a *chosen-plaintext attack*. Being able to encrypt plaintexts with the same algorithm and key that is used by the communicating parties makes forward search attacks possible. For this reason, the RSA cryptosystem adds random values to plaintexts as padding to make short messages secure against forward search attacks.

7.4.2 Schneier's Law

> *"Anyone, from the most clueless amateur to the best cryptographer, can create an algorithm that he himself can't break. It's not even hard. What is hard is creating an algorithm that no one else can break, even after years of analysis. And the only way to prove that is to subject the algorithm to years of analysis by the best cryptographers around." - Bruce Schneier*

History demonstrates that it is difficult to create a secure cryptosystem. It seems inevitable that every cryptosystem, outside of the one-time pad, will eventually be broken. In fact, most will be broken quickly. *Schneier's law*, proposed by cryptographer *Bruce Schneier*, states that *anybody can create a cryptosystem that he himself cannot break*. The trick of course is to create a cryptosystem that nobody else can break! Intellectual pride is the cryptographer's downfall.

Tempting both Kerckhoffs's principle and Schneier's law, many cryptographic algorithms have been designed and implemented in secret, hoping that the underlying algorithms would not be revealed. History has shown they are dissected and broken due to weaknesses soon discovered in their designs. Therefore, the best practice in modern cryptography is to publish the full details of a proposed cryptosystem before it is implemented so that it can be peer reviewed. Only after a lengthy and thorough review process, assuming no cracks in the design are found, should the system be considered secure enough to implement for real-world purposes. The most secure cryptosystems are the tried and true ones that have been in use for a long time. New cryptosystems should always be treated with suspicion until time proves them secure.

7.4.3 Simplicity and Security

"Make things as simple as possible, but no simpler." - Albert Einstein

Cryptography inevitably becomes more complex over time as past schemes are broken and new schemes are devised to replace them. We saw this play out in classic cryptography as we went from the Caesar cipher to alphabetic shift ciphers to the Vigenère cipher to one-time pads—each evolution was stronger than the previous but required more work to implement. Although we did not go into the details of computer cryptography, it is far more complex than classic cryptography because it has to withstand sophisticated cryptanalytic attacks aided by computers.

This creates a tension because more substituting and transposing can always be performed in an attempt to make the cryptography stronger, but it also adds time to encrypting and decrypting messages, and it adds extra complexity which invites more vulnerabilities. Cryptosystems should be as simple as possible but no simpler. Simpler cryptosystems are easier to understand. The fewer lines of code required to implement a cryptographic algorithm the better. Simpler cryptosystems are also easier to use. The more steps that people are required to perform to create secure messages, the more likely they are to make mistakes, and the more opportunities for adversaries to undermine the process. For cybersecurity in general and cryptography in particular, simplicity and security go hand-in-hand. In the same way that good knots are strong and also easy to tie and untie, good cryptosystems are easy to understand and efficient (see Figure 7.18). The best cryptographic algorithms are marked by an elegant simplicity.

Figure 7.18

A complex tangled knot of dubious strength and a simple and elegant strong knot.

Also due to the increasing complexity of cryptography, care must be taken to implement cryptographic processes correctly. Even when using libraries like OpenSSL that encapsulate the cryptographic algorithms, it is easy to make mistakes that result in cryptography that is not secure. For example, when using public key cryptography, people have been known to publish their private key instead of their public key! When using symmetric key cryptography, errors can be made that forfeit the bullet-proof security that people assume they are gaining. Mistakes like these are due to the complexity of computer cryptography and also ignorance of how the process is supposed to work. Care must be taken to learn the proper way of doing things in order to be confident that one's messages are in fact secure. For example, Figure 7.8 shows a short message being encrypted with AES 256. Note the many arguments used in the encryption command—failure to understand what they do can result in a less secure encrypted message. Even though AES 256 is a highly secure encryption algorithm, this encrypted message is not safe from prying eyes because it is encrypted with the key *SECRET* instead of a random 256-bit string that must be used as the key with AES 256 to get the security properties it promises. Therefore, this ciphertext is vulnerable to a simple forward search attack using a dictionary as a wordlist.

Fortunately, most of the cryptography users depend on to keep them secure is embedded behind the scenes in software written and vetted by experts, so they do not have to worry about things like command line flags. Many people do not even realize that their Internet traffic is being encrypted and decrypted by their web browsers—but it is! Of course, combining the complexity of computing with the complexity of cryptography ensures that no implemented cryptosystem is provably perfectly secure. Any claim to the contrary is misleading at best.

7.5 Conclusion

The history and evolution of cryptography is fascinating. In the past its use was mostly limited to governments and militaries, but today it is used by nearly everybody on the planet. Without computer cryptography, the modern Internet would not be possible. There would be no way to protect messages propagating over untrusted channels. Cryptography is also used locally on computers to protect the privacy of individuals and organizations. It is believed that no entity, no matter how big and powerful, can read messages properly encrypted by the standard and freely available cryptosystems in use today. Later in Section 9.2.2 of this text we will cover some more of the practical uses of cryptography.

Cryptography is the bedrock of cybersecurity. It is the primary way data confidentiality and integrity are achieved in cyberspace, and it is also used for an important component of access control: authentication. The next chapter explores access control in depth.

Chapter 8

8. The Means of Cybersecurity: Access Control

"Whence and what art thou, execrable shape! / That darest, though grim and terrible, advance / Thy miscreated front athwart my way / To yonder gate? Through them I mean to pass— / That be assured—without leave asked of thee."
- Satan to Hell's gatekeeper Death in Paradise Lost *by John Milton*

"Unauthorized access" is the key phrase used in most definitions of cybersecurity and in cybercrime laws. Cybersecurity is primarily concerned with controlling access to computers and data. Therefore, *access control*, or *monitoring and controlling access to computer systems and data*, is the means of cybersecurity. Access control mechanisms are designed to make sure that the appropriate people, and only the appropriate people, have access to the resources they need. Failures of access control occur when people obtain access to computing resources they are not authorized to use, view, or modify.

Access control is made up of three interrelated components: authentication, authorization, and accounting. These are typically referred to by the acronym *AAA*—the second most important acronym in cybersecurity behind CIA. Authentication asks the question, "Are you who you say you are?" Authorization asks the question, "Are you allowed to do that?" And accounting is a record of who did what when. These three components work together to control and monitor access to computer systems and data. This chapter covers each part of access control in turn.

8.1 Authentication

"Who in the world am I? Ah, that's the great puzzle." - Alice's Adventures in Wonderland *by Lewis Carroll*

Authentication is verifying an identity. In the real world, people know one another through personal interactions. We recognize each other based on the way we look, speak, and act. When we need to be authenticated to strangers, like at the airport, identification cards (IDs) issued by trusted authorities that include the person's name, photograph, and sometimes signature are used.

The goal of authentication is to get the correct answer to the question, "Are you who you say you are?" In cyberspace, computers, not humans, perform the work of verifying identities. Authentication is controlled by an authentication mechanism which mediates access to the requested computing resource. The resource could be a computer system, a computer program, or data. The authentication mechanism is the gatekeeper, and is analogous to a security guard who verifies identities before permitting people to enter a secure space.

The authentication question is asked and answered frequently in cyberspace, like when logging into a computer and logging onto websites. Email is a good example. Email providers have many users, and when a person checks their email, the email provider needs to know which email records to serve—people should have access to only their own email messages. Therefore, email providers must authenticate users before granting them access to emails.

In cyberspace, all interactions are digital. Because all digital artifacts can be perfectly replicated, verifying identities in cyberspace is a tricky problem and impersonation is a real threat. While cyberspace seems like an anonymous world, most interactions involve some form of authentication. Even when a person is browsing the web anonymously, they likely had to first log in to their smartphone or laptop. Plus, while surfing, their browser authenticates the web servers they connect to even if the user is not authenticated in return. Authenticating web servers is important for protecting users because it ensures they are interacting with real, not imposter, websites. We saw how this works in Chapter 7 with the public key infrastructure (PKI). How did Alice know whether the public key that was sent to her belonged to Bob or someone else? This is the same authentication problem web browsers face when connecting to web servers. How can they be sure they are connected to the real news, ecommerce, or social media server and not to a fake website? PKI solves this problem by using certificate authorities (CAs) to verify identities. This is similar to how IDs are used in physical space.

For any authentication system, two opposite errors are possible. The first is the *false positive error*. A false positive error is *when the wrong person passes authentication*. The other is the *false negative error*. A false negative error is *when the right person fails authentication*. Both errors are damaging. False positives allow impersonation and unauthorized access which is a breakdown of cybersecurity. False negatives prevent the right people from accessing their resources which is a breakdown of the availability goal of the CIA triad. These errors are in tension because attempts to minimize one likely increases the odds of the other. How to balance the two is determined by the relative priority of security and availability—another security trade-off like we saw in Chapter 5.

Authentication is composed of two phases: the *enrollment phase* and the *recognition phase*. During the enrollment phase, *the user's access credentials are registered and stored in an authentication database*. The user either creates a username or one is assigned. The system makes sure that the username is unique among all the registered users on the system. Email addresses are sometimes used as usernames because they connect users to the control of an external account that can be verified. Usernames may or may not be associated with a person's legal name.

During the enrollment phase the user is also asked to provide or is given an *authentication token*. An authentication token is *an artifact that uniquely identifies a user*. The username and token are stored together in the authentication database. Users need to go through the enrollment phase only once. Accounts are registered the first time a person uses a computer or visits a website. Figure 8.1 shows a minimal account registration page.

Figure 8.1

An account registration page—the enrollment phase for password-based authentication.

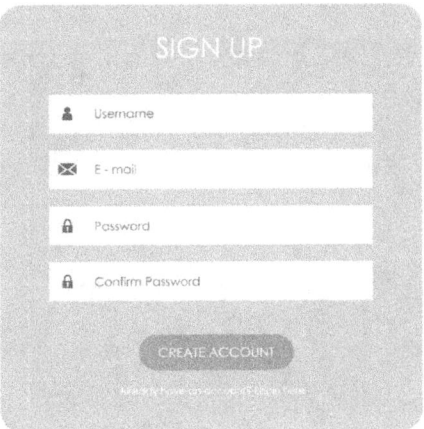

After enrollment, users go through the recognition phase when they need to access their account. During the recognition phase, *the user's identity is validated*. Unlike the enrollment phase, the recognition phase is repeated frequently. This phase starts with the identification step. Users are asked to identify themselves by inputting their username. Usernames are not necessarily secret, and there is nothing that would prevent a user from inputting another person's username. Once the username is entered, the computer asks for proof, in effect saying, "How do I know you are who you claim to be?" This is the authentication question. At this point the user provides their authentication token. This token is compared to the value stored in the authentication database, and if it matches, the user is authenticated.

There are three main types of authentication tokens. They are based on something the user knows, something the user is, or something the user has. We will explore each of these methods in the following subsections.

8.1.1 Something You Know

> *"And since there's only him and me in the whole of the universe what knows about those signals, sir, he'll open up for me without any doubt and without calling out.'"* - The Brothers Karamazov *by Fyodor Dostoyevsky*

Something you know is the most popular type of authentication token. During the recognition phase, the computer in effect asks the user, "If you are who you say you are, prove it by telling me something that only the real [*username*] would know." We refer to this secret piece of information as a *password*. The assumption is that only the real user knows his password, so if he is able to input it into the computer, then he must be the user he claims to be. The familiar login screen is the recognition phase for password-based authentication schemes (see Figure 8.2).

Figure 8.2

A login page—the recognition phase for password-based authentication.

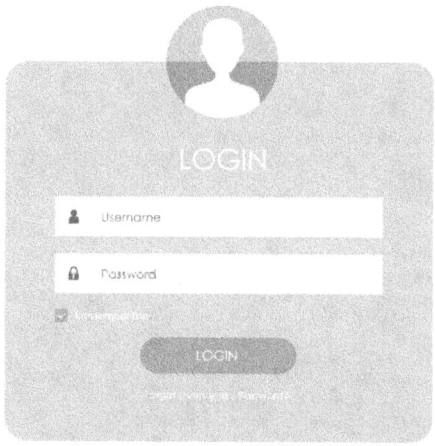

8.1.1.1 Passwords

Password based authentication is popular because it is inexpensive to implement and maintain. The overhead costs of assigning passwords, storing them, and verifying users during the recognition phase is minimal. No extra hardware is required and there are minimal storage and processing costs. Also, in theory, it should work really well—after all, only the correct user should be able to provide his secret piece of information.

However, as we saw in Section 4.1.2.3 on credential stealing, there are numerous ways that password-based authentication fails. False positive errors—an imposter authenticating as a valid user—are common. The assumption that only one person, the rightful account owner, knows the password does not hold. Passwords are a great example of the practical realities that confront cybersecurity mechanisms. Many ideas that work well in theory do not work well in practice. The price of failures are cybersecurity incidents. This has definitely been the case with passwords.

False negatives—users being denied access to their own account—are also an issue with password-based authentication schemes because people sometimes forget their passwords. To try and prevent impersonation attacks, passwords should be long, random, and complex. All these factors make it difficult for users to remember their passwords. When users forget their password, they are unable to be authenticated, causing frustration and loss of work. Most password-based authentication schemes include a convenient password reset mechanism. This allows users to create a new password. Unfortunately, password reset mechanisms are themselves targets for hackers and have their own vulnerabilities. The fundamental problem is that it is difficult in the anonymous world of cyberspace to accurately and inexpensively authenticate users in a scalable way. Passwords scale well, but they do not provide strong cybersecurity.

The burden passwords place on users is significant. When they register an account users are asked to create a password they can remember yet have nothing to do with anything they care about (to prevent password guessing attacks), that are long and use several different character sets (to prevent password cracking attacks), and that are different from all of their other accounts (to prevent credential stuffing attacks). This is unrealistic, especially for users who see passwords as a burden for them as opposed as an obstacle for hackers. It is the rare user that can come up with strong passwords that they can keep track of.

To make matters worse, it used to be a best practice in cybersecurity to set password expiration dates, forcing users to update their passwords periodically. This practice is designed to limit *surreptitious account access*. Surreptitious account access is when attackers spy on victims by logging into their accounts. For example, an attacker might access a victim's email account to read their messages. If the compromised password expires and is reset by the account owner, then the attacker would lose access to the account, stopping the attack. While still employed in some workplace environments, password expirations are no longer considered a standard cybersecurity best practice. Surreptitious account access is relatively low risk—it is rare, and even when it happens, the impact is usually not severe. In the majority of cases when a user's password is compromised, the attacker uses his unauthorized access right away to accomplish his objectives, not worrying that this will reveal the password compromise—by then, the damage has been done. Even if an attacker is interested in maintaining surreptitious access, he may inadvertently make his presence known leading to a password reset at that time. Meanwhile, the practice of password expirations does not always eliminate surreptitious account access. If the attacker was able to obtain a user's password in the first place, he may be able to do it again the same way after the password update—for example, by repeating a shoulder surfing attack. Additionally, as we saw in Chapter 5, cybersecurity practices come at a cost and tradeoffs must be considered. One of the costs of password expirations is the burden it places on users because it results in even more passwords for them to track. In some work environments that enforce regular password expirations, exasperated users resort to writing down their passwords and keeping them near their desks. In those cases, password expiration policies backfire by increasing the likelihood of password compromises.

As we become more dependent as a society on a secure cyberspace, sheer password-based authentication is on the decline. It has long been known that passwords are a major cybersecurity vulnerability. However, because of their convenience, they will not be going away completely anytime soon, so we must educate ourselves and others on the risks of passwords and how to choose and manage passwords wisely. Chapter 9 provides practical guidance on creating and managing passwords.

8.1.1.2 PINs

In addition to passwords, *personal identification numbers* (*PINs*) fall under the something you know category. PINs are designed to be easily remembered. PINs are short passwords, and despite the name, are not always numeric. Because PINs are short, they should never be used as the sole means of authentication, and they should always be used with an authentication mechanism that limits login attempts. PINs have traditionally been (and many still are) four numeric digits. There are only 10,000 unique PINs in this range—the values ranging from 0000 to 9999. Computers can attempt every PIN in a tiny fraction of a second, and even humans can attempt every PIN manually in only a couple of hours. For this reason, users usually only get a few attempts to enter their PIN. After this, the account is either locked, requiring a different form of authentication to unlock it, or a login timeout period is imposed to slow down further login attempts. Some smartphones have a setting that deletes all the phone's data after too many failed PIN attempts. The assumption behind PINs and passwords is the same—it is a piece of information that only the genuine user knows, and like passwords, they can be attacked.

8.1.1.3 Security Questions

One last category of "something you know" is usually referred to as *security questions*. These are sometimes compiled during the account registration process, and are used as either an additional or alternative authentication token. Typical questions are the name of the elementary school you went to, your mother's maiden name, or the name of your first pet. Unfortunately, security questions almost always fail the "only you know" assumption because they are based on real world facts that others may also know or be able to find out. Basing them on facts is by design so that users are more likely to remember the answers since they are based on their real life history.

Security questions are sometimes used as part of the password reset process. Usually multiple security questions are asked to make it more difficult for attackers to impersonate their target. Sarah Palin was the Republican United States Vice Presidential candidate in 2008, and her personal Yahoo email account was famously hacked and exposed in the run-up to the election. The hacker bragged that it took him only a few minutes to gain access to her account. First, he found her Yahoo email address—this was not difficult since email addresses are not usually secret and can sometimes be guessed or seen posted online. Then, he successfully reset her account password by answering three security questions. He did not know Sarah Palin so he had no personal insight into her background, but since she was a famous public figure, he was able to find the answers to all her questions through online searches!

While there are clever security questions that people can remember yet are not likely to be published anywhere or known by others, this is a weak type of authentication token and should be used with caution. Some security-conscious people provide bogus answers to security questions to make it difficult for them to be guessed by an attacker, but this can also make it difficult for them to remember their own answers! Some take the shortcut of always just using the same bogus answer (e.g, "I'm not telling") to every question everywhere, but this exposes them to credential stuffing attacks, and is not an ideal solution either.

8.1.1.4 Summary

The main issue with passwords is that they put the onus of security on the user. Most users do not understand the risks they are taking and are not prepared for the burden of creating their own authentication token. When users are asked to come up with a password, they think more about convenience than security. We saw this tradeoff in Chapter 5 on cyber risk management. Users prioritize convenience and choose poor passwords they can remember and type in quickly. Unfortunately, these passwords are easy for attackers to guess or at least crack. This is not really a failure of the user as much as it is of the cybersecurity industry. The reason passwords are so commonly used is not because they work exceptionally well, but because they work somewhat well and are inexpensive to use and maintain.

8.1.2 Something You Are

> *"The Gileadites captured the fords of the Jordan leading to Ephraim, and whenever a survivor of Ephraim said, 'Let me cross over,' the men of Gilead asked him, 'Are you an Ephraimite?' If he replied, 'No,' they said, 'All right, say "Shibboleth."' If he said, 'Sibboleth,' because he could not pronounce the word correctly, they seized him and killed him at the fords of the Jordan." - Judges 12:5-6*

In the real world authentication is based on physical characteristics. We use our human senses, such as our senses of sight and hearing, to authenticate one another. When computers use physical characteristics to authenticate people it is called *biometric authentication*. Biometric authentication requires a hardware sensor on the computing device, such as a camera or scanner. During the enrollment phase, users provide a biometric sample via this hardware sensor that is digitized and stored in the authentication database. Then in the recognition phase, users provide a fresh sample which is digitized and compared to the sample on file—if they match, the user is authenticated. This type of authentication is based on *something you are*—some physical characteristic of the person.

Numerous biometric candidates have been explored in the cyberspace era. The most common ones are facial recognition and fingerprint recognition. Call centers sometimes use voice recognition to help ensure they are talking with the correct customer and not to an identity thief. In biblical times, "accent recognition" was used to distinguish Gileadites from Ephraimites (see quote at the beginning of this subsection). Many other biometrics exist such as gait recognition—how a person walks. In this section we will explore what features make for a good biometric.

8.1.2.1 Distinguishable

The most important feature of a biometric is how well it distinguishes people from one another. For example, height would not make a good biometric because most people fall into a relatively narrow range of heights. For biometric authentication to work reliably, there need to be many more measurement possibilities than there are users to make it improbable that any two people share similar measurements. With height, even if measurements could be taken to a fraction of an inch (which would be prohibitively difficult to do), multiple people would still share the same measurement making it impossible to distinguish between them. For this and other reasons, height is a poor biometric candidate.

A biometric that rates much higher in distinguishability is fingerprints. It is well-known that no two people have the same fingerprints, not even identical twins even though they share the same DNA. And people's fingerprints differ widely enough that it does not require microscopic measurements to distinguish them from one another. Historically fingerprints were mostly used for identification, like in crime scene investigations, not for authentication. Identification compares a single fingerprint to many different ones to try to identify a person among many possible people. Authentication compares a single fingerprint to a fingerprint on record to see if it is the same person. In fingerprint authentication in computer systems, the fingerprint on record is collected during the enrollment phase, and many fingerprint samples are collected over time during the recognition phase to be compared with the fingerprint on record.

Regardless of the biometric, distinguishability is affected by the precision of the measurements taken. The more precise the measurements the more measurement possibilities and the bigger average distance between measurements. However, there are limits to the level of detail that sensors can collect and their costs increase as they become more sensitive.

For biometrics the probability of false positive authentications is called the *fraud rate* because it permits fraud to occur—an imposter is authenticated as a legitimate user. The probability of false negatives is called the *insult rate* because the authentication mechanism blocks the right person—thus "insulting" them by not believing they are who they say they are. These two rates are in tension because they are based on how closely the recognition sample is compared to the enrollment sample. A person's recognition samples will never exactly match his enrollment sample, but they should fall within a small range. If the biometric system's acceptable range is too narrow, this increases the insult rate. On the other hand, large acceptable ranges means that more nearby measurements are accepted, maybe even including other people's recognition samples, leading to a higher fraud rate.

For a given biometric the level of specificity where the fraud rate and the error rate are set equal to one another is called the *equal error rate*. If the equal error rate is .1% for a given system, this means one out of a thousand times the right person will be denied access (i.e., insulted), and one out of a thousand times an imposter will be granted access (i.e., fraud will occur). This measure is a good way to make an apples-to-apples comparison between competing biometric systems—the smaller the equal error rate, the better that system performs in the distinguishability category. For a given biometric installation, an organization may not calibrate to the equal error rate given their risk tolerance. For example, if they believe that the loss of availability would cost more than the risk of fraud, they may decrease the insult rate to .01% (one in ten thousand), which might raise the fraud rate to .2% (one in five hundred).

8.1.2.2 Permanent
Permanence is another important category for biometric authentication. Ideally, users should only have to be enrolled once. If the biometric characteristic being measured changes over time, then people would need to be re-enrolled to collect updated samples. This is one reason why weight would be a poor biometric! A person's weight fluctuates with age and even with the seasons. Therefore, weight measurements would become outdated and enrollment samples would need to be updated regularly.

Faces change over time, but there are facial features that are stable for long periods of time. Good facial recognition algorithms home in on these stable characteristics. By focusing on specific dimensions such as the distance between facial features, facial recognition systems can tolerate changes in weight and facial hair, and can even work when the subject is wearing glasses or a hat. Facial recognition systems have become highly accurate and do a better job than humans in recognizing faces. They are even capable of telling identical twins apart.

Some other biometrics rate higher on the permanence scale than facial recognition. For example, iris patterns in the human eye do not typically change from adolescence through old age. This makes iris recognition a good candidate for a biometric. In theory, children could be enrolled and they could still be authenticated as an elderly person—that would not be true for facial recognition or many other biometrics. The most permanent biometric of all is DNA—it is set at the moment of conception and never changes! However, DNA does not make for a good biometric for other reasons, as discussed below.

8.1.2.3 Universal

The biometric should also be universal. This makes hair color a poor biometric because not everybody has hair. Similarly, gait recognition would not work for people in wheelchairs. Biometrics should exclude as few people as possible.

8.1.2.4 Costs

Another concern is costs. What is the initial investment and what are the ongoing costs of a biometric system? Some types of scanners are more expensive to buy, install, and maintain than others. This is ultimately a cyber risk management decision because it involves weighing risks against costs taking into account relative security benefits.

8.1.2.5 Collectible

Another practical consideration for a biometric is collectability. The recognition phase is repeated frequently, therefore, the easier and quicker it is to capture the recognition sample, the better. Facial recognition on smartphones scores highly in this category because front-facing cameras can scan faces when users look at their smartphones, requiring no deliberate effort on the part of the user in order to be authenticated. Fingerprints are captured by a simple scan in a fraction of a second and put little burden on users. Iris scans are more onerous since they require a higher resolution image of the eye. DNA makes for a great biometric when considering the permanence factor, but it rates poorly on collectability. DNA samples could be collected through cheek swabs, blood draws, or hair samples, but those can be time consuming, tedious, and invasive.

8.1.2.6 Secure

Lastly, biometric schemes should be rated for security. This is another area where adversarial thinking is important for cybersecurity. How easy would it be to hack the biometric? What are some approaches adversaries might take? In some early facial recognition systems, a photograph of a person could be used to login as them. Some fingerprint recognition systems were susceptible to fingerprints lifted from a glass cup using scotch tape. Biometric authentication systems, like all computers, are potentially hackable and need to be tested not just for performance and reliability but also for security.

8.1.2.7 Summary

Table 8.1 illustrates how these categories can be used to rate biometric candidates. In this illustration, facial recognition comes out on top, but in reality, specific implementations would need to be compared with one another, including multiple different candidates in the same category. Plus, organizations may give more weight to some factors over others based on their cyber risk management profile.

"Something you are" authentication tokens are becoming more popular because their performance has been improving and their costs have been decreasing. They place less of a burden on users than passwords because most people prefer to scan a fingerprint over having to remember and type a password. Also, they are less susceptible to some of the low-cost attacks against passwords. For example, shoulder surfing attacks are devastating against passwords, but they are ineffective against biometrics—observing a person scanning a fingerprint does nothing to help a hacker.

Table 8.1

Comparisons of biometrics rated on a scale of 1 (low) to 5 (high)—these ratings are for illustration purposes only and are not necessarily accurate.

Biometric	Distinguishable	Permanent	Universal	Costs	Collectible	Secure
Face	4	4	5	4	5	4
Fingerprints	4	4	5	5	4	3
Iris	5	5	4	3	3	4
DNA	4	5	5	1	1	4
Gait	3	3	3	3	2	4
Weight	1	1	5	3	3	3
Height	1	2	5	3	3	3

One advantage of passwords, however, is that they cannot be "extracted" without the user's cooperation. This is not true of biometrics. People can potentially be forced to provide a recognition sample, for example, by holding their phone up to their faces. Attackers could also trick a person into providing a biometric sample without their knowledge—this would be much more difficult to do with passwords.

Another potential downside of biometrics compared to passwords is that they cannot be reset. If a password is compromised, the user can always just create a new password. However, biometrics cannot be modified, so a compromised biometric sample might disqualify that particular biometric for future use.

8.1.3 Something You Have

The last category of authentication token is *something you have.* Here, the authentication mechanism verifies that the user is in possession of a unique device. The device is registered during the enrollment phase. The device could be a smartphone, a smart card, a USB stick, or another physical token.

8.1.3.1 Smartphone

"Something you have" authentication tokens work by verifying that the user has a device. For smartphones, one method is to send the user a text message with a one-time passcode. The user checks his phone, sees the passcode, and types it in. Passcodes seem similar to passwords, but they serve a different purpose. The passcode is randomly generated, only used once, and has a short expiration. The assumption is that if the user can type in the passcode, then he must have the device it was sent to. Since the passcode is only valid for a short period of time, this ensures that the user is in possession of the device at the time he logs in. The longer that passcodes are valid, the more they resemble passwords because a user (or an eavesdropper) could type in an old passcode even if he does not have the device in his possession at login time.

Text messages do not technically verify the person is in possession of a smartphone, but rather, a *phone number*. This is an important distinction because it opens up an attack vector called *SIM swapping*. SIM stands for *subscriber identity module*. It is a unique ID used by mobile carriers to identify customers when they change phones. If a hacker is able to deceive a mobile carrier into switching a target's SIM to a SIM that the hacker controls, then all of the victim's calls and text messages will be routed to the hacker's phone. If text messages are used to send passcodes, then the hacker will receive the text message and could be authenticated as the victim. There have been several high-profile SIM swapping attacks, including one conducted against the former CEO of Twitter, Jack Dorsey, that allowed a hacker to control the Twitter account of the Twitter CEO!

Many smartphone-based authentication tokens rely on smartphone apps instead of text messages. When a user tries to log in, a notification is pushed from the authentication server to an app on the user's smartphone, and the user is prompted to either accept or block the login attempt. With a simple click of the accept button, the user proves he has the device and is authenticated. This method is not susceptible to SIM swapping attacks because mobile apps are not tied to SIMs. In general, it is a much more secure method because smartphone apps and authentication servers are able to authenticate one another cryptographically. Push notifications are still hackable, however, due to user error. If a user clicks accept on accident, out of confusion, or through deception, an attacker could still be authenticated as him.

A nice advantage of smartphone-based authentication is that if a user receives a text message or a push notification when he is *not* attempting to login, this alerts him that his account may be under attack.

8.1.3.2 Smart card

Another example of something you have is a *smart card*. A smart card is *a plastic ID card with an embedded integrated circuit that can perform cryptographically-secure authentication*. During the enrollment phase, each user is assigned a unique smart card. Since they are the same size as a credit card, smart cards are easy to carry around in a wallet or a purse. They also frequently double as an always-visible physical ID badge and include the name and picture of the user. At many workplaces, employees attach their smart cards to lanyards and wear them around their neck at all times. This is a security measure that helps to control physical access to buildings and rooms.

Smart card readers can be purchased inexpensively and added to a computer and are sometimes even built-in to laptops and keyboards. When a user needs to be authenticated, he must insert or scan his smart card. An attacker would need to gain possession of the target's smart card in order to be authenticated as him. That would be difficult because smart cards are protected closely and are handled with care. If one is lost or stolen, it must be reported as soon as possible, and the lost smart card can be deactivated and a new one issued.

Debit cards are similar to smart cards. When a bank customer uses an automated teller machine (ATM) to withdraw cash from his account, he must insert his debit card. The machine reads the debit card to determine the account number. ATMs also require a PIN number (i.e., "something you know") for logging in. This is true with most smart card-based authentication mechanisms—another authentication token is usually required.

8.1.3.3 Security Key

A similar type of token is a *security key*. A security key is *a USB stick that can perform cryptographically-secure authentication*. When a user tries to login to a computer or a website, if his security key is plugged into the computer, then the user can be authenticated. If the user does not have their security key, then they cannot be authenticated.

Some security keys include a built-in fingerprint reader to collect a "something you are" token as well at login time. Because the fingerprint is also needed, users can leave their security key plugged into their computer without too much risk of an imposter logging in as them.

8.1.3.4 Keychain Pseudo-Random Number Generator

Another hardware-based system uses keychain-sized battery-powered tokens with a built-in number display (see Figure 8.3). The token runs a simple *pseudo-random number generator* (*PRNG*) to generate an endless stream of unpredictable numbers that change every thirty seconds or so. At login time, the user is prompted to enter the number displayed on the token at that time, proving he has the device. Even though the token is offline and has no way of communicating with the authentication server, it is synchronized with the authentication server during the enrollment phase, so the server can always calculate what number is displayed on the device at any given time. Every physical token uses a different seed for the PRNG so they can be uniquely identified. Having access to one device will not help a person hack into someone else's account. However, these types of tokens are hackable. If an attacker discovers the seed for a device or cracks the PRNG algorithm—which is possible since it is deterministic—he may be able to determine the numbers on the device just like the authentication server does. For this reason, these devices should always be used with another authentication token like a password.

8.1.3.5 Summary

Using a physical token for authentication is effective, but it can be an inconvenience. If a user needs to login but does not have his token, then he cannot be authenticated. It could be that he lost it or that his token is inaccessible—maybe he is at home but left it at work—or just that the token is in another room—he is upstairs but the token is downstairs. Foreseeing circumstances like these, many users will push back on adopting "something you have" tokens fearing it will be an inconvenience and cost them time—a clear security versus cost trade-off. This is one reason why smartphones are good candidates for "something you have" tokens. Many people carry their phones with them at all times, even around their house, guard them closely, and do not think of them as an extra burden they have to track.

Figure 8.3

A physical authentication token that displays a series of random numbers.

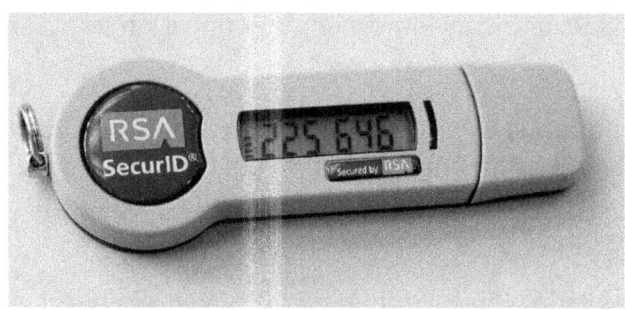

"Something you have" tokens are much more susceptible to compromise than "something you know" and "something you are" tokens because they are physical devices that can be misplaced, "borrowed," or stolen. Therefore, they should always be used in conjunction with other forms of authentication. This is called *multi-factor authentication* and is explained in the next subsection.

8.1.4 Multi-factor Authentication

Multi-factor authentication (MFA) is authentication based on tokens from two or more different categories. For example, inserting a debit card into an ATM and typing in a PIN number to withdraw cash is an example of MFA because it uses something you have (the debit card) and something you know (the PIN) tokens. *Two-factor authentication* is when exactly two authentication tokens are used. Most MFA is two-factor, but in some circumstances more than two factors are used.

MFA must draw from different categories in order to qualify as multi-factor. An authentication mechanism that forces users to enter a password, a PIN, and answer a secret question is not MFA because all three of those tokens are from the "something you know" category.

MFA is a well-established best practice in cybersecurity. It comes at a relatively low cost and provides substantial security benefits. Hackers may be able to compromise a token in one category, but it is much more difficult to compromise multiple categories of tokens. For example, conducting a keylogging attack on a public computer to compromise passwords is effective and relatively simple to perform. However, the compromised passwords are useless if users are also authenticated with a smartphone app. Similarly, a security key may be lost or stolen, but it cannot be used to login without the user's fingerprint.

MFA does not provide 100% security. There are many cases where it has been defeated by hackers, such as by using a password cracking attack and then a SIM swapping attack against a target, but the level of sophistication and work required to defeat MFA poses a significant hurdle for hackers. Unless it is a targeted attack, hackers are likely to just move onto an easier target and take lower hanging fruit.

8.1.5 Network-based Authentication

Authentication servers can be accessed over computer networks to perform *identity and access management* (*IAM*) services. This is a common practice in organizations. Even when logging in to a laptop computer, the login tokens can be sent to the authentication server to verify the user's identity and access credentials—authentication does not have to be done locally on the laptop. In order for this to work, the laptop needs to be online, but most computers are online all the time so this is not an issue. When a user is offline, such as on an airplane, then he can still be authenticated locally on the laptop as a backup option.

Remote authentication allows access information to be updated in real time from a central location. For example, a user's account access can be granted, modified, or removed by an administrator and the changes will take place immediately across all the devices the user accesses. Microsoft's *Active Directory* (*AD*) is a commonly used platform that performs IAM (and other) services. The user's access credentials are stored in the *domain controller* (*DC*). The DC is queried when a user authenticates within the AD network. When a user changes his password, it is changed on the DC so his new login credentials will be in effect everywhere. This is a major convenience at places like universities where students may need access to various computers across campus. Because the computers connect to the centralized authentication server, the student's one set of access credentials work on all the campus computers. The authentication tokens are sent securely over the network. A common protocol used by AD and other similar platforms to securely send authentication tokens is called *Lightweight Directory Access Protocol* (*LDAP*).

Single sign-on (SSO) is an authentication scheme that allows a user to sign-in once, be granted an authentication token, and then use that token to be automatically logged-in to other websites. Many SSO implementations use third parties to authenticate users, easing the burden that managing authentication can place on an organization. For example, instead of having to store and manage access credentials for their userbase, a startup can authenticate users through Google or some other well-known company. The startup's users likely already have a Google account so it is convenient for them to not have to create a new username and password, and Google is trusted to make sure that the access credentials are managed securely. When the user navigates to the startup's website, if they are already signed into Google, then they can be signed into the website without having to type in a username and password. Like we saw in Chapter 4, access credentials are a major target for hackers, so this is a way to mitigate risk for an organization. One downside of SSO for users is if their credentials are compromised by a hacker, then the hacker would have access to all their accounts that rely on SSO. This is the *keys to the kingdom dilemma*. It is convenient for users to have fewer keys, but if those keys open more doors, then they become bigger targets and do more damage if compromised.

8.1.6 Authentication on the Web

Local authentication does not need cryptography, but authentication over the network clearly does. Since the Internet has an open infrastructure and all messages are assumed to pass though many "untrusted hands" between endpoints, passwords cannot be sent in plaintext. This problem cannot be fixed with password hashes. If a password is hashed locally and then the hash is sent over the network to the server, then this just makes the hash the authentication token instead of the password. The hash can just as easily be observed by a hacker and used in a credential stealing attack.

Therefore, every time a person logs into a website by typing in a username and password, this information must flow over an already-established cryptographic channel to prevent eavesdropping attacks. These cryptographic channels are established between web browsers and web servers over HTTPS connections. The *S* stands for *secure* and it means that an end-to-end encrypted channel is created. Multiple messages are passed between the web browser and web server as part of the *connection handshake*, and the handshake typically establishes web server authentication, message integrity, and message confidentiality for the data to follow.

Authentication is common when browsing the web. Social media sites, email accounts, work and school-related accounts, and ecommerce websites all require repeated authentication while browsing. Because the HTTP protocol is stateless, every page visit is a brand new request from the web server's perspective. Stateless means that the web server does not retain any information about previous interactions. It is as if the web server has amnesia—every time a web server answers a request for a new webpage, it treats it as a first-time encounter with a never-before-seen user. For example, on amazon.com, every page visited while searching for products, adding items to the shopping cart, and then during checkout requires the user to be authenticated to Amazon—a quick and simple shopping session could generate dozens of unique webpage requests. However, the user is typically only asked to login once at the beginning of the browsing session, and sometimes not even then if the website is set to "remember" the user. How do web servers authenticate users on pages where no login information is provided?

Web browsing authentication is handled transparently in the background by web browsers using *authentication cookies*. An authentication cookie is *a text string assigned to a web browser by a web server for the purposes of authenticating a user*. When a user successfully logs into a website by providing his authentication token (e.g., username and password), the web server creates an authentication cookie in its database and sends a copy to the web browser (see Figure 8.4). The web browser stores the cookie and sends it back to the web server with every new page request for that website. Web servers inspect webpage requests for authentication cookies so they can "remember" users and serve them the appropriate webpages. In this way, web servers put the onus of maintaining state onto web browsers. The cookie acts exactly like a username and password.

Figure 8.4

Authentication cookies stored in a web browser for a website.

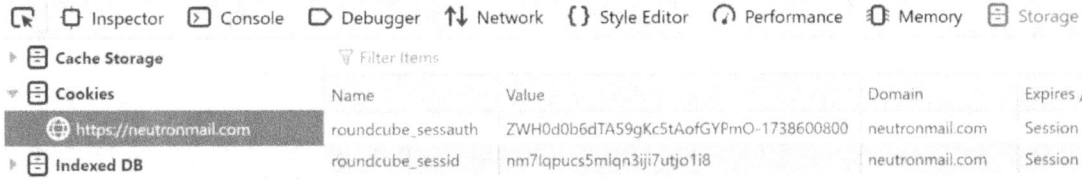

Authentication cookies are a kind of "something you have" token similar to passcodes used in text-message based authentication. Like passcodes, they are randomly generated so they cannot be guessed and are sent to users by the authentication server. They also have an expiration date, but not necessarily as short as one-time passcodes. Most banking websites expire authentication tokens after only a few minutes, but many websites, like webmail sites, set much longer-term expiration periods.

Authentication cookies, like passwords, are sent from web browsers to web servers over an end-to-end encrypted channel to prevent eavesdropping attacks. However, they still pose a security vulnerability similar to other "something you have" tokens. A cookie stealing attack is when a hacker copies a user's cookies, enabling the hacker to be authenticated as the user. For example, if Mallory steals Bob's Amazon cookies, then when Mallory visits Amazon's website, her browser will send Bob's cookies, and Amazon will think that Mallory is Bob! This will allow Mallory to see Bob's shopping cart and account information. Mallory does not need to know Bob's password in order for this attack to work—she skips the login page and is authenticated as Bob on the basis of sending his cookies.

This type of attack requires the hacker to have access to the cookies on the target's machine. This could be accomplished through obtaining physical access to the target's computer through an evil maid attack or by installing malware on the target machine and copying the cookies over the network. However, even if an attacker obtains valid cookies, the attack still may be thwarted because the web server may recognize that the request is coming from an unusual IP address or web browser—a protection against cookie stealing attacks employed by some web servers. Also, to continue the Amazon example, if Mallory were to attempt to do something sensitive, like make a purchase or change Bob's password, Amazon may prompt her to login just to be sure that the user really is Bob. This is another defense web servers employ against cookie stealing and other types of unauthorized access attacks. It poses a minor inconvenience for users since they have to login again, but it limits the amount of damage that an attacker can do.

8.2 Authorization

Once a person has been authenticated, the system is able to associate the user with data and activities. This makes authorization possible. *Authorization is permitting or denying access to a resource.* Authorization asks the question, "Now that I know who you are, are you allowed to do that?" For example, when a user tries to open a file, the operating system (OS) needs to verify the access that the user has for that specific file, and then permit or deny the request.

8.2.1 Operating Systems

On a computer system, permissions are set and enforced by the OS. The person who owns a computer typically has the highest level of user privileges available. For Windows machines, this is called *administrator access*, and for Linux and Mac machines, it is called *root access*. Users with this level of access are able to take any available user action. Typical actions on a computer system include creating, updating, and deleting files and users, installing, running, and deleting programs, modifying system configuration settings, and assigning permissions to other users. Users issued computers from their employers for work typically do not have administrator access, preventing them from changing certain settings or installing programs for security reasons.

Some computers, like the computers in a computer lab, have multiple users. Different users might have different permissions on the computers. For example, Alice might have full permissions over files in her home directory but be unable to view the files in Bob's home directory, and vice versa.

Most OSs that run on personal computers implement *discretionary access control* (*DAC*). In these systems the OS allows users to assign permissions to other users. For example, a user that creates a file could allow or restrict other users from viewing or modifying the file.

Users can be associated with other users through *groups*. A group is *a collection of users*. Groups can be assigned permissions just like individual users, and all of the users in a group share the same permissions. For example, for a university's introduction to programming class, all of the students could be given user accounts on a Linux computer, allowing them to write, compile, and run their programs. Rather than setting their permissions one-by-one, the instructor could add all the students to a *student* group. The *student* group could then be given permission to view all of the files in the instructor's *labs* directory, but not permissions to modify the files. When a student clicks on a lab file, the OS checks the permission, sees that the user is in the *student* group, and serves the file to the student as *read-only*—this means the student can read but not modify the file. Instructors for the class could be added to a *teacher* group. The teacher group could have read and write privileges for the *lab* directory as well as the *answer key* directory. A failure of access control would occur if a student is able to modify a lab or view an answer key file.

To implement authorization, OSs are organized around *subjects* and *objects*. Subjects are the actors in a computer system. They include users, groups, and processes. Objects are the resources on a computer, including files,directories, and programs. Objects are acted upon by subjects. Non-human users, called *system users*, are needed to take actions on behalf of the OS. System users are not assigned a password and are not for logging in to the computer. They are managed by the OS. For example, a system user might trigger an antivirus program to run on a set schedule to scan for malware. In most OSs, when a program runs, it runs by default with the permission of the user that started it—this is called *delegation*. So, if the antivirus program needs to read all of the files in the system, then the system user that starts it must have permissions to read all the files, too. In some OSs it is also possible to override delegation and have a program run with the permissions of the user that owns the program instead of the permissions of the user that started the program.

All of the permissions in an OS can be recorded in a *permissions matrix* (see Table 8.2). A permissions matrix captures all the permissions subjects have on objects. If every subject and object is listed in the matrix, then it would detail every possible interaction. The columns in a permissions matrix are the subjects and the rows are the objects. The intersection of a row and column records what permission the subject has on that object.

Table 8.2

Example Linux-style permissions matrix.

	alice (H)	bob (H)	student (G)	teacher (G)	system (S)
/bin/gedit (P)	RX	-	RX	RX	RWX
/bin/gcc (P)	-	-	RX	RX	RWX
/home/alice/lab1 (T)	RW	-	-	R	RW
/home/bob/sort.cpp (T)	R	RW	-	RW	RW
/home/instructor/labs (D)	-	-	R	RW	RW
/home/instructor/labs/lab1 (T)	-	-	R	RW	RW

Key: H=human user, G=group, S=system user, P=program, D=directory, T=text file
R=read permission, W=write permission, X=execute permission

Linux OSs employ a basic *read, write,* and *execute* permissions model (abbreviated RWX, respectively). Read is *the view permission*, write is *the modify permission*, and execute is *the permission to run a program or script.* It does not make sense to have an execute permission on an ordinary text file because it cannot be run, but users do need execute permissions for programs and scripts. Permissions have different meanings for files and directories. *A directory is a container for files* and is sometimes called a *folder* in the Windows OS since it is represented by a folder icon. The read permission for a directory means a user can see the files and subdirectories within it, the write permission means the user can create files and subdirectories within it, and the execute permission means a user can change directories into it with the *cd* command.

Besides these basic permissions, OSs can define many other permissions, including the permission to assign permissions, and more granular permissions, such as a modify permission that permits a file to be modified but not deleted. Some OSs, such as Windows, also include negative permissions. A negative permission is an explicit denial of an action (see Figure 8.5).

Figure 8.5

Windows OS file permissions showing positive (Allow) and negative (Deny) permissions.

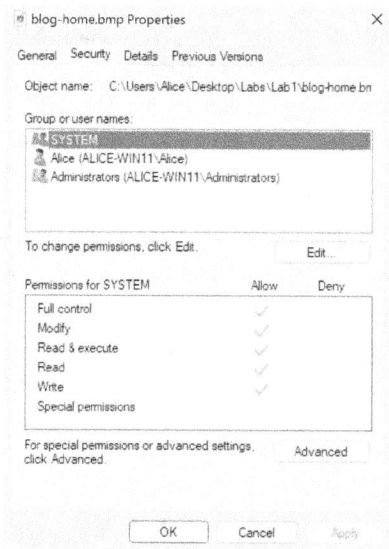

The objects in the first column of Table 8.2 are *fully qualified*. This means their full directory path is provided, uniquely identifying them in the *file system*. The file system is *the organization of the files in an OS*. The beginning forward slash / is the root directory. The root directory contains files and subdirectories, and those subdirectories in turn may contain files and additional subdirectories, and so on. The result is a file tree structure with branches and levels (see Figure 8.6).

Each directory creates a *namespace*. Within a namespace, there can be no *name collisions*. A name collision is when two objects have the same name. It is possible for two files to have the same name in a file system as long as they are not in the same directory. For example, in Table 8.2, there are two files named *lab1*. Presumably Alice created a copy of the *lab1* document from the instructor *labs* folder. Her copy is in her home directory, and since she created the copy, she has read and write permissions on her copy.

In Table 8.2, the user *alice* has the permission to read and execute (i.e., run) *gedit*, but *bob* does not. (Gedit is a text editor for Linux similar to Notepad in Windows.) The *student* group also has permission to read and execute *gedit*. If the user *bob* is in the *student* group, then his privileges for *gedit* are ambiguous. Do the group permissions trump the user permissions? This begins to illustrate some of the complexities that can arise with permissions in OSs. An OS's documentation would explain how conflicting permissions are resolved, and users can perform experiments to see for themselves how different situations are handled. In this case, on a Windows and Linux OS, *bob's* permissions are the union of his individual permissions and the permissions for all the groups he is in. Therefore, he would have permission to run *gedit*. One exception would be on a Windows system if *bob* had or was in a group with a negative permission that denied him access. Negative permissions take precedence over positive permissions when they are in conflict.

Figure 8.6

Linux OS file tree starting from bob's home directory.

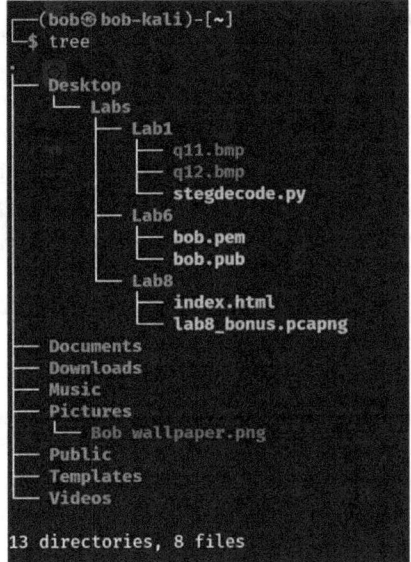

Another complexity that arises with permissions in OSs are inheritance rules. For example, if a user is given permission to *read* a directory, then by inheritance, he can view all the files and directories under that directory. Similarly, if he is given permission to *write* a directory, then he can create subdirectories underneath that directory.

OSs do not implement a data structure like the full permission matrix illustrated in Table 8.2. Instead, they track and enforce authorization by associating lists of user permissions with files. In other words, files "carry around" their sets of permissions. This method of managing permissions is called *access control lists* (*ACLs*). Figure 8.7 shows the ACL for a Python script in a Linux OS called *stegdecode.py*. It details that the user *bob* has *RWX* permissions on the file. This means that *bob* (*user* in Figure 8.7) can read the script, edit the script, and run the script. Meanwhile, all users in the *student* group can read and run the script, but they cannot edit it. All other users (*other*) have no permissions on the script, so they cannot view it, modify it, or run it. Windows also uses ACLs. Figure 8.5 above shows an ACL for the *blog-home.bmp* file. It shows that the *SYSTEM* user has full control over the file, and that *Alice* and the *Administrators* group also have been assigned certain permissions for the file.

Figure 8.7

The access control list for a Python script in a Linux OS.

```
┌─(bob⬤bob-kali)-[~/Desktop/Labs/Lab1]
└─$ getfacl stegdecode.py
# file: stegdecode.py
# owner: bob
# group: student
user :: rwx
group :: r-x
other :: ──
```

8.2.2 Multi-level Security

Most OSs use DAC, but for some highly sensitive information environments, *mandatory access control* (*MAC*) is preferred. In a MAC system, permissions are managed exclusively by administrators. This makes it much easier to monitor who has access to what resources because users are not able to assign permissions to other users.

The United States Department of Defense (DOD) uses MAC to protect national security information. They control access using a multi-level security (MLS) scheme that assigns permissions based on information sensitivity levels. The levels they use are *Unclassified* (*U*), *Confidential* (*C*), *Secret* (*S*), and *Top Secret* (*TS*). The levels are called *clearances* for subjects and *classifications* for objects.

Data is classified based on the damage it could "reasonably be expected to cause" if it were disclosed: U poses no risk, C could cause "identifiable damage," S could cause "serious damage," and TS could cause "exceptionally grave damage" to national security.[1] Examples of TS information might be the names and locations of spies abroad and the details of a military or intelligence operation. As we saw in Chapter 3, *cyber operations are intelligence and military operations that take place in and through cyberspace.* Information surrounding cyber operations are highly likely to be classified TS.

People are provided clearances based on what they need to know in order to do their jobs. Since it is a MAC system, all authorizations are approved and managed by administrators. The higher the clearance, the more vetting necessary for the subject. TS security clearances require an extensive background investigation that can take months to complete. The subject must disclose the places he has lived, the foreign countries he has visited, and the jobs he has held. Investigators interview people that know the subject to gather additional information and to verify facts. Investigators look for ties to foreign countries, any illegal activities the subject may have engaged in, and anything in his background that could make him susceptible to blackmail, such as significant financial debt, a history of

---

[1] From the United States Code of Federal Regulations Title 41 105-62.101

gambling or drug use, etc. Some special TS clearances also require the subject to undergo a psychiatric evaluation and a polygraph examination (i.e., a lie detector test). Most cyber operators need a TS clearance in order to do their jobs.

Once cleared, subjects have access to information on a *need-to-know* basis. This means they can only access information relevant to their job duties—they do not automatically have access to all the information at their level. This is a control that mitigates the risk of *data leaks*.

There are two competing and contradictory models for how access to information in a MLS ideally should be managed. Both have non-intuitive aspects. The *Bell-LaPadula model* (named after the two researchers that defined it) is concerned with the *confidentiality* of information. It states that subjects should only be able to read information at their clearance level and below. This makes sense. A subject with an S clearance should be able to read information classified U, C, and S, but not TS. However, it also stipulates that a subject should not be able to write information below his level of clearance. The reason is because if a TS subject composes a S document, some TS information might "leak" because the subject's mind contains a jumble of sensitive information that cannot be neatly parsed into TS and S. Therefore, in the Bell-LaPadula model, a subject can only write documents at or above his clearance level. This is known as the *high-water mark principle*. The highest level of information that a subject is exposed to sets the bar, and no writing is permitted beneath that level to prevent information leakage—a confidentiality concern. The Bell-LaPadula model is sometimes summarized as, "No read up, no write down."

A researcher named Biba devised an alternative MLS model that comes to opposite conclusions. The *Biba model* is concerned with the *integrity* of information. This model states that a subject should not be able to write information above his classification level. This makes sense because he is not authorized to know information at that level, so his ability to compose it is suspect. The Biba model also stipulates that a subject should not be able to read any information below his classification level. The reason for this is that the lowest level of information that a subject is exposed to "pollutes" him with potentially misleading or incomplete information. This is known as the *low-water mark principle*. Therefore, to prevent polluting information at higher levels—an integrity concern—he is not permitted to *write* above the lowest level he has been exposed to. The Biba model is sometimes summarized as, "No write up, no read down."

Table 8.3 summarizes both MLS models. While it makes for an interesting academic debate, it is impossible to satisfy both concerns simultaneously, so in the real world compromises are made to appropriately handle sensitive data.

Table 8.3

A summary of the Bell-LaPadula and Biba models for MLS schemes.

	Read up?	Read down?	Write up?	Write down?
Bell-LaPadula	No	Yes	Yes	No
Biba	Yes	No	No	Yes

Companies may also control access to data using a classification system. For example, they may visibly *watermark* documents using categories such as *sensitive, confidential, proprietary,* or *not for distribution*. A visible watermark is *a conspicuous marking in the background of a document* (see Figure 8.8). Organizations can develop processes around marking and handling documents. Seeing the marking reminds employees to be careful when distributing the information in digital and printed form.

Figure 8.8

A corporate document watermarked confidential.

ACME CORP

LONG-TERM
STRATEGIC PLAN

8.2.3 Authorization in Applications

Authorization is also a concern for many desktop, web-based, and smartphone applications. Most applications need some kind of authorization functionality for the services and data they provide. Many use a simple scheme that classifies users into a small number of roles such as *viewer* and *admin*. Then, based on their role, users are able to perform actions and access data. This method of managing authorization is called *role-based access control (RBAC)*. For example, in a web application used by the human resources division of a company, the web server could serve a "Manage Employee Salaries" webpage to users in the *admin* role, but not to users in the *viewer* role. If an employee in the *viewer* role tried to access the webpage, the web server's answer to the "Are you allowed to do that?" question would be "No!" and the user would receive a permissions error.

8.2.4 Firewalls

In computer networks, authorization is enforced by *firewalls*. As we saw in Chapter 5, a firewall is a software application or hardware appliance that allows or denies network traffic based on a set of rules. Firewalls act as a gatekeeper, inspecting all incoming and outbound traffic. Rules are created to either allow or deny packets based on their contents. Different types of firewalls inspect packets at different levels. A *packet filter firewall* focuses only on metadata in the TCP and IP headers of individual packets. For example, certain source IP addresses or destination TCP ports could be blocked. A *stateful firewall* maintains a memory of inbound and outbound packets for a window of time and uses that context to determine a packet's fate. For example, if a computer from inside the network initiates a connection outside the network, the firewall might permit the response packet to come back into the network. However, if an outside computer tries to initiate a connection to a computer inside the network unsolicited, the firewall might block it. An *application firewall* inspects not only packet metadata, but also the payload of packets. The payload of a packet is the application layer data that it carries. By inspecting the payload, an application firewall can scan incoming packets for dangerous executables and other malware and prevent them from coming into the network. These types of firewalls can also scan for sensitive and proprietary information in outbound packets and block them from leaving the network. This practice is known as *data loss prevention* and can help prevent a data leak.

Firewalls are installed on network appliances such as routers and also on *endpoints*. Endpoints are *the computers, smartphones, and other devices on the network*. Windows computers come configured with Windows *Defender Firewall* (see Figure 8.9). The Windows firewall can be configured to block applications from communicating over the network.

Firewalls can also block access to websites, preventing users from visiting them. If a user tries to visit a blocked website, they may see a generic HTTP 404 "Page Not Found" error or a message from the firewall informing the user that the site was blocked. There are two main approaches to blocking websites: *whitelists* and *blacklists*.

A whitelist is *a list of explicitly approved resources*. Any resource not on the list is denied. This approach is highly restrictive because, by default, all websites are blocked. A website must be explicitly allowed in order for a user to visit it. Whitelists prioritize security over costs. They ensure a high degree of security because websites are vetted and approved before being added to the whitelist. They come with a cost, however, because it is difficult to maintain a whitelist and make sure that all the websites a user needs are available. There are many websites, with new ones coming online everyday, and users may need access to different websites across time in order to do their jobs. If they need access to a website that is not on the whitelist, then the user must wait for the site to be vetted and approved before he is able to visit it, reducing his productivity.

Figure 8.9

Windows Defender Firewall.

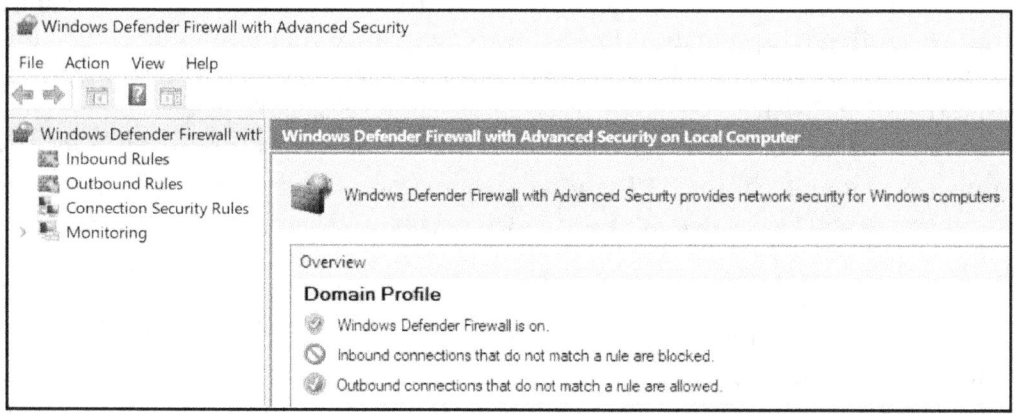

A blacklist is *a list of explicitly denied resources*. Any site not on the list is approved. This approach identifies unsafe websites and adds them to the blacklist. Blacklists prioritize costs over security. Because only known unsafe or undesirable websites are added to the list, users never need to visit these sites in order to do their jobs. However, it is likely that there are many bad websites that are not on the list. This opens the door to a user visiting a malicious website and causing a security incident. After the first visit, the website might be added to a list of websites to be vetted, and if so, it could be added to the blacklist to prevent users from navigating to it in the future, but by that time the damage may already have been done.

Some smartphone users use whitelists or blacklists to control who can call their phones. If a whitelist is used, only known good callers are allowed through, such as friends and family in their contact list; all other callers are sent straight to voicemail. This eliminates *spam* phone calls. Spam is *an unsolicited and unwanted communication*. The downside of this approach is that some important phone calls might be missed. Some callers not in a users' contacts might have legitimate reasons for calling, such as a company calling to schedule a job interview. If a blacklist is used instead, only certain numbers are sent straight to voicemail and everyone else is allowed through. This can prevent harassing calls from the same number, but since spam phone calls often come from different numbers, adding them to a blacklist is like playing a game of Whac-A-Mole: whenever one number is knocked down, a new number pops up!

8.3 Accounting

Authentication and authorization mechanisms work together with accounting. *Accounting is recording who did what when*. As actions are processed, they can be recorded. This is also known as *logging*. Logging is *recording cyberspace events*. Because actions in cyberspace are composed of digital signals processed at incomprehensible speeds (see

Chapter 2), everything that happens in cyberspace is ephemeral. If actions are not logged, there is no way to know who did what when.

Many types of activities can be logged. Some are more obviously relevant for cybersecurity than others. For example, logs recording when a malicious program was installed and by which user would be helpful for investigating how a cyber incident occurred. Other types of activities, like a system log recording when a background process was started by the OS, is less obviously helpful, although it could still be relevant for some types of cybersecurity incidents. As we covered in Chapter 3, cyber forensics is the analysis of cyberspace evidence. In general, the more events that are logged, the more potential clues could be found by cyber forensics experts to determine what happened. It is also the case that the more information that is logged and the longer it is kept, the more data that could be exposed in a data breach—some logging data might be valued by attackers. This is yet another cybersecurity tradeoff that must be weighed carefully.

Logging is a setting that can be configured by administrative users in an OS. Some events are logged by default, but not every event. Users are able to manage the events they want to track. While it would be nice to track everything all the time "just in case," there is a cost to logging. Logs occupy space that must be managed. Also, the more logs that are kept, the more "noise" to sift through to find events of interest. Log files accumulate and grow over time; they never shrink. Logging systems can be configured to only keep logs going back a set period of time. Older log records are deleted to make space for new logs. Alternatively, logs can be limited by size instead of time. An example would be to keep only the most recent one hundred MBs of log files for tracking certain types of activities. When the limit is reached, the oldest logs are deleted to make room for newly recorded events. This approach caps the size of the logs but leaves the time period they cover variable. Instead of deleting them, old logs can also be saved on removable media and stored offline. This would allow an organization to retain records of old events, but it may be more trouble than it is worth. The usefulness of logs decreases exponentially with time, and meanwhile, copying logs onto removable media and storing them can be time consuming. Plus, as the data accumulates over time and is not easily accessible, information may be difficult to locate even if it could be useful if found.

Logs are saved as OS system files and are protected. They are meant to be created only by the OS and never to be modified. As we saw in Chapter 4, however, hackers typically want to maintain access to systems and cover their tracks. They have discovered ways to subvert OS authorization protections for log files and are able to sometimes modify or delete them. Deleting logs, though, indicates malicious activity in and of itself. If hundreds of logs exist per minute every minute except for a five-minute period in the middle of the night, this indicates a hacker might be present on the system. Deleting logs is an event that also can be logged, although hackers have devised ways to delete logs without creating a log record. To prevent attacks on logs, it can be helpful to mirror log files to another system over the network in real time. This makes it much more difficult for hackers to track down logs and alter them.

The Windows OS provides the *Event Viewer* to configure log settings and to view logs (see Figure 8.10). Windows classifies logs into three main groups: *application*, *security*, and *system* logs. Application logs record events associated with programs installed on the OS, such as when they are installed, modified, or deleted, and when they are used. Security logs record information related to user accounts such as logins and logouts, file accesses, and other security-related events such as password changes. System logs record actions taken by the OS such as running background processes, and also issues related to the operating system like hardware failures and system shutdowns.

Figure 8.10

Windows Event Viewer security logs.

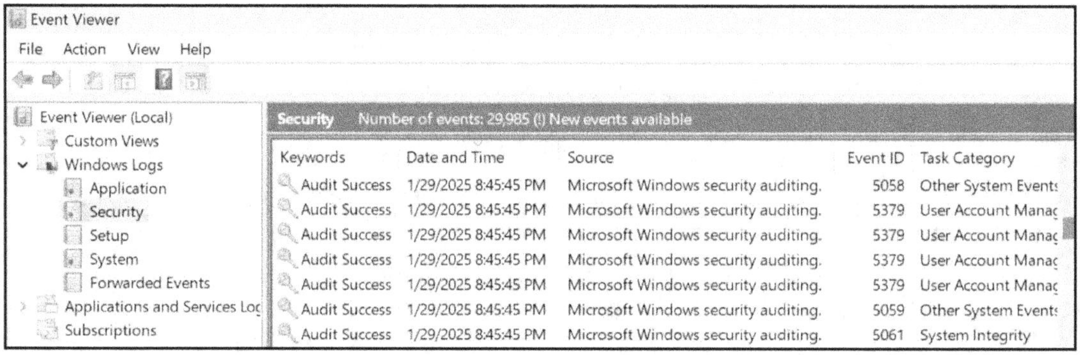

Logs label events they record with attributes such as *timestamp, event ID*, and *level*. The level attribute records the importance of the event. Example levels are *information, audit, error, warning*, and *critical*. Items marked critical need to be reviewed. For example, if a user is locked out of his account due to too many failed login attempts, this could be logged as critical because it may be an issue that needs to be fixed or investigated.

In addition to the OS, many applications keep their own logs. These logs can be helpful for diagnosing issues with the application, such as why a user action failed. Applications typically show users a generic error message, but they record the technical details of what went wrong in the logs. Showing a user too much diagnostic information might raise more questions than it answers, and it can also be a security risk because it could reveal sensitive information about the inner-workings of the application. If a user encounters an error, systems administrators with proper permissions can review the application logs to diagnose and address the root issue.

Applications also record information that acts as implicit logging. For example, web browsers record the sites visited, keep local copies of web pages in the cache, and store cookies. This information is not logs per se, but it can be helpful for accounting (i.e., determining who did what when). Similarly, OSs record when a document was created and when it was last modified. The Windows OS saves deleted files in a trash folder for a period of time. These records can also be valuable sources of accounting information.

On a local area network, network appliances such as routers and firewalls log information about packets sent and received. These logs might contain information about websites visited, computers logged into, and data accessed. This data can be associated with users through the IP addresses of the devices the user signed into. Packet sniffers like Wireshark are also a type of logging. They record packets sent and received in a *packet capture* (*PCAP*) file. Packet sniffers record entire packets so PCAP files can grow large quickly. They can be used with filters to only record packets meeting specific criteria to narrow their scope. For logging purposes, packet sniffers are likely only used for a set period of time and for a particular purpose.

Network logging is important for network maintenance and support because it provides information about the resources being used. For example, logs can record user connections to a wireless network access point (AP). If an AP can only support 100 connections at once and it is commonly handling 90 or more, this is an indication that the wireless network infrastructure might need to be upgraded. Similarly, printers, copiers, and voice-over-IP (VOIP) phones can keep logs of all their activity. These logs are important for IT support purposes, but they can also be useful for accounting.

Logs are not only kept and controlled by users and systems and network administrators on local devices and networks. Logs are also kept in cyberspace by third parties. These logs are invisible to users and are outside of their control. More on this in Section 9.2.5.4.

8.3.1 Analyzing Logs
Logging is a key part of accounting, but accounting also includes the processing and analysis of logs. Logs can be analyzed to create detailed pictures of cyberspace activity. This is vital for cybersecurity because bad actors leave behind breadcrumbs in the form of explicit and implicit logs. Logs can be analyzed actively for *indicators of compromise* (*IOCs*) to determine that a cyber incident has occurred, and they can also be analyzed after a known incident by cyber forensic experts to determine how it happened and what the hacker did with his unauthorized access. After an incident logs can reveal what dates hackers were active and what data they accessed, modified, deleted, and exfiltrated. Logs can also be used to uncover the real-life identity of hackers so that they can be arrested, and logs can serve as evidence in a court of law to help convict bad actors.

Using logs to identify if an incident has occurred or if hackers are probing an organization is an active and ongoing process. Experienced hackers use tradecraft to remain as quiet as possible and to make it difficult for their activity to be noticed and flagged, while inexperienced hackers may "make a lot of noise" as they bump around within the cyberspace of an organization. Logs need to be analyzed to discover all kinds of suspicious activity, but the massive volume of data generated by logs is mostly filled with normal events. How does one find the proverbial needles in the haystack that might need to be investigated?

Security information and event management (SIEM) is the process of aggregating logs and analyzing them for suspicious activity. Because of the vast volume of logs that are collected, software takes the first pass at analyzing logs looking for anomalous activity. If it finds anything suspicious, it can create an alert that can be investigated by a human being. Anomalous activity is defined by a baseline of normal activity. Therefore, this type of software might require a period of training and fine tuning. For example, if it is normal for users to login to the network from foreign countries, then those types of logins will be ignored—otherwise, they would be highly suspicious and trigger an alert.

False positives and *false negatives* are two different errors that can occur in alerting systems. In this context, a false positive is *an alert that turns out to be normal activity.* A false negative is *when there is malicious activity but no alerts are generated.* These two errors are in tension because attempts to reduce one increase the other. False positives are inefficient because they waste time. Plus, they may have the adverse effect of causing *alert fatigue.* The more alerts that are investigated that end up being normal activity, the more likely future alerts are to be ignored, like the boy who cried wolf in the fable. On the other hand, false negatives are also clearly a problem. If a hacker has been inside an organization for months and all of the evidence of his presence was in the logs but went unnoticed, then the organization's processes around analyzing logs and creating alerts was clearly deficient. AI promises to improve the ability of SIEMs to reduce false positive and false negative alerts and is an important area of cyber research and development.

SIEM systems aggregate logs from multiple different sources. This provides significant advantages. For one, because the logs are centralized, they can all be viewed in one place using the same platform and functionality. Logging platforms differ widely in how they store and present data for searching and viewing. Learning many different platforms and switching and translating between them is inefficient. For example, different logging systems may use different scales for alert levels and different categories for similar events. Logs are vendor specific. But with a SIEM platform, all logs are brought under one umbrella and converted so that they fit under a common standard. Plus, SIEM interfaces are designed to be intuitive to use and include advanced features for analyzing logs. They make it easy to search for and view logs and to sift through them to find events of interest.

Another benefit of log aggregation is that it can expose wider patterns that indicate nefarious activity that would go unnoticed if logs were siloed. For example, a single logging system might reveal that the user Mallory tried to access one of her manager's files. That isolated event is not likely to trigger an alert—it could be that Mallory accidentally clicked on a wrong file. However, a centralized logging system might reveal that Mallory tried to access files belonging to many different managers across the organization from several different computers. This could indicate that Mallory might be an insider threat or that her account may have been compromised.

8.4 Conclusion

Access control is the means of cybersecurity because cybersecurity is primarily concerned with preventing unauthorized access to computer systems and data. Hackers strive to disclose and alter data and to deny access to authorized users. A prerequisite to many of their objectives is gaining unauthorized access. Hackers can do this by defeating and subverting access control mechanisms.

The access control acronym *AAA* stands for *authentication, authorization*, and *accounting*. The goal of authentication is to make sure that users are who they say they are. The goal of authorization is to make sure that computer systems and data are only used in ways that are permitted. The goal of accounting is to record events so that it is possible to see who did what when. As we have seen in this chapter, there are many detailed aspects of each area.

None of these three systems work perfectly—all of them have weaknesses. But having well designed access control mechanisms in each of the three areas is a vital component of a sound cybersecurity posture.

Chapter 9

9. The Application of Cybersecurity: Principles and Practices

"If you think technology can solve your security problems, then you don't understand the problems and you don't understand the technology."
- Bruce Schneier

As we have seen throughout this text, managing cybersecurity is complex. There are many facets that need to be understood including how cyberspace works, cyber adversaries and their tactics, cyber risk management, cryptography, and access control. Implementing cybersecurity takes all of this knowledge and more. It involves understanding the nature of the problem and the variety of possible solutions. It involves making shrewd choices and wise investments. It involves technology-based solutions as well as people, process, and facility-based solutions.

This chapter reviews basic principles of cybersecurity and then several best practices. In theory, every practice is rooted in a principle of cybersecurity. Principles are akin to rules of thumb and practices are specific actions that implement principles. Following sound principles and best practices leads to strong cybersecurity. This chapter outlines the top ten principles and several areas of best practices—it is this text's most practical chapter in terms of implementing cybersecurity on the ground.

9.1 Cybersecurity Principles

"Effective people lead their lives and manage their relationships around principles; ineffective people attempt to manage their time around priorities and their tasks around goals." - Principle-Centered Leadership *by Stephen Covey*

Principles are high-level guidelines that inform daily priorities and decisions. Wise principles lead to good decisions and positive results. The converse is also true: faulty principles lead to poor results. It is not necessary to be able to trace specific causes-and-effects in order for principles to work their magic. Identifying precise cause and effect relationships is difficult to do because of the real world's numerous confounding variables. In the short term, no results may be apparent, or the initial results might even be negative. Therefore, acting on principle requires faith. One makes decisions in line with his principles, trusting that they will lead to good results over time. At times there are temptations to take shortcuts believing that negative consequences will not materialize. But in the end, one reaps what he sows. Sooner or later, poor decisions lead to bad results. It takes discipline and a measure of faith to consistently live out good principles.

One example of a wise principle for life and relationships is the Golden Rule. The Golden Rule states, "Treat others how you would like to be treated." While it is a high-level concept, living by this rule impacts words and actions on a daily basis. The results of living this way may not be immediately apparent, but they accumulate over time and lead to flourishing relationships and positive well-being in the long run. There are many other well-known principles for life, including principles covering personal finances (e.g., "live below your means") and communicating in the digital age (e.g., "never put anything online you would not want your parents to see").

Figure 9.1

The top ten principles for cybersecurity.

In cybersecurity there are several principles that can help guide the countless decisions organizations must make on a daily basis. Making choices that align with wise principles is important. This section outlines ten of the most helpful high-level principles for cybersecurity (see Figure 9.1). They describe ways of thinking and seeing the world that will impact decisions all across an organization.

9.1.1 Adversarial Thinking

"It helps you to react better if you are thinking about what the other person is going to be thinking about—how he is going to react to your reactions—[rather] than just assuming that he is going to be a complete idiot." - Anonymous Cedarville Intro to Cybersecurity student

Cybersecurity is only necessary because of the existence of people who deliberately attack computer systems and networks. Therefore, adversarial thinking is central to cybersecurity. Chapter 6 focused on this principle as the fundamental skill of cybersecurity.

The principle of adversarial thinking states that one must never forget about the existence of intelligent human hackers. Maintaining constant awareness of their technological capabilities, unique perspectives, and strategic reasoning skills results in more vigilant and better cybersecurity practice. Forgetting or underestimating the adversary results in naive choices that create vulnerabilities that can be exploited by hackers. Cyber defenders must be able to think like a hacker.

Every security context, including cybersecurity, contains three fundamental components: bounty, barriers, and bad guys. The bad guys must breach the barriers in order to get their hands on the bounty. There is a temptation to focus on the value of the bounty and on bolstering the barriers while ignoring the bad guys. But, if we get so focused on any one cybersecurity best practice, technology, tool, etc., that we forget about the bad guys, we do so at our own peril. We need to frequently lift our eyes from the barriers and put ourselves in the shoes of the bad guys who are sizing up our defenses and looking for cracks.

This principle does not downplay the importance of best practices—they are absolutely necessary—it just stresses that we should always do so while remembering the reason it all exists: hackers. Following the principle of adversarial thinking should shape everything an organization does. Onboarding new employees? They must be made aware of the cyber threat and how they can help keep the organization safe. Purchasing new technology? This increases the organization's cyber footprint, so it must be installed with appropriate cybersecurity parameters and monitored for malicious activity. Writing new software? It must be designed and tested with cybersecurity in mind. The list goes on and on. Cyber vigilance naturally follows from keeping the cyber adversary at the forefront of one's mind.

9.1.2 Depth Wins

> *"We [being cyber attackers] put the time in ...to know [the network] better than the people who designed it and the people who are securing it. You know the technologies you intended to use in that network. We know the technologies that are actually in use in that network." - Rob Joyce, former head of NSA's Tailored Access Operations group*

In 2016 Rob Joyce, once known as the nation's Chief Hacker, gave a rare public presentation in which he "gave away" the secrets of nation state hacking. His main message was that cyber attackers succeed because they know more than the cyber defenders—more about the target network and how the technologies work on that network. This is the *depth wins principle.* This principle states that *the success of cyber attacks and cyber defense often comes down to who knows more.*

People have different levels of understanding of technology. If an attacker understands technology at a deep level, he can use that to his advantage in attacking his target. For example, think of a stereotypical, computer-illiterate granny (see Figure 9.2). Since granny has no real understanding of how technology, computers, and software work, her computer is a black box to her—a complete mystery. When she sees security-related prompts, she is not sure what they mean and does not know what to do. She can be fooled into clicking things she should not click on and installing things she should not install. Further, she will not recognize even obvious indications of compromise because she is not sure what is normal and what is not—hackers can hide in plain sight. In short, granny is an easy target because her knowledge is so limited.

Figure 9.2

The depth wins principle.

The depth wins principle generalizes to all people, not just to hacking grannies! Many students reading this textbook are tech-savvy digital natives who are highly competent computer users—not at all like granny. They are power users on their computers, and may even be comfortable modifying configurations to make their systems more secure. They are several layers deeper in knowledge than granny. But, what they may not realize is that elite hackers are to them what they are to granny! In other words, elite hackers are several layers deeper than they are. This makes even tech-savvy digital natives an easy target for elite hackers. Elite hackers can trick them into clicking things they should not click on and installing things they should not install. Elite hackers can even hide "in plain sight" on their computers without them even noticing anything amiss.

The depth wins principle works because computers are complex. There are layers and layers of depth. As Bruce Schneier reminds us, "Complexity is the worst enemy of security." Hackers can find vulnerabilities in this complexity, and can hide in black boxes—or places in the operation of computers that are mysterious and not understood by users. For example, Figure 9.3 shows just a few of the many processes running in the background of a typical Windows computer. Are these all normal processes that are supposed to be running? It takes expertise to be able to differentiate benign from malicious behavior. In general, he who knows and understands more wins. The less you know, the easier you are to hack. Success in hacking often comes from burrowing beneath the technological level of understanding of the defender. The hacker can operate in these deeper technology layers unnoticed because they are blindspots to the defender. He cannot differentiate normal from suspect activity in these layers. The attacker is invisible to the defender, and meanwhile, going deeper opens up new vistas and promising opportunities for hackers.

Figure 9.3

A partial listing of the background processes running on a Windows computer.

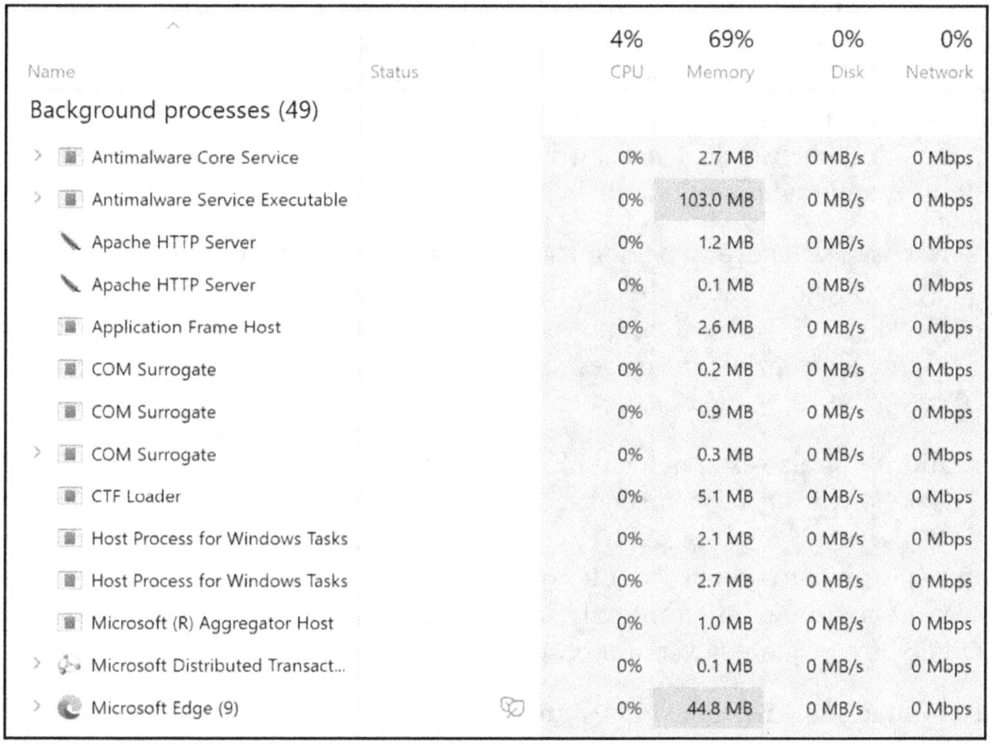

Name	Status	4% CPU	69% Memory	0% Disk	0% Network
Background processes (49)					
> 🔲 Antimalware Core Service		0%	2.7 MB	0 MB/s	0 Mbps
> 🔲 Antimalware Service Executable		0%	103.0 MB	0 MB/s	0 Mbps
❋ Apache HTTP Server		0%	1.2 MB	0 MB/s	0 Mbps
❋ Apache HTTP Server		0%	0.1 MB	0 MB/s	0 Mbps
🔲 Application Frame Host		0%	2.6 MB	0 MB/s	0 Mbps
🔲 COM Surrogate		0%	0.2 MB	0 MB/s	0 Mbps
🔲 COM Surrogate		0%	0.9 MB	0 MB/s	0 Mbps
> 🔲 COM Surrogate		0%	0.3 MB	0 MB/s	0 Mbps
🔲 CTF Loader		0%	5.1 MB	0 MB/s	0 Mbps
🔲 Host Process for Windows Tasks		0%	2.1 MB	0 MB/s	0 Mbps
🔲 Host Process for Windows Tasks		0%	2.7 MB	0 MB/s	0 Mbps
🔲 Microsoft (R) Aggregator Host		0%	1.0 MB	0 MB/s	0 Mbps
> ⟳ Microsoft Distributed Transact...		0%	0.1 MB	0 MB/s	0 Mbps
> ◉ Microsoft Edge (9)	🗐	0%	44.8 MB	0 MB/s	0 Mbps

In cyber, attacks often come from below. The depth wins principle teaches that time invested in understanding technology, especially the technologies that are in use on one's own computer and network, is time well spent. Ignorance is not bliss. The failure to pry into black boxes to understand what is inside them is a major security vulnerability. Investing in learning and gaining expertise confers substantial advantages.

9.1.3 Trusting Trust

> *"To what extent should one trust a statement that a program is free of Trojan horses? Perhaps it is more important to trust the people who wrote the software."*
> *- Ken Thompson in "Reflections on Trusting Trust"*

In Section 4.1.2.4 on supply chain attacks we mentioned the famous Turing Award lecture written by Ken Thompson called "Reflections on Trusting Trust." Thompson's point is that trust is inherent in computer security and can never be fully eliminated, but it should at least be acknowledged for the risk it poses. *The principle of trusting trust states that trust relationships should be explicitly identified and examined so that they can be managed appropriately.* The default, and easier, alternative is to ignore or downplay the levels of trust placed in others, but this leads to significant unacknowledged and unmonitored cybersecurity risks.

Organizations depend on numerous trust relationships. They need to trust employees, customers, contractors, vendors, software, and hardware. This principle does not imply that these trust relationships are inherently bad, but only that they should be scrutinized and managed well. If there are ways to limit trust, similar to the principle of least privilege (see below), then those ways should be pursued. Measures should also be taken to verify trust—in other words, trust but verify. This means to the extent possible proper accountability and reviews should be in place to make sure that trust is not being abused.

Trust is exercised whenever a person uses somebody else's computer. As we saw in Section 4.1.2.3, keylogger attacks are easy to conduct. Could the owner of the computer be running a keylogger? If the computer is in a public space, is the organization responsible for it taking appropriate measures to prevent hardware-based keyloggers from being installed?

Trust relationships also exist in technology on networks. In computer networks computers and devices typically have special access to other computers and devices on the same network. This trust can be abused by hackers because by gaining access to one computer, they can easily pivot to other computers and devices. For this reason, trust relationships on networks should be limited to only what is necessary, and additional authentication steps should be considered even if it poses an inconvenience.

After examining some relationships, it may be the case that the trust relationship should be eliminated. For example, the United States has had a complicated trust relationship with computer hardware and software originating from China, an adversary of the United States. The United States government at different times has warned citizens not to trust certain devices or apps because they could be used to spy on Americans. China denies any wrongdoing, in essence saying, "trust us." But because they have a motive to surveil United States citizens, and it is not generally possible to verify what they are and are not doing, one solution is to just reject this trust relationship, and the United States government has done just that on occasion.

Individual computer users also maintain trust relationships that should be scrutinized. When online, users need to be careful about divulging personal information and details. When we provide personal information, we are trusting the person or organization we are sharing them with. Is this trust well-placed? Will they safeguard our information from hackers? Is it possible that they could use the information themselves for nefarious purposes, like identity theft, blackmail, or password guessing? We should avoid unnecessary trust relationships and oversharing.

Trusting trust has important ramifications for AI. Machine learning is an AI technology that performs layers of calculations on enormous data sets. Due to its complexity, the outputs it produces are not easily proven or explainable, therefore, some measure of trust is required to evaluate AI-driven recommendations. If we start seeing AI as a superior form of intelligence, even if we are skeptical about its conclusions, we may be inclined to trust it over our own instincts. For example, if some day in the future the military began using AI for war planning as a substitute for the judgement of generals, as a society we would be placing a tremendous amount of trust in technology. Some might argue this trust is well-placed compared to human judgement which history has proven to be liable to error, biases, and vanity. Others might argue that hidden in its complexity is the potential for bugs and malicious hacking by our adversaries, making it less trustworthy. As a society we are steadily marching towards a future full of these kinds of dilemmas in the fields of medicine, law, and the military. One thing is for sure: as we become more dependent on technology for more consequential decisions, cybersecurity will become that much more important.

The principle of trusting trust boils down to examining trust relationships and then deciding what course or action to take based on the perceived risks and benefits.

9.1.4 Simplicity

"'Tis the gift to be simple." - from the Shaker hymn "Simple Gifts"

Several times in this book we have highlighted how complexity is the enemy of security. The more complex something is, the more difficult it is to understand—in other words, there is less light and more darkness. In this darkness there are potential vulnerabilities and places for attackers to hide. When it comes to security, the more light (both literally and figuratively), the better.

The principle of simplicity states that simplicity should always be pursued. There are a couple of straightforward implications of this principle. One, because simple systems are easier to secure than complex systems, and the more features a system has the more complex it is, unnecessary features should be eliminated from systems. This advice is counterintuitive because it is generally understood that more is better. If a person is trying to decide which app to download from an app store, and one app has just the core feature needed and the other has the core feature plus bonus features, it is tempting to believe that the app with more features is better. Even if a person is not sure if or how he would use some of the bonus features, he might imagine that those extra features could come in handy someday. This reasoning, however, does not take into account the added risk that these unused extra features pose. The app with more features is more complex, and the principle of simplicity would caution against automatically assuming it is the better choice. More cyber risk is assumed by choosing it.

The other issue with complexity is that it is difficult to correctly implement complicated things, and bad implementations are potential security vulnerabilities. For example, security policies should be as simple as possible. Every extra step and extra clause in a policy is an opportunity for errors to creep into the process. Section 2.9 of the ACM Code of Ethics even treats this as an ethical mandate in stating, "Computing professionals should discourage security precautions that are too confusing." The same is true for technology tools and software. Simple and straightforward tools are more likely to be secure than complicated and unclear ones. We also saw this principle in Section 7.4.3 on cryptography. The best cryptographic algorithms are marked by an elegant simplicity. They are easier to understand, implement, and use. As a developer of tools this principle argues for secure defaults because users that do not understand all the options and their implications are more likely to accept the defaults.

Simplicity also has implications for *data retention*. Data retention is *the practice of storing data*. On the one hand, the more data the better—one never knows when it might come in handy someday, and data storage is cheap, so why not just hang onto as much as possible for as long as possible? But more data means more complexity. It requires more resources to monitor and safeguard. It also means a bigger fallout if a data breach or doxxing attack occurs. The principle of simplicity would suggest that data storage should be minimized. An organization should only keep the data they know they will use and only for as long as it is useful. Older data should be purged. Like all cybersecurity decisions, this comes at a cost, but it limits the exposure data poses.

Complexity is inherent in cyberspace, but simplicity should always be pursued to the extent possible.

9.1.5 Weakest Link

"A chain is only as strong as its weakest link." - Popular saying

There is an asymmetry in cybersecurity that benefits the attacker: cyber defenders must protect everywhere at all times, but cyber attackers only need to find one opening at one point in time in order to succeed. *The weakest link principle states that hackers will take the easiest path towards accomplishing their objectives.* As they survey an organization, they might notice multiple angles of attack, but they will always pursue the easiest one. This is not because attackers are lazy; it is because they are smart. They will not take more risks and work harder than necessary to accomplish their goal.

The weakest link principle derives its name from the picture of a chain with multiple chain-links. As more and more weight is added to a chain, it will eventually break and always at the point of its weakest link. This principle is helpful for cybersecurity because it can be used to prioritize defensive measures. It is more efficient to invest in identifying and reinforcing the least secure aspects of an infrastructure than to just add security where it is most convenient to do so. When it comes to a chain, reinforcing the strongest link has no impact on its overall performance and is a waste of resources. In security, making the walls thicker and higher will not contribute anything as long as there is an unlocked door that attackers can just walk right through.

As we saw in Section 4.1.2.1 there is saying in cybersecurity about humans being the weakest link in cybersecurity. This is why social engineering is one of the most pursued and effective attack vectors. Attackers are not going to spend countless hours crafting a complex exploit to gain unauthorized access if they can just compose a *spear phishing* email (i.e., a highly targeted phishing attack) to obtain login credentials to the network. According to this principle, it may be tempting to spend resources on the latest and greatest security technology, but it is likely better for cybersecurity to spend those resources on cyber awareness training and on security processes for employees.

The weakest link principle helps cyber defenders to scrutinize where attacks may come, and to spend their limited resources bolstering the most cost-effective barriers.

9.1.6 Least Privilege

"A man's got to know his limitations." - Clint Eastwood as Dirty Harry

As we have seen throughout this textbook, cybersecurity is concerned with preventing unauthorized access to computer systems and data, making access control the means of cybersecurity. In computer systems, access control assigns subjects permissions to objects—permissions such as read, write, and execute. *The principle of least privilege states that permissions should be granted only up to the level needed and only for as long as necessary.* In other words, users and other resources should be given the *least privilege* necessary to perform their functions.

This principle seems like common sense, but it is often ignored because it is easier to assign permissions liberally. Managing fine-grained access control can be tedious and time consuming. For example, when a new employee begins working at a company, the specific accesses he needs may not be fully known. The temptation is to just give him every permission he *may* end up needing because this will save time and hassle later and make the organization more efficient. However, doing this would be a violation of the least privilege principle. What it gains in work efficiency it costs in cybersecurity exposure. A new employee would never be given a master key to every building and room at an organization just in case he might need it—this would be taking obviously undue risk. In cyberspace, the danger in giving a user administrator access to computer systems and networks is not as readily apparent, but it is similarly risky.

Another temptation is to give permissions to a resource when they are only needed temporarily, but then not have a process in place to rescind the permissions later when they are no longer needed. This results in people being given access for longer than they need—another violation of this principle. For example, if an unusual circumstance arises and a non-IT employee needs access to the server room, one option would be to just give him a key to the room and assume that he will return it later. But a better option would be to escort him to the room, unlock the door, and wait while he completes his task. The second option reduces the amount of time he has access to the bare minimum.

An implementation of least privilege in multi-level security systems where people handle classified information is *need-to-know* (see Chapter 8). This limits the exposure of classified information. Least privilege is also implemented in Linux operating systems with the *sudo* command. Sudo is short for "substitute user, do" and it allows users to run a command as another user. By default, sudo runs commands as the *root* user. Figure 9.4 illustrates Bob trying to view a sensitive file for which he does not have read permissions, therefore, he receives a "Permission denied" error. Then Bob runs the same command using sudo and is able to view the contents of the file. The screenshot also shows that Bob is re-authenticated at the point when he requests access to the file—he must type in his password—as an extra security precaution. Sudo can be used in this way to temporarily escalate a user's permissions in line with the principle of least privilege. Obviously, because sudo allows users to run commands as root whenever they want, most users are not given sudo privileges, so just typing the word *sudo* in front of a command you want to run does not guarantee it will work!

Figure 9.4

Using sudo to view a sensitive file on a Linux machine.

```
┌──(bob㉿bob-kali)-[~]
└─$ whoami
bob

┌──(bob㉿bob-kali)-[~]
└─$ cat /etc/shadow
cat: /etc/shadow: Permission denied

┌──(bob㉿bob-kali)-[~]
└─$ sudo cat /etc/shadow
[sudo] password for bob:
root:!:19879:0:99999:7:::
daemon:*:19879:0:99999:7:::
bin:*:19879:0:99999:7:::
sys:*:19879:0:99999:7:::
```

Following the principle of least privilege has several benefits. For one, it limits the damage that a malicious insider can cause. For example, if all the employees at a company were given access to the entire database of customers, then any employee could cause a wide-scale data breach. Least privilege also limits the damage that can occur by external threats who, as we have seen, often compromise user accounts. After compromising an account, a hacker has the same access as the compromised user. For this reason, compromising an administrator's account is a major coup for the hacker, so the number of users who have admin access should be small. Hackers also exploit vulnerabilities in computer processes, and when they do this, they assume the permissions of that process. If the process is running with root privilege, then the hackers have access to the entire system. Systems administrators must take care when assigning access to processes and computer programs, and they must be highly vigilant in protecting their own credentials.

Perfectly implementing the principle of least privilege is not possible. It would consume an enormous amount of resources and would not be worth the cost. However, that does not mean that the principle is not helpful. It should be aspired to and never flagrantly violated for expediency's sake.

9.1.7 Defense in Depth

> *"So the principles of warfare are: do not depend on the enemy not coming, but depend on our readiness against him. Do not depend on the enemy not attacking, but depend on our position that cannot be attacked."* - The Art of War *by Sun Tzu*

The principle of defense in depth states that security should be implemented in layers. The idea is to put in place multiple barriers for an attacker, forcing him to overcome all of them in order to accomplish his objectives. A well-fortified medieval castle is a good illustration of defense in depth (see Figure 9.5). Picture a castle built on elevated ground and surrounded by a moat. The castle has thick, high walls with guards stationed on top. The bridge is the only way to get to the big iron gate, and then the gate must be open in order to enter the castle grounds. At the gate guards interrogate and inspect everyone who enters and leaves. Inside the castle grounds, the crown jewels are hidden behind lock and key in an interior room where more guards roam the halls. The crown jewels themselves are surrounded by booby traps. In order for a thief to get away with this castle's bounty, he would need to overcome every single one of these obstacles both on the way in and on the way out!

In the realm of cybersecurity, the closest thing we have to castle-like defenses may be data centers. Data centers are extremely high-value targets, therefore, they typically implement layers and layers of security. They are surrounded by prison-like fences forcing all cars to enter through the gate which is continually monitored by a security guard. Entering the building requires another round of authentication involving multiple factors, including biometrics. Security cameras are everywhere. Once inside the building, sensitive areas are protected by even more security measures. And these are just the physical space facility-based security measures!

The idea of defense in depth is to make a hacker have to exert tremendous effort to overcome a barrier, only then to be confronted with another, even higher barrier. After that barrier is yet another barrier, etc. Each barrier drives up the costs of a successful attack.

The higher the costs, the less likely the hacker will succeed and the more likely he will be caught. If a would-be attacker perceives that the costs will be too high to attack a target, he will choose a different, easier target.

Figure 9.5

A medieval castle illustrating the defense in depth principle.

This principle reminds us that implementing a single defense and then thinking everything will be fine is hubris. One defensive measure is good, but two are even better, three are better still, etc. Ideally, the layers of defense are independent of one another, so they cannot fall like dominoes. Each layer takes a different creative approach to overcome. In cybersecurity the layers include many different controls and categories of security, such as physical security, access control, network security, system security, alarm systems, and more, implemented all across the people, processes, technology, and facilities of an organization.

9.1.8 Compartmentalization

> *"I cannot imagine any condition which would cause a ship to founder. I cannot conceive of any vital disaster happening to this vessel. Modern shipbuilding has gone beyond that." - Captain Smith, Commander of the Titanic*

The Titanic was believed to be an unsinkable ship because its hull was made up of several watertight compartments (see Figure 9.6). The theory was that if the ship was ever in a collision, only the compartments directly involved in the impact would flood, and the other compartments would stay dry and provide enough buoyancy to keep the ship afloat. Tragically and ironically, on its very first voyage the Titanic struck an iceberg and sank. For various reasons, the theory did not hold up in practice. Even though it is actually a counterexample, the Titanic provides a memorable illustration of the principle of compartmentalization. *The principle of compartmentalization states that access to resources should be segmented*. There are two parallel benefits of compartmentalization: to limit exposure and to avoid intermingling resources that could compromise access control.

Sandboxes, like the virtual machine (VM) pods used for cybersecurity labs, are a good example of using compartmentalization to both limit exposure and avoid the intermingling of resources. VMs run on top of the host operating system and provide a barrier for the processes that run within the VM—segmenting them from the host. Malware running in a VM cannot access files and processes belonging to the host. Like the Titanic, the design is not full-proof, and sometimes it is possible for malware to escape from a VM, but the principle holds. Operating systems implement compartmentalization in a similar way. They enforce memory space barriers between running processes, making it more difficult for a malicious process to gain unauthorized access to another process's resources (access control), and if a process crashes, the damage is limited to just that one process (limiting exposure).

Figure 9.6

A diagram of the Titanic illustrating the compartmentalization principle.

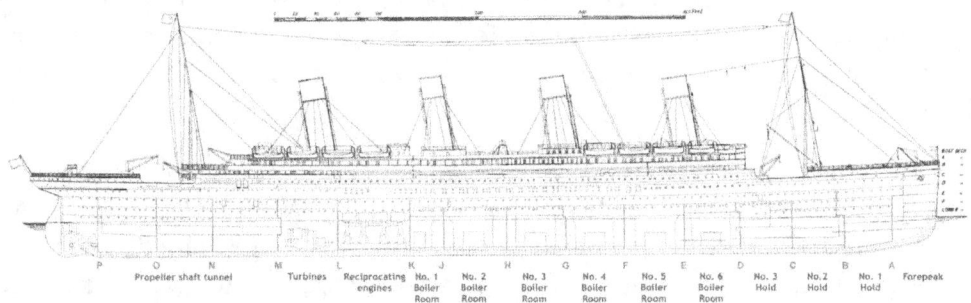

Compartmentalization limits *exposure*. As we saw in Chapter 5, exposure is *the potential losses that could result from an incident*. An organization's computer network should be segmented to limit exposure, creating barriers between different computing resources on the network. One such implementation of this is called a *DMZ* for *demilitarized zone*. A DMZ is *a segmented portion of a computer network that contains Internet facing servers*. These servers are directly accessible from the Internet and are most vulnerable to attack. Therefore, they are susceptible to compromise, and if that were to occur, the network is designed to confine the compromise to the DMZ network only. This type of network design is analogous to a safe room in a home—if a home invasion occurs, the safe room is an interior, fortified and secured compartment where a family can retreat while they wait for help to come. The rest of the house is like a DMZ, and the safe room is like the core, internal network, much more difficult to breach.

Compartmentalization is also important for avoiding the intermingling of resources that could compromise access control. For example, in Section 4.1.2.2 we learned about injection attacks. In these attacks user inputted data is executed as code. This is a failure of compartmentalization—inputted data should never be interpreted as code to be executed. An injection attack intermingles code and data, and it is ultimately a failure of access control—the attacker is able to run unauthorized code. Another famous example of the failure of compartmentalization is the vulnerability in the old landline telephone system that enabled phone phreakers to abuse the system (see Section 3.2.5). By inputting certain frequencies such as 2600 Hz into a phone, hackers could gain command control over the phone system. The system failed to isolate data (human speech) from code (control frequencies). Interestingly, this same vulnerability is built into most generative AI systems. The same prompt used for asking questions (data) can also be used to input commands and source data (code). *A prompt injection attack is an attack where malicious prompts are fed into large language models to manipulate their behavior.*

Compartmentalization also applies for access to information. In classified environments, information is segmented into *compartments*. A compartment *is a category of sensitive information*. Examples of compartments within the United States Department of Defense (DOD) classification system might include "nuclear weapons," "terrorist threats," and "Chinese intelligence." The information within the compartments still carry classifications, but even people with Top Secret (TS) clearances may not be able to access information in certain compartments (this is similar to need-to-know). There is a special type of clearance called TS/SCI. SCI stands for *sensitive compartmented information*. This is an even higher bar than a TS clearance and gives people access to more categories of sensitive information.

In the DOD multi-level security system compartmentalization is also enforced with facilities. People with clearances often work in a special environment called a *SCIF*. A SCIF is a *sensitive compartmented information facility*. SCIFs are specially designed to contain and isolate classified information. SCIFs have their own isolated computer network, their walls are designed to block radio signals, and computing devices are carefully vetted before being allowed in or out. Most cyber operations-related work takes place in SCIFs. Unvetted computers and technology, including personal smartphones and smartwatches, are not allowed in SCIFs because they could introduce data into the compartment or exfiltrate data out of it. SCIFs are designed to prevent unauthorized personnel, data, and resources from being mixed with authorized personnel, data, and resources.

Compartmentalization is a key security principle. It limits exposure and prevents compromises of access control by creating barriers around resources.

9.1.9 Security as a Process

"Security is a process, not a product." - Bruce Schneier

As we have seen throughout this book, cybersecurity is implemented in various ways across organizations through people, processes, technology, and facilities. It is naive to think of cybersecurity as a one-and-done checkbox, a set-it-and-forget-it solution, or a technology product that can be purchased. We have also seen that there is no such thing as 100% security, and that there is always room for improvement. Furthermore, organizations, technologies, and the threat landscape are ever evolving. *The principle of security as a process states that cybersecurity must permeate all aspects of an organization and be continually monitored and improved.*

On the one hand this principle is difficult to bear, because it makes it clear that there is no rest for the weary cyber defenders. The job is never done and constant vigilance is required. On the other hand, it is reassuring because everybody has to start somewhere. Imagine being put in charge of cybersecurity for a small organization—maybe a non-profit that does not have the budget to hire a seasoned expert and has virtually no cyber program in place. The task is daunting because the threat is real and there is so much to do. The principle of security as a process provides some breathing room. Not everything has to be completed on day one. Pick a place to start such as inventorying the organization's cyber assets, implementing a new security process, or installing a piece of technology, and come back the next day and do a little more. Each day the goal is to be more secure than the day before. A strong cybersecurity posture is achieved over time through incremental improvements.

This principle is also humbling in a helpful way. When it comes to cybersecurity, one never arrives. As organizations, technologies, and cyber attackers change, cybersecurity must adapt. Last year's product or process may not be the best choice this year. Even best practices change over time as we learn more about what makes for effective cyber defense. Pride comes before the fall, and this principle inhibits pride. There is always more to learn and more to do. For example, new cyber vulnerabilities are discovered every day. Several free and publicly available resources are continually updated so that organizations can learn about the latest vulnerabilities (see Table 9.1). Technology such as antivirus software and vulnerability scanners rely on these resources to keep their databases up-to-date.

Table 9.1

Catalogs and bulletins for cybersecurity vulnerability awareness.

Title	Oversight	Description
Common Vulnerabilities and Exposures (CVE)	MITRE and CISA	The main classifier and catalog of cybersecurity vulnerabilities.
National Vulnerabilities Database (NVD)	NIST	A catalog that provides additional guidance for CVEs including criticality scores and remediation.
Known Exploited Vulnerabilities (KEV)	CISA	A catalog of CVEs that have been exploited by threat actors.
Microsoft Security Bulletins	Microsoft	A notice of vulnerabilities discovered in Microsoft software.

The *Common Vulnerabilities and Exposures (CVE)* catalog was established in 1999 and has become the primary source of information for cyber vulnerabilities. CVEs are published every day, describing newly found vulnerabilities in every kind of software product. They are the standard by which vulnerabilities are named and categorized—they keep everybody in the cybersecurity community on the same page. Without CVEs people would have different ways of referring to the same issues, preventing collaboration, dissemination, and the efficient flow of information. Some CVEs are famous because they became front-page news. CVE-2014-6271 (the 6271st CVE added in 2014) was coined *ShellShock* because it disclosed a vulnerability in *Bash*, a command-line interpreter used in many Linux systems (see Figure 9.7). It was easy to exploit, could lead to arbitrary code execution, and nearly every Linux system in the world was vulnerable. CVE-2017-0144 is a Windows known as *EternalBlue*, and was disclosed within a cache of cyber weapons purportedly belonging to the NSA that were dumped on the Internet in 2017. Later that year, both the *WannaCry* and *NotPetya* malware exploited the EternalBlue vulnerability to wreak havoc across the globe.

Figure 9.7

The CVE describing the ShellShock vulnerability.

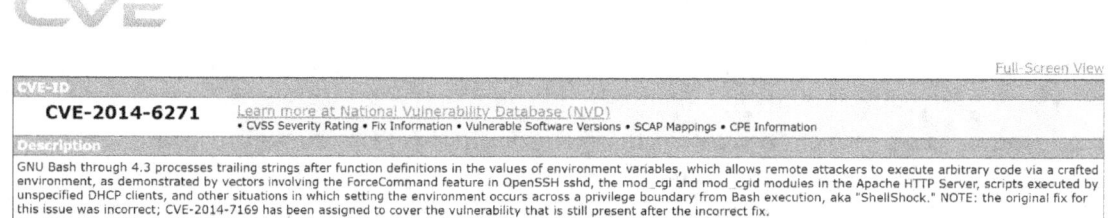

In addition to CVEs, major software companies such as Microsoft and Adobe regularly publish security bulletins to disclose vulnerabilities found in their products. In 2003 Microsoft started the tradition of using the second Tuesday of every month to publish their monthly bulletins. Since then, other vendors have followed suit, and this day has become known in the cybersecurity industry as *patch Tuesday*.

Finally, the principle of security as a process is a reminder that just because things seem to be operating normally does not mean that an incident has not occurred. Cyber professionals must diligently monitor logs and look for indicators of compromise. Typically, a significant amount of time elapses between the initial compromise and when hackers are able to achieve their actions on objectives because they need time to orient and position themselves on the network. *Dwell time is the amount of time that an unauthorized actor remains undetected on a system or network.* Encouragingly, statistics show that average dwell time has been decreasing over the past decade from months to weeks to days. This is partly due to organizations being more proactive in searching for clues of a cyber compromise rather than just assuming that if no obvious damage has occurred, there must not be a problem. As we saw in Chapter 4, there are multiple links in the cyber kill chain after initial exploitation that defenders can interrupt to prevent attackers from achieving their ultimate objectives. The sooner the bad guys are detected, the less likely they will be to accomplish their goals.

The principle of security as a process helps organizations to invest in continual improvement and to stay vigilant at all times.

9.1.10 Planning for Failures

"Those who fail to plan for a security incident are planning for failure."
- The Art of Deception *by Kevin Mitnick*

As we saw in Chapter 5 on risk management, disaster planning is essential for cybersecurity because it is understood that there will be failures. Organizations must be prepared. *The principle of planning for failures states that organizations must assume that cyber incidents will occur.* This principle is not to be taken fatalistically, as in, "We are going to fail and there is nothing we can do about it, so why try?" Rather, following this principle provides a healthy dose of reality that can help organizations properly prepare and stay vigilant, ultimately improving their cybersecurity posture. It is also an encouragement that a future exists after an incident and recovery is possible. Cyber incidents are an existential threat for some organizations, but proper planning makes it much more likely that an organization can recover from an incident and move forward.

A helpful way to prepare for and hopefully prevent cyber incidents is to perform a *pre-mortem.* A pre-mortem is *a thought experiment where one imagines a failure has occurred and explores how and why it could have happened.* This is similar to what we did in Section 5.4 on disaster recovery planning. Exercises like these motivate the planning process and provide helpful insights. They can lead to actions that make incidents less likely to occur. For example, a pre-mortem might imagine that a *business email compromise* has occurred within an organization. A business email compromise is when an employee is tricked into making a fraudulent funds transfer. Often in an attack like this, a cyber criminal is able to impersonate a company executive and insist that a payment be made immediately. By imagining that this has taken place, and then asking the question, "How could this have happened?," an organization might be able to purchase a technology, create a process, or raise awareness among employees that would make a business email compromise much less likely.

Planning for failure also involves implementing appropriate detective and corrective controls. Detective controls can sometimes prevent incidents altogether or at least limit their damage. For example, creating a process around reviewing log files can shrink the dwell time of cyber adversaries. Corrective controls help an organization recover from an incident. A control like data backups can provide an organization with options in the wake of a ransomware attack.

Zero trust is an implementation of the planning for failures principle. Zero trust is *a security strategy that assumes internal systems may be compromised.* Therefore, it requires that all requests be verified, even ones from trusted systems, instead of just assuming they are legitimate. This is similar to the trusting trust principle covered above, and it comes from the mindset of planning for failures. It is a somewhat pessimistic outlook, but, as history has demonstrated, it is also realistic.

The principle of planning for failures leads to better cyber vigilance and preparedness across an organization, and it helps to minimize the damages when failures do occur.

9.2 Practices

Cyber principles lead to daily practices that improve cybersecurity. This section explores a few of the most impactful cybersecurity best practices. All of them can be mapped to one or more of the principles above. Ignoring best practices invites cybersecurity incidents.

While most of this textbook focuses on organizational cybersecurity, this section homes more in on personal cybersecurity. These practices also apply to organizations, but many organizational practices do not apply to individual persons or families. The motivation of this section is to provide practical and actionable steps that readers can take in their own lives. More so than the rest of the book, this section risks becoming outdated.

9.2.1 Manage Authentication Credentials

Section 4.1.2.3 explored credential stealing, one of the most efficient and effective ways that cyber attackers gain unauthorized access to networks and computer systems. This section explores a few best practices for managing user credentials.

9.2.1.1 Use Strong Passwords

> *"'Does it make any difference whether he lies there for ever or walks the quadrillion kilometers? It would take a billion years to walk it?' 'Much more than that. I haven't got a pencil and paper or I could work it out.'" - the devil answers Ivan in* The Brothers Karamazov *by Fyodor Dostoyevsky*

As we learned in Section 8.1, the most common authentication credential is a username and password combination. Password-based authentication is based on "something you know" and assumes that only the rightful user knows his password. In reality, accounts are compromised routinely because this assumption fails. There are multiple ways that hackers can discover a user's password. However, there are things that users can do to make it less likely that their passwords will be compromised.

As we saw in Section 7.2.3.3, authentication databases store password hashes, not passwords. Even though password hash files are closely guarded, *hash dumps* (collections of password hashes) routinely fall into the hands of hackers. When a user creates a password for an account, he needs to anticipate that this could happen. As we have seen, hashes are one-way so they are attacked through a forward search attack called password cracking. The mathematics of forward searches can be analyzed to make password cracking attacks practically impossible.

The United States National Institute of Standards (NIST) released password guidance in the early 2000s that has since been repudiated and replaced with much better guidance. Unfortunately, the old guidance took root during an era of rapid Internet expansion, and it is still widely practiced and promoted. The guidance required that passwords be at least eight characters long and have at least one uppercase, one lowercase, and one non-alphabetic character. Password math (similar to keyspace math) can be used to determine the number of passwords in this space. Password math takes the length of the password and the character set to compute the total number of possibilities. There are twenty-six uppercase letters [A-Z], twenty-six lowercase letters [a-z], ten digits [0-9], and thirty-three other typable non-alphabetic characters (e.g., *!, @, #, $*, etc.). Therefore, for every place in a password there are ninety-five possible choices. This means that for an eight character password, the total number of possible passwords is:

$$95 \times 95 \times 95 \times 95 \times 95 \times 95 \times 95 \times 95 = 95^8 \approx 6 \times 10^{15}$$

Converted to a power of two, this number is approximately 2^{53} (see Section 2.1.3 for the base two-base ten conversion rule). If we assume that a sophisticated hacker could compute 2^{40} hashes per second (one trillion per second), to crack any password in this space would take:

$$2^{53} \text{ passwords} / 2^{40} \text{ hashes per second} / 2 = 2^{12} \text{ seconds}$$

This is a little over an hour to crack any one of these passwords (see Section 7.1.1 for the brute-force keyspace attack math).[1] If we assume a common hacker could compute 2^{30} hashes per second (one billion per second), it would take:

$$2^{53} \text{ passwords} / 2^{30} \text{ hashes per second} / 2 = 2^{22} \text{ seconds}$$

This is around 48 days—still not very much time. Clearly an eight digit password is not long enough.

But the reality is actually much worse than this. Users have adopted NIST's guidance in predictable ways. For example, typical passwords following these guidelines have a structure similar to this: *base word + digit or symbol*. Examples of passwords matching this structure are *Password1* and *Password!* In other words, passwords are not drawn randomly from the password space—instead they occupy only a tiny fraction of the potential passwords (see Figure 9.8). The password math for passwords matching this structure is the number of base words times the number of digits and symbols. If we assume there are one million English words and forty-three digits and non-alphabetic characters, this means there are forty-three million passwords in this space, or around 2^{26} passwords—far fewer than the possible 2^{53}. If a common hacker can compute 2^{30} hashes per second (one billion per second), then a formulaic password like this could be cracked in:

$$2^{26} \text{ passwords} / 2^{30} \text{ hashes per second} / 2 = .03125 \text{ seconds}$$

This equals just three hundredths of a second!

---

[1] The password cracking math in this section assumes salt is used but not key stretching (see Section 7.2.3.3).

It is true that not all user passwords following NIST's old guidance fit this simple structure—some combine two short words, modify capitalization in other ways, insert symbols in multiple places, and are a little longer. For example, *p@ssword11*, *#passWORD23*, *!PA55word1*, etc. These examples probably look more like the passwords a typical reader of this book might use, but the main point still applies. Any formulaic password around the length of eight to ten characters is likely to be cracked quickly with a *dictionary attack*. A dictionary attack is *a password hashing attack that draws base words from a wordlist (e.g., a dictionary) and applies string mangling*. String mangling is *modifying base words in formulaic ways by changing capitalization, using character substitutions, and adding prefixes and postfixes*. Free password cracking programs such as *John the Ripper* make these attacks easy to perform.

Figure 9.8

The difference between random and user-selected passwords for an eight character password (not drawn to scale).

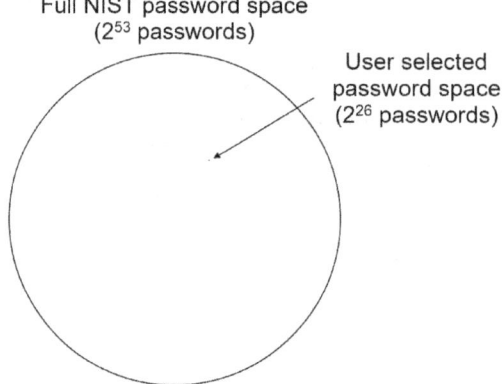

Full NIST password space
(2^{53} passwords)

User selected
password space
(2^{26} passwords)

The password math for the old-style approach works in a hacker's favor. To turn the math in the user's favor, passwords should be long and complex. They should not be built upon a single base word and then modified in a predictable way—passwords need to have some randomness. The problem with randomness is that it is difficult to remember. However, there are tricks to make it easier to remember a password while it still appears to be random. One way is to take parts of letters in a phrase to build a quasi-acronym. Here are a couple of strong passwords that look random but are still relatively easy to remember because they are based on a memorable phrase: *cyIZ4ullof$+@dv* (phrase: "cybersecurity is full of fun and adventure") *@L&BobRbstfr1D$* (phrase: "alice and bob are best friends") The letter selections and character substitutions to build the quasi-acronym are chosen arbitrarily, but after typing them a few times, they start to feel natural and become memorable.

For fifteen character passwords, the number of passwords in this space is 95^{15} and this equals approximately $4 \times 10^{29} \approx 2^{98}$ possible combinations. Importantly, because of their semi-randomness, they are relatively well distributed throughout the entire space as opposed to being concentrated in a small area like in Figure 9.8. If we assume that a nation-state hacker could compute 2^{50} hashes per second (one quadrillion per second), it would take:

$$2^{98} \text{ passwords} / 2^{50} \text{ hashes per second} / 2 = 2^{47} \text{ seconds}$$

This many seconds is more than four million years, and this assumes a highly advanced adversary.

Hopefully this password math convincingly demonstrates the need to use long and complex passwords. But even the best passwords are susceptible to keystroke logging and shoulder surfing attacks, so following this advice does not turn passwords into a perfectly safe authentication token. However, choosing strong passwords definitely helps protect users from account compromises.

9.2.1.2 Password Management

Passwords should not only be long and complex, but users should have different passwords for different sites to protect against credential stuffing attacks. This way the damage of an account compromise is limited to a single account. The problem is that even using the quasi-acronym technique outlined above, it is not possible to remember unique passwords for all the different accounts that a user needs to maintain. Therefore, a best practice for managing user credentials is a *password manager*. A password manager is *a software solution that stores user credentials in an encrypted file (AKA vault)*. Vaults are unlocked with a master password. Once unlocked, all of the user's passwords are accessible. Because they are encrypted and protected by a master password, password managers are far superior to storing usernames and passwords in an ordinary file such as a spreadsheet or text document. If a hacker gains access to a computer, he is likely to find such password files through string searching and pattern matching even if they are disguised with an innocuous name or hidden.

Password managers can be online or offline. There are pros and cons to each approach. Online password managers can be accessed by signing in from any Internet-connected computer. Once signed in, the password manager software can automatically fill in passwords on websites on any device, and user credentials can be added and updated from anywhere. Unfortunately, being online also means that the password managers can be attacked from any Internet-connected computer. Attackers could steal a user's password manager credentials, thereby gaining access to all of the victim's accounts. Plus, password vaults are a high-value target for hackers. If a password manager company is hacked and their password vaults are breached, the attackers could try to crack them via brute-force password attacks. In addition to trying to brute-force crack password vaults, it is possible that shortcut attacks exist via cryptanalysis or through a backdoor or vulnerability in the password manager software.

This is a reminder that using a password manager, whether online or offline, places a significant amount of trust in the vendor. Is their software secure? Are they honest? Password manager companies likely use *closed-source software*. Closed-source means the software's source code is not published. This makes it difficult to verify the security of their software. The alternative is *open-source software*. Open-source software means that the software's source code is published. Therefore, open-source software can be verified by anybody willing and able to go through it line-by-line. Most people believe that vulnerabilities such as bugs and backdoors are more prevalent in closed-source software than open-source, although there have been cases of these types of vulnerabilities "hiding in plain sight" in open-source software, too.

For online password managers, another question of trust arises: is the vendor invested in maintaining strong cybersecurity? A company named LastPass was an early entrant into the online password manager space, and in 2022, they suffered a data breach. This allegedly resulted in some LastPass customers' vaults being cracked, and some victims claimed they lost millions of dollars in cryptocurrency because their crypto keys were stored in their LastPass vaults.

An alternative to online password managers is offline desktop software. In this model, the password vault is stored locally, either on a device's hard drive or on portable storage, and it can only be opened by the password manager software. It is the end-user's responsibility to manage software updates, backups, and security. A downside of this approach is accessibility. The password vault may not be available when it is needed. If the vault is stored on portable storage such as a thumb drive, physical security is a major concern. If the thumb drive is lost or stolen, even if a back-up exists, the lost vault is forever outside of the user's control and could be subjected to brute-force and cryptanalytic attacks.

A hybrid approach between online and offline is also possible. An offline password manager can be used but the password vault can be stored in the cloud just like other cloud-synced files. The vault can then be accessed from different devices that have the password manager software installed on them through file syncing or through the cloud storage provider's web interface. The vault could also be shared between multiple systems, but overwrites are possible when edits are made. This approach could be seen as either a "best of both worlds" or "worst of both worlds approach." The end-user has more control but the vault is online and potentially vulnerable.

Web browsers can act as a type of password manager by remembering passwords. They can then conveniently pre-fill passwords on login pages. If the browser is linked to a cloud account, then users can access their passwords from any online device once signed-in.

Table 9.2

Comparison of online and offline password managers.

Type	Costs	Storage	Data Backups	Accessibility	Shareable	Security
Online	Free or subscription	Online	Company responsibility	Any online device	Yes	Company responsibility
Offline	Free or one-time	Local	End-user responsibility	Limited to physical storage	No	End-user responsibility
Hybrid	Free, subscription, or one-time	Online and local	Cloud storage provider responsibility	Any online or synced device	Yes, but overwrites are possible	End-user and cloud storage provider responsibility

This is a type of password manager solution, but it may not be as secure as a full-featured password manager. There may not be a master password protecting the passwords, plus, the browser's password security may be more vulnerable to attack.

The bottomline on password managers is that they represent yet another security tradeoff. They make it feasible and convenient for people to use unique and uncrackable passwords for every website. This is a major security gain. However, they also suffer from the keys to the kingdom dilemma. If a single password is compromised (the master password), then all of a user's passwords are compromised. The master password must be a strong password because password vaults are subject to cracking attacks just like password hashes. Another issue is that if the auto-fill feature is used with a password manager or a web browser, evil maid attacks are much more devastating. Once access to the browser is gained, then all of a victim's online accounts could be compromised. This is one reason why password managers should only be opened when they are needed and closed immediately afterwards.

An alternative to password managers is to write passwords down on paper. This is a valid security model because it eliminates the threat of the online hacker. It heightens the in-person threat, but for most people, the risk of physical theft or snooping is small compared to the online threat. However, it is also much less convenient—passwords are difficult to maintain and update on paper—and care needs to be taken to "backup" and safeguard the paper-based copy.

9.2.1.3 Use Multi-factor Authentication

This text has mentioned multiple times the importance of multi-factor authentication. It heightens a user's security profile substantially without requiring a significant amount of extra work. This is one area where the security versus cost tradeoff is clear—the costs are definitely worth it. Especially for important online accounts, for example email, social media, and any account involving financial information or health records, users should adopt a second factor. For most users the most convenient second factor is a push-based smartphone app. Another option is investing in a security key (see Section 8.1.3.3).

9.2.2 Use Cryptography

Cryptography is the bedrock of cybersecurity. It is how cyberspace information is protected from theft and manipulation. Data needs to be protected in storage (at rest) and during transmission over computer networks (in transit). This section will highlight some best practices in both areas.

9.2.2.1 Protect Data at Rest

Data at rest is data stored on a computer's hard drive and on removable media. This includes all kinds of data such as personal documents, pictures, and videos. If the data is not encrypted, then it can be viewed by anyone that gains physical access to the storage medium. For example, if data is copied onto a thumb drive as a backup or for physical transport, and if that thumb drive is lost, stolen, or "borrowed" for a period of time, then the information on it is not secure by default. Anybody who comes in possession of the drive can plug it into a computer and access all of the information on it. The same holds true for a laptop or smartphone. A person with full physical access to a computing device can bypass operating system-based access controls by directly reading the data stored on the device's hard drive.

A better approach for disk storage is to use *full disk encryption*. Full disk encryption stores data in encrypted form and decrypts and encrypts data transparently as needed. When full disk encryption is used, if an attacker gains physical access to a device, its data is encrypted and is of no value to the attacker. Because it works transparently in the background, the user experience is not affected—encryption and decryption occurs automatically. The data is unlocked with a master key, typically at log-in time. Windows has a built-in disk encryption utility called *Bitlocker*. Bitlocker can be used to encrypt an entire hard drive and removable media devices. For removable media, if data is transported from one computer to another it requires that Bitlocker be installed on the destination computer. In addition to Bitlocker, there are a variety of other disk encryption utilities—some cost money and others are free.

Data at rest can also be encrypted in a one-off fashion with encryption software. *OpenSSL is a free command line utility that performs a large variety of cryptographic operations.* OpenSSL includes several different encryption algorithms. It comes installed by default on Linux and macOS systems and can be added to Windows devices. It can be used for symmetric key and public key cryptography, for creating hashes and message authentication codes, and for many other purposes. If used properly, OpenSSL can encrypt files that no person or government could ever hope to decrypt—the highest standards of cryptography available. Figure 9.9 shows a file named *secret* being encrypted as *secret.crypt* with the Advanced Encryption Standard (AES) cipher and a user-supplied password as the key. In this example, the security of the encryption resides in the password, so as long as it is a strong password, the file will remain safe. Many similar cryptographic tools are also available, both free and paid, with some featuring more user-friendly graphical user interfaces.

Data at rest also includes data stored in the cloud. Cloud data is encrypted while it is being uploaded and downloaded over the network, but it is likely stored unencrypted on the cloud servers. This means that cloud storage providers can read their customer's data. Depending on the service agreement and the trustworthiness of the storage provider, the data could be mined for marketing and other purposes, or shared with others, including law enforcement and government officials under certain circumstances. Plus, it is possible that it could be accessed by hackers either through the end-user interface or by compromising the cloud service provider's servers. To protect against this, users can encrypt their data before uploading it to the cloud. This would make their data useless for data mining and sharing and to hackers if a data breach occurs. It is more work for the user, but it provides better security. For example, if a person wants to share a sensitive file with a friend using a file hosting provider, he can encrypt it before uploading it. This will protect the file from abuse and theft, and the friend can decrypt it once it is downloaded (assuming the key has been securely shared with him).

Figure 9.9

An *OpenSSL encryption example using a password.*

```
┌──(bob㉿bob-kali)-[~/Desktop]
└─$ cat secret
this is a secret

┌──(bob㉿bob-kali)-[~/Desktop]
└─$ openssl enc -aes256 -in secret -out secret.crypt -pbkdf2
enter AES-256-CBC encryption password:
Verifying - enter AES-256-CBC encryption password:

┌──(bob㉿bob-kali)-[~/Desktop]
└─$ cat secret.crypt
Salted__◆'◆◆◆◆j◆◆◆◆◆r|◆JDzg◆◆za♀◆◆t◆W◆◆\D
```

One caution when using encryption for data at rest is key security and management. Keys can be stolen or lost. If a key is lost, then the encrypted data is forever locked and can never be recovered. Also, if a poor key is chosen or a weak password from which a key is derived, then the encrypted data is vulnerable to brute-force attacks.

9.2.2.2 Protect Data in Transit

"Gentlemen don't read each other's mail."
- United States Secretary of State Henry Stimson

When data is sent over a network, it passes through untrusted servers. Therefore, protecting data in transit from eavesdropping and manipulation is critical.

Many of the older networking protocols do not use encryption. These include Telnet, File Transfer Protocol (FTP), and Hypertext Transfer Protocol (HTTP)—these protocols send data in plaintext so anybody sniffing network traffic, and all the intermediary servers between endpoints, can read the data. Therefore, these protocols should not be used. Email is another old protocol without built-in encryption. Email messages traverse between email servers like snail mail letters traverse between post offices. One of the primary protocols used for transmitting emails over the Internet is called *Simple Mail Transfer Protocol (SMTP)*. SMTP traffic was originally not encrypted, but it has been updated to use encryption to protect email messages as they travel over the Internet. However, email messages stored on servers are not encrypted. In other words, email providers can read their users' email and could mine them for various purposes and share their contents with third parties. It is also easy for end-users to forward sensitive emails to others, either by accident or intentionally. Hackers may also be able to access emails, either through compromising a user account or an email provider's servers. For these reasons, email is not considered a secure means of communication. Encrypted email software does exist and encrypted attachments can be sent over regular email, but these solutions require key distribution and management as well as additional steps, and are rarely used.

Figure 9.10

A website protected with HTTPS.

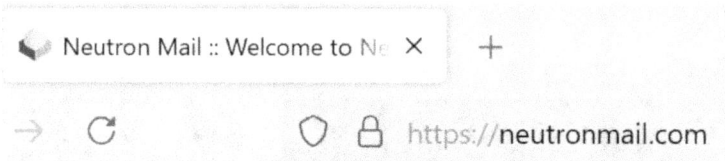

In Section 8.1.6 we briefly covered the HTTPS protocol that is used to make web browser connections end-to-end encrypted—it is the secure replacement for HTTP (see Figure 9.10). When browsing the Internet, users need to be cautious on sites that are not served over HTTPS. With HTTP, the web server has not been authenticated as genuine, and no encryption is used. Most traffic is HTTPS encrypted by default on today's Internet. However, on local area networks and when writing or using custom networking software, users need to be aware of the threat of eavesdropping and take appropriate precautions. This may include encrypting data before it is sent over the network. It should be assumed that traffic sent over the network will be sniffed, and therefore, one should investigate to be sure that it is protected. For example, on an organization's network, are Voice Over Internet Protocol (VOIP) phone calls encrypted? How about data sent to copiers and printers? These questions are worth asking. Below in Section 9.2.4.2 we examine wireless networks and how that data can be protected.

9.2.3 Harden Systems

If there is one thing this textbook has made clear it is that cyberspace is rife with vulnerabilities. Computers of all kinds, including laptops, smartphones, routers, and smart devices, need to be *hardened*. Hardened means *made secure*. This section outlines some steps that users can take to make their devices more secure.

9.2.3.1 Patch Systems

Section 4.1.2.2 explored technical vulnerabilities and explained the importance of software patching. As every cyber attacker knows, n-day exploits are effective because cyber defenders do not always apply patches in a timely manner. Operating systems, desktop applications, and smartphone apps need to be patched. Cloud-based applications, like Google Docs, are patched by the cloud provider since the software runs on their servers. However, web browsers such as Google Chrome and Mozilla Firefox run locally on computers. Similarly, smartphone apps run on end-users smartphones. Therefore, it is the end-user's responsibility to patch their operating systems and the software that runs locally on their devices.

Because patching is so important, software vendors have made it easier over time for users to apply patches—it usually happens automatically and sometimes without the user even knowing. Patches are typically pushed out to users as part of software updates. The updates may include new features in addition to security fixes. If a user is prompted to apply an update, he should do so as soon as possible. Applying patches can be inconvenient because it may require a system restart, but it is essential that time be taken to do so.

Other types of computers, like routers and smart devices, should also be patched. Patching these systems may not always happen automatically, and in some cases, it may not even be possible. Many devices have an admin page that is accessible at their IP address on the local area network through a web browser. Users should login to this page periodically to check for updates and apply them when identified. The admin page typically also enables the user to change the device's default password which is another best practice.

It is also recommended to occasionally restart computers, smartphones, and devices. Many computing devices run continuously for days on end. The normal restart procedure may trigger updates to be installed. Plus, the longer a computer runs and the more processing that it does, the more inefficient it can become. Restarting a computer creates a fresh, clean slate for the memory, and this may result in better performance. Also, some malware is resident only in memory and can be eliminated by a system restart. After a restart, as long as the user does not take the same action that resulted in the malware being installed in the first place, then his computer or smartphone will be secure again.

9.2.3.2 Be a Minimalist

In cybersecurity, neatness counts. In pursuit of neatness and simplicity, users should keep app installs and the number of running processes to a minimum. Programs should only be open when they are being used. Once a user is finished, the program should be terminated. If an app is no longer needed, it should be uninstalled. *Bloatware is a pejorative term for applications that come pre-installed on devices.* If possible, users should delete bloatware from their computers and smartphones. Users should also scrutinize which applications are launched at system startup, and deactivate all applications that are not necessary (see Figure 9.11). It is also helpful to minimize the number of running background services and processes, but this takes a higher level of expertise (see Figure 9.3 above). Users should also keep active browser tabs to a minimum to reduce the complexity of their computing environment. In addition, browsers should be closed and reopened periodically to create a fresh slate in memory.

Users should also limit the number of users on their systems. If a user account is no longer needed, it should be deleted. Users should be cautious about sharing their computer or smartphone with others and when using peripheral devices such as removable media. Portable storage devices should only be used if they come from trusted sources. If a Bluetooth or wireless network connection is no longer needed, the connection should be forgotten.

Only the bare minimum number of permissions should be given to apps. Permissions to access photos and contacts should be allowed only when absolutely necessary. Many apps ask for permissions that they do not actually need to perform their core function, such as constantly sending location data to their company servers.

Figure 9.11

Configuring Windows startup apps.

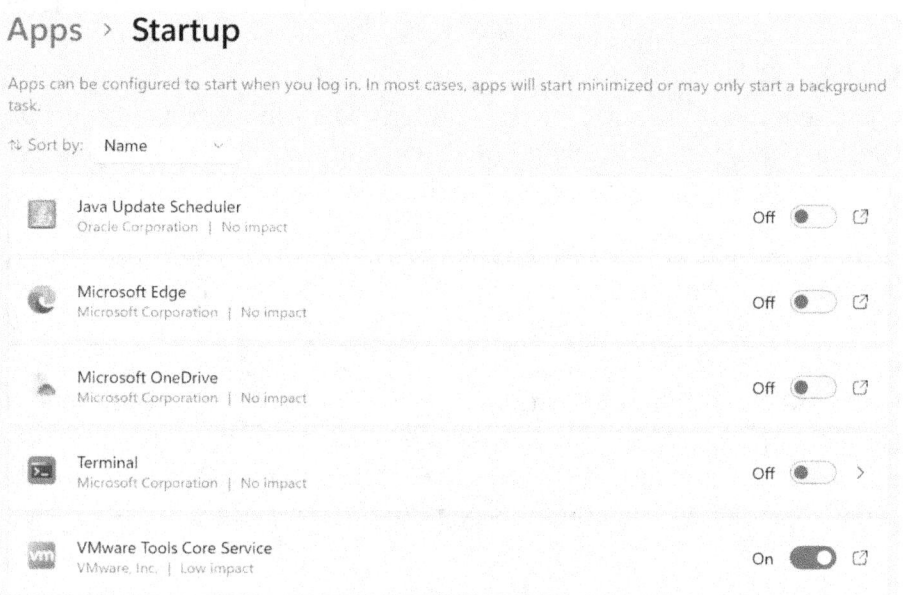

Computers can do many things at once, but people cannot. Users should strive to create a simple computing environment where they can maintain a better grasp of all the activities being performed by their computers.

9.2.3.3 Run Antivirus Software

It is disconcerting when a person's computer starts behaving weirdly, and he begins to fear that he may have been hacked. Maybe the computer seems slower than normal, programs start running randomly on their own, or some system settings have apparently been changed. These could be indicators of compromise, but they could also be due to hardware issues, software bugs, or software bloat that has built up over time. It can be difficult to determine the root cause and to rule out the possibility that malware has been installed or other malicious activity has occurred, and without knowing for sure, users can feel anxious. Running *antivirus software* can help users have peace of mind and not jump to the conclusion that they have been hacked. Antivirus software is *a program that scans files to identify malware.* Windows ships with Microsoft Defender Antivirus and macOS ships with XProtect. These included antivirus programs are pre-configured and run by default and meet typical home computing needs. They can be turned off, but this is not advisable. Before antivirus software was built into operating systems, there was a major market for either free or paid antivirus programs. The market for these types of products is diminished since they provide at most only an incremental benefit over the pre-installed versions. However, Linux systems do not ship with built-in antivirus software, so Linux users should consider installing either a free or paid for product. Users should only change the default behavior of their antivirus programs if they are confident they know what they are doing and why a change is necessary. Any warning or alert that an antivirus program raises needs to be taken seriously and reviewed before proceeding.

Antivirus software uses *signature detection.* Signature detection scans software looking for *malware signatures—a specific sequence of 1s and 0s in known malware.* False positives are rare due to the improbability of two different files having the same signature. Some users may install a program that triggers a match on purpose, maybe for the purpose of cybersecurity testing, but this does not really count as a false positive because the malware was accurately identified. In these instances, a user could ignore the warning if he is confident that he understands the risk. False negatives, on the other hand, are relatively common for signature detection. A false negative is when malware is able to slip past the antivirus program undetected. If a piece of malware has not been cataloged by the antivirus vendor, then it will evade detection. This is true of all novel malware, but is also true of known malware that has been modified. A best practice in hacking is to subtly transform malware so that its signature changes but not its functionality. This makes it more likely that antivirus programs will not identify it. Signature detection also suffers from the problem of an ever-increasing catalog of signatures. There is a limit to the number of samples that software can be compared against. Therefore, malware scanners prioritize some signatures over others, and this sometimes means that even known malware can evade detection.

A more advanced type of protection system is called an *intrusion detection and prevention system* (*IDPS*). These systems can perform *anomaly detection* in addition to signature detection. Anomaly detection *monitors the behavior of software looking for unusual or suspicious behavior*. For example, if a program attempts to modify certain operating system settings or tries to create a network connection to a server, the IDPS program can detect and prevent the behavior and trigger an alert. It can also "learn" over time to identify normal behavior by monitoring the activity on a computer system. Once a baseline is established, it can more accurately identify unusual events. Anomaly detection is more likely than signature detection to suffer from false positives. Too many false alarms results in alert fatigue, so anomaly detection needs to be tuned down to an acceptable level. One of the big advantages of anomaly detection is that it could potentially identify and prevent never-before-seen malware—this is not possible for signature detection.

These protection techniques are illustrative of the general arms race between cyber defenders and cyber attackers. As cyber defenders get better at identifying malicious software, hackers adapt and find new creative ways to evade detection. AI promises to improve both signature and anomaly detection, but it will also be used by hackers to improve evasion. AI could also be used to find vulnerabilities in software. This type of functionality could be used by both hackers to attack systems and defenders to patch systems. There is an ongoing debate whether AI will fundamentally change the balance of power in cybersecurity or whether it will favor both sides such that the status quo will be maintained.

9.2.3.4 Use a System Firewall

System firewalls are similar to antivirus programs, but they focus solely on inbound and outbound network connections and traffic. Windows comes with Microsoft Defender Firewall preconfigured. It can help protect systems and alert users to suspicious behavior and should not be turned off unless a user understands the risk. When alerts pop-up, users should try to understand what is being communicated and should consider searching online for more context before allowing exceptions. Alerts that pop-up not in response to a specific user action are indicators of system compromise. Linux and macOS operating systems come with a system firewall but they are not configured and activated by default.

Smartphones do not need antivirus software or a firewall because they are much more constrained than personal computers. Smartphone owners do not have true administrative or root access on their phones—this helps to make them more secure by default. The security of smartphones mostly boils down to the apps that users deliberately install. It is a risk to install apps acquired from outside of the approved stores because they have not been vetted. They are more likely to be malicious. Even apps from the approved stores could be problematic—users should scrutinize the permissions that apps require during the installation process. The default should be to deny apps permissions unless they are absolutely necessary.

9.2.4 Secure Networks

Cyberspace is synonymous with computer networks and the Internet. Therefore, network security is a primary concern of cybersecurity. This section briefly covers some basic network security measures.

9.2.4.1 Use a Network Firewall

Network firewalls perform the same function as system firewalls except that they can protect multiple computers because they act as a chokepoint for all traffic entering and leaving the network. Most home networks use a router provided by their Internet service provider (ISP). These routers likely have firewall features that the homeowner can configure. Wireless network routers that users purchase and add to their home networks typically also have a built-in firewall. Firewalls can be configured with blacklists or whitelists on a per device basis (see Section 8.2.4).

Network firewalls, like system firewalls, can prevent unsolicited connections from coming into the network and malicious connections from being made to outside servers. Firewalls can also log network activity so it can be determined who did what when on the network. Firewalls also include features to limit the amount of time a device can be online and the amount of data that a device can use, and they can set time windows for when devices are able to access the Internet. Advanced firewall rules can be created to allow applications to only access certain computers on the network.

Routers can also be configured to provide firewall-like functionality through a service called *DNS filtering*. DNS filtering prevents access to blacklisted websites and is enabled by pointing a router's DNS servers to a DNS filtering service (see Section 2.4.3 for more info on DNS). When computers on the network request the IP address of a blacklisted site, the DNS server will not respond with the correct IP address. Therefore, the computer cannot initiate the connection to the website because it is unable to determine what "number to dial." It is as-if the website does not exist. DNS filtering is a free service provided by third parties, and it is a cheap and easy way to perform content filtering on a home network. DNS filters are helpful for blocking known malicious websites and offensive content such as pornography, but they are not perfect systems. Some sites that should be blocked may not be, and determined users can find workarounds to access forbidden websites.

9.2.4.2 Secure Wireless Networks

The *SSID* (*service set identifier*) of a wireless network is the network name seen by users when they scan for available wireless networks. Wireless routers come with a default SSID that typically includes the device manufacturer's name. The default SSID should be changed to something unique, but one that does not reveal any personally identifiable or sensitive information such as an address or name. Many wireless routers allow users to create separate SSIDs for guests and this is a best practice. It creates a barrier between the wireless network that guests use and the devices in the home. Both the home and guest network need to be protected with strong passwords because wireless networks are susceptible to brute-force password guessing attacks. Depending on how many guests use the guest network and how frequently, it may be prudent to change the guest network password periodically to prevent previous guests from continuing to connect to the network without authorization.

Once connected to a wireless network, users with login credentials can gain administrative access to the router using their web browser via the default gateway's IP address. As illustrated in Figure 9.12, this IP address is readily available to anybody that is connected to the network. It used to be common for wireless routers to use a model-wide default username and password, and lists of these default credentials were widely available on the Internet. Fortunately, most wireless routers today ship with unique default administrator passwords. However, they are typically printed on the bottom of the router and could be observed by bad actors. Therefore, either way, people that connect to a wireless network may be able to determine the default password of the router and gain administrative access. Therefore, it is important to change the default administrator password to a unique strong password—definitely one that is different from the wireless network password!

Figure 9.12

Wireless network connection details including the default gateway IP.

```
Wireless LAN adapter Wi-Fi:

   Connection-specific DNS Suffix   . :
   Description . . . . . . . . . . .: Intel(R) Wi-Fi 6 AX201 160MHz
   Physical Address. . . . . . . . .: BE-CA-02-1A-E3-43
   DHCP Enabled. . . . . . . . . . .: Yes
   Autoconfiguration Enabled . . . .: Yes
   Link-local IPv6 Address . . . . .: fe80::8a97:edb0:3a51:28cd%12(Preferred)
   IPv4 Address. . . . . . . . . . .: 192.168.0.38(Preferred)
   Subnet Mask . . . . . . . . . . .: 255.255.255.0
   Lease Obtained. . . . . . . . . .: Saturday, March 29, 2025 5:49:14 PM
   Lease Expires . . . . . . . . . .: Thursday, April 3, 2025 12:53:54 AM
   Default Gateway . . . . . . . . .: 192.168.0.1
   DHCP Server . . . . . . . . . . .: 192.168.0.1
   DHCPv6 IAID . . . . . . . . . . .: 213830146
   DHCPv6 Client DUID. . . . . . . .: 00-03-00-01-BE-CA-02-1A-E3-43
   DNS Servers . . . . . . . . . . .: 192.168.0.1
   NetBIOS over Tcpip. . . . . . . .: Enabled
```

When using wireless networks away from home, users should be especially wary of networks that are not protected with a password. Anybody can join these networks anonymously, and other users of the network can potentially eavesdrop on the metadata being exchanged between the router and other clients. It may also be possible for an attacker to impersonate the router by injecting wireless traffic into the network, and this could trick users into visiting spoofed websites.

Users of public wireless networks are also vulnerable to *evil twin attacks*. An evil twin is a fraudulent wireless network that appears to be legitimate. Evil twins typically broadcast a similar SSID or even the same SSID as the official one at a place of business. Therefore, people may join the fake network without realizing it. It is likely that they will still have Internet access and will see no obvious indicators of compromise. This is a man-in-the-middle attack, and it puts the hacker that is running the evil twin in a place of authority. They can monitor the IP addresses the victim is visiting, read all unencrypted web traffic, and serve spoofed websites. Some devices can be tricked into connecting to an evil twin without the victim needing to take any action. Laptops and smartphones record past wireless credentials and will automatically reconnect to wireless networks that are remembered when they come within range. If an attacker knows the credentials of a remembered wireless network for a target, he may be able to spoof that wireless network and trick the victim into connecting to it. One important thing to note, evil twin attacks cannot be conducted over the Internet. They are local attacks because the hacker's wireless signal must be in the physical proximity of the victims. This mitigates the risk compared to other cyber attacks that can originate from anyone on the Internet anywhere in the world.

A best practice to bolster security when using a public wireless network, whether it is password protected or not, is to sign-in to a *VPN* (*virtual private network*) immediately after connecting to the network. VPNs create an authenticated and encrypted channel between the user and the VPN server, and the VPN server becomes the launching-off point for all web browsing. All traffic that flows to a VPN must still first go through the untrusted wireless router, but the traffic cannot be read or undetectably tampered with by the router. When using a VPN, the user shifts trust from the local Internet gateway to the VPN provider, mitigating the risks of attacks like the evil twin attack (more on VPN best practices below). An even better option than using a VPN is connecting to a personal mobile hotspot instead of public wireless networks. However, this requires the user to have access to a mobile hotspot through a cellular provider which can be expensive, and in some locations a cellular signal may not be available.

9.2.5 Online Safety

"I am sending you out like sheep among wolves. Therefore be as shrewd as snakes and as innocent as doves." - Matthew 10:16

Most cyber threats are invited in from online. People can be lulled into a sense of safety and anonymity online and forget to be vigilant. This section covers some basic online safety measures.

9.2.5.1 Avoid the Dark Alleys of the Internet

In physical space, some places are scarier than others. Dark alleys are places where a person could be robbed or assaulted. Most people instinctively avoid dark alleys, sensing the potential danger. The dark alleys of the Internet are not so obviously dangerous because the visceral sense of being physically vulnerable is not present. Examples of such online places include websites promoting illegal or immoral activity such as those offering free downloads of copyrighted software like games, textbooks, music, and movies. Visitors of these sites are already morally compromised, and this makes them more vulnerable to being victimized. Cyber criminals use sites like these to entice people to click on links to draw them further into danger and to download files that come laced with viruses and remote access trojans. Users of these sites may later notice unusual behavior on their computers but be reluctant to seek help for fear that their browsing history may be exposed. This reluctance can provide hackers more time to compromise their victims.

Users may also enter into relationships with strangers they meet online. Bad actors assume fake identities that they use to cultivate trust relationships with unsuspecting victims. *Catphishing is a social engineering attack where a bad actor creates trust through online interactions and then manipulates victims financially, emotionally, or otherwise.* Attackers may deceive a person into sharing private information, pictures, or videos and then turn around and use that information to blackmail the person by threatening to share it with their friends, families, or co-workers. This could result in further humiliation and harm at the hands of the attacker. Attackers can be ruthless, threatening physical harm to the victim and their loved ones. Although these threats are empty, they can scare the victim into silence and prolong the abuse. If a person has been victimized in one of these attacks, they must ignore the threats and seek help by telling someone what happened—this is the only path to freedom.

Internet dark alleys should be avoided at all costs because they pose real danger. The risk is significant that a user of these sites might inadvertently invite a hacker or cyber criminal right into his computer. They also open users up to associations that could result in being harassed, blackmailed, or even victimized in physical space.

9.2.5.2 Caveat Emptor

"If you didn't buy the product, you are the product." - Anonymous Internet saying

Caveat emptor is a Latin phrase that translates to "let the buyer beware." It means that *a buyer accepts the risk for a purchase.* In the Internet age, many online services do not cost anything. However, caveat emptor still applies, because there are hidden costs to using free services. It may be the case that the user's data is being monetized by the online service by selling it to third parties who may then use it for targeted marketing or other purposes. Most users do not realize that they are providing consent for their data to be used in this way, but it is likely stated in the *end-user license agreement (EULA)* they clicked "Accept" to when signing up. People rarely read these documents but that does not make them invalid.

The social media website Facebook faced public criticism in 2018 when it became known that they were selling their users' data to third parties. Since people share so much of their lives on Facebook's website, Facebook has the ability to create detailed dossiers of their users, and this information has substantial value. Users did not realize that the information they shared on Facebook could be used by Facebook and third parties to manipulate them and shape their opinions. For example, Facebook users could be shown targeted propaganda designed to subtly shift their political views further to the right or the left. Facebook (since renamed Meta) did not violate any laws, but their founder and president, *Mark Zuckerberg*, was forced to testify before Congress about their privacy practices, and many people believe that what they did was unethical. This is a good illustration that there is no such thing as a free lunch. When online, if you are not paying for the product, you are the product!

9.2.5.3 Exercise Caution

In addition to the dark alleys of the Internet, users can be exposed to danger anytime they are online—they are only one click away from compromise. When users download files and click on links, they are exercising trust and need to be cautious. Downloading files whether from a website or in an email is a risk because any file could contain a virus. Installing programs from unknown or unvetted sources is clearly dangerous, but merely opening files, including documents, pictures, and videos, can also pose a risk.

Phishing emails are a prime example of how a user can be attacked online (see Section 4.1.2.1). Links in emails can sometimes be deceiving. A user should always hover over a link to verify the URL before clicking—the link text may not always match the actual URL (see Section 2.4.3). Subdomains are not officially registered so domain name owners can choose any text they want for a subdomain, and bad actors have no problem including trademarked names. If the primary domain looks suspicious, then a user should not click on the link. Phishing emails will sometimes include an official-looking subdomain to trick users into clicking. For example:

> www.amazon.customerhelp.com/feedback

connects to a customerhelp.com server, not an Amazon server (i.e., the primary domain is not amazon.com). URLs with such a deceptive structure are indicative of phishing attacks.

Copying and pasting from online can also be dangerous, especially if copying and pasting computer code or commands. It is common for programmers (especially novices) to search online or query AI for code or a command to accomplish a task. While many of the results are helpful and legitimate, it is possible that a hacker could have planted a backdoor in code or inserted a malicious command. It is important that users trust the source of the information, and if they do copy and paste code or commands, that they have reviewed the text and understand how it works.

When in doubt when it comes to downloading, clicking, or copy and pasting, users can take these actions from within a sandboxed environment such as a VM. This will likely confine any fallout to the VM, and if a VM is compromised, it can easily be destroyed and recreated. Once a host computer is compromised, on the other hand, it is difficult to ever verify with certainty that the threat has been completely eliminated.

Users must also exercise caution when uploading files and typing information into web-pages. This certainly includes anything posted to a public website such as a social media page or a web forum, but it is not limited to only these types of websites. Once data has been uploaded or input online, even if it is not posted publicly, it is shared with the web server. If a bad actor runs the web server or if it is compromised, that data can potentially be used for nefarious purposes. Company employees sometimes unwittingly share pro-prietary information or source code online while searching for answers to questions, dia-loguing on forums, and when using AI tools. Although they may not realize it, they have caused a data breach because they have sent sensitive information out of the company's network. Once out, that data cannot be retrieved. Even non-nefarious AI tools might store user inputted data and later output it in response to future user queries.

Being cautious about posting online goes beyond the risk of sharing sensitive or propri-etary company data. In general, users should never put anything online or send anything through cyberspace, including text, pictures, and videos, that they would not want their parents to see. Following this principle promises to save future pain and regret. This includes pictures and videos that are synced to the cloud, like most of the data on smart-phones is by default, and also data sent via messaging apps. Once something enters cy-berspace, it is forever out of the user's ability to completely control. Even if a user deletes something online, it is impossible to be sure it is really permanently deleted.

9.2.5.4 Be Discerning

As we saw in Section 8.3, logs are kept by systems and network administrators on local devices and networks. Logs are also kept in cyberspace by service providers. These logs are invisible to users and are outside of their control. However, users need to be aware that this data exists and should think twice before typing or clicking. The Internet is not as anonymous as it appears. Being aware of what is possible will help users make wise choices.

When a user is logged in to a website, the actions taken by that user can be logged by the web server. For example, Google might log what searches a person makes and what results they click on along with the date and time and the user's location. Online word processing applications such as Google Docs record all the changes made to a document over time, including deleted sections of text. Social media companies can also log activity including who viewed what when, what they clicked on, and any information they posted, including content that was later deleted or modified.

Some companies are able to log actions taken by users across the Internet as they browse different websites through the use of *tracking cookies*. A tracking cookie is *a text string assigned to a web browser by a web server for the purposes of uniquely identifying a user*. For example, advertisers partner with many different websites and aggregate in-formation across their advertising network using tracking cookies. If a user searches for gold coins on one website, then that user might be shown advertisements for gold coins on other seemingly unrelated websites. Tracking cookies are a powerful tool that can be used to create a detailed picture of a person's life and interests.

Cellular providers log information about the phones in their network. It is unlikely they record conversations, but they do record *metadata*. Metadata is *the attributes of an item of data*. For phone calls, metadata includes the phone numbers of the caller and the callee, the date and time of the call, and the duration of the call. Cellular providers can log text messages in their entirety, including pictures that are sent and received. Cellular providers can also log a phone's physical location at all times, whether the person is using the phone or not, and sometimes even when the phone is supposedly turned off. Since we carry our phones wherever we go throughout the day, this means cellular providers know everywhere a person has been. Cellular providers can also determine all phones that were near a specific location at a particular point in time. Needless to say, cell phone records can easily reveal a person's habits, hobbies, associations, and much more.

Smartphone apps can also log the actions of their users, including their location. Some apps track users even when the app is not actively being used. Android and iPhone OSs force apps to ask users for permission before they are able to collect sensitive data such as this. Some app developers claim they delete the information they collect, but such claims should be viewed with suspicion because they cannot be verified. If a company has access to data, they may or may not collect and store it. Some "anonymous" messaging apps have been exposed in the past for storing messages even though they claimed all messages were ephemeral, anonymous, and immediately deleted.

Digital assistants such as Amazon's Alexa, Apple's Siri, and Hey Google also log activities. They are always listening in the background in case their voice prompt is spoken. In addition to the information they process while they are being used, they could conceivably record and upload to their company servers everything they overhear at all times. So even in the privacy of one's home, in a car, or during a hike in the woods, if a person has a digital assistant or is wearing a smartwatch or carrying a smartphone, it is at least technically possible that his conversations could be overheard and even recorded.

ISPs log website connections made by their customers. Because ISPs are their customers' gateway to the Internet, all web browsing must go through them before reaching the wider Internet. Cellular providers are the ISPs for their smartphone customers. Most traffic is end-to-end encrypted with HTTPS and is opaque to ISPs, but even for HTTPS connections, metadata is not encrypted. For example, the source and destination IP addresses in HTTPS traffic are sent in plaintext (see Figure 9.13). IP addresses can be tied to websites, revealing which websites a person visited. For example, it would be possible for an ISP to determine that a person accessing the Internet at a particular residence visited a website at a specific time, such as google.com. The ISP could not see the search query the person typed nor the search results they received, but they could determine the next web server the person visited after Google, revealing some information about a likely query.

Figure 9.13

Alice browsing the Internet—her ISP can see the IP addresses of the websites Alice is visiting.

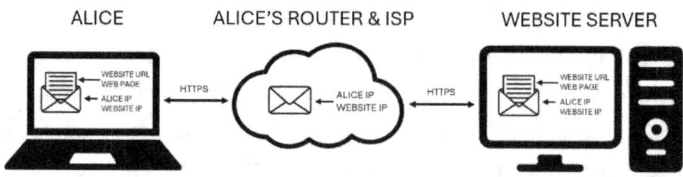

Some people use VPNs so that their ISP is unable to perform this type of data collection. VPNs create an encrypted channel between the user and the VPN, and the VPN becomes the launching-off point for all web browsing. All traffic that flows to a VPN still must first go to the ISP, but because of the encrypted channel that is created, the ISP is only able to see that the person is using a VPN and nothing else—not even the metadata of the HTTPS connections (see Figure 9.14). However, the tradeoff is that now the VPN can see all the websites a person visits. In some cases this can be more invasive to a person's privacy, because if he always uses the same VPN even when connecting to the Internet on different devices and through different ISPs, then the VPN can track *all* of the websites that he ever visits.

Figure 9.14

Alice browsing the Internet while using a VPN—her ISP only sees that Alice is using a VPN, but the VPN provider sees the IP addresses of the websites Alice is visiting.

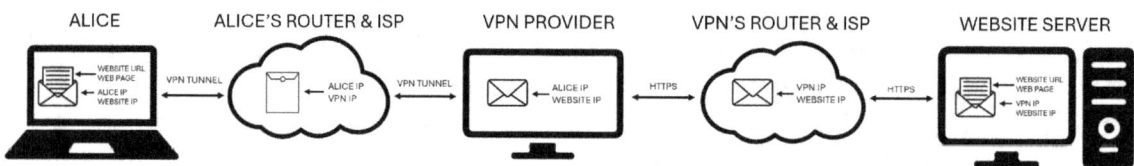

VPNs are often used to provide local area network resources to off network users. For example, businesses often restrict access to some of their resources (e.g., file servers) to employees who are at work and connected to the business's local area network. Employees working from home, therefore, would be unable to access those resources. However, if an employee at home were to connect to his employer's VPN server, then he would gain access to those restricted resources. The VPN server becomes the point of origin for his requests and allows the user to browse resources as if he were at work. The worker is at home physically, but he is at work *virtually*. Similarly, when a person uses a VPN for privacy, the webservers he visits see his requests as coming from the VPN server, not from the user's actual IP address (look closely at the IP address the website server sees in Figure 9.14). This creates another layer of privacy for users.

Most companies and ISPs are honest and are not looking to abuse their customers, but it is still helpful to know what access to information they have. If they retain data, they are legally obligated to turn it over to law enforcement if there is a signed warrant for the information—more on this in Section 10.2.3. Also, if a trustworthy company is hacked, the data they collect on their customers can fall into the hands of bad actors who may use it for evil purposes. Hopefully this section has made it clear that cyberspace is not as anonymous as it appears!

9.3 Conclusion

This chapter has outlined several principles and best practices for cybersecurity. The best practices are tied to principles in different ways. Following these principles and best practices does not guarantee perfect cybersecurity, but they will result in fewer incidents and make the user a much more difficult target for cyber attackers.

 **Chapter 10**

10. The Boundaries of Cybersecurity: Ethics, Rights, and Laws

"Do not remove a fence until you know the reason why it was put up in the first place."
- G.K. Chesterton

In Chapter 3 we defined ethical hacking as *behaving ethically at all times, respecting the rights of all citizens, and obeying all applicable laws and legal authorities.* In Chapter 5 we learned that understanding ethics, rights, and laws is a vital part of cybersecurity governance. In this chapter we will explore these boundaries of cybersecurity and ethical hacking in more detail.

Ethical gray areas occur relatively frequently in cyberspace due to the uniqueness of the domain, its intrinsic attributes (e.g., quasi-anonymity, lack of accountability, social distance from actions, etc.), and the fact it is a new area of human exploration. People who work in cybersecurity need to understand the ethical implications of their actions so that they can confidently and successfully navigate these gray areas. Section 10.1 focuses on the field of ethics and explains how to reason about moral decision making with the goal of providing the necessary clarity to behave ethically at all times.

As we have seen, cybersecurity is about protecting and respecting the rights of every individual and organization in cyberspace. Because so much of society and the economy revolves around cyberspace, it is essential that cybersecurity personnel understand what these rights are, where they come from, and exactly what they need to protect. Section 10.2 focuses on the United States Constitution and the rights that it guarantees to every citizen so that cybersecurity workers can respect the rights of all citizens in cyberspace just like they are in physical space.

Because the stakes are high, the laws that pertain to cyberspace have become increasingly important. In order to protect themselves from criminal prosecution and possible litigation, cybersecurity experts and ethical hackers must understand the law and the legal authorities and how they apply to their activities. Section 10.3 focuses on federal statutes and international law so that workers in the field of cybersecurity can know and abide by the law at all times.

Section 10.4 concludes the chapter with a brief look at how ethics, rights, and laws intersect and how they could potentially come into conflict.

10.1 Ethics

"These, then, are the two points I wanted to make. First, that human beings, all over the earth, have this curious idea that they ought to behave in a certain way, and cannot really get rid of it. Secondly, that they do not in fact behave in that way." - Mere Christianity *by C.S. Lewis*

Cyberspace is a frontier world where clearly defined norms and rules of behavior are still emerging. Furthermore, many of the inherent dynamics of cyberspace contribute to a minimum of accountability, creating the temptation to take ethical shortcuts. Cybersecurity is about protecting the rights of individuals and organizations in cyberspace, and it is vital that practitioners act with integrity at all times, even when there is little fear of repercussions. This is a responsibility the field owes to society in return for the trust placed in cybersecurity experts. Studying ethics provides clarity for discerning right from wrong and conviction for operating with confidence.

10.1.1 Ethical Analysis

"Doing ethics is not like finding the maximum element in a list...there are no algorithms that enable you to 'solve' a moral problem as neatly as you can construct a binary search tree." - Ethics for the Information Age *by Michael Quinn*

Cybersecurity professionals must develop the clarity to determine the difference between right and wrong and the character to make the right choice, even when nobody will probably ever know the difference. But how can one distinguish between right and wrong in the gray areas of cyberspace?

In an ethical debate, appealing to intuition or divine inspiration are conversation stoppers because those bases for right and wrong cannot be meaningfully engaged with by others who do not share the same intuition or religious beliefs. Equally invalid are relativistic arguments such as, "It is true for you, but it is not true for me." A subjective basis for right and wrong makes debate pointless. Without agreement on an objective reality, two people can come to opposite conclusions yet still both be "correct" in their own moral dimensions.

Religious convictions provide the ethical standard for those who hold them, but one may not be able to persuade others in a debate by appealing to a religious authority alone. For example, the argument, "The Bible says it is wrong so it is wrong," assumes the Bible is the mutually understood final authority. If that is not the case, this statement moves the discussion away from the topic at hand to other debates such as evidence for the existence of God and divine revelation. Appealing to a religious text can have a meaningful place in a debate. Therefore, "the Bible says it is wrong" can be an appropriate point to make in support of an ethical argument, but it is not always persuasive as a stand alone argument.

Throughout history, philosophers have sought to develop objective standards based on reason by which the morality of actions can be meaningfully explored. Table 10.1 lists four such approaches that have carried weight in societies across time: *utilitarianism*, *social contract theory*, *Kantian ethics*, and *virtue ethics*. Each takes for granted the goals of human and societal flourishing and subjugates individual choices to these higher ends, and each emphasizes different approaches to realizing these goals. All of the approaches are based on reason and each method has pros and cons. Therefore, a practical approach to evaluating an ethical grey area is to consider each philosophical perspective and see how the arguments stack up overall. They may not all agree on the conclusion, but each will help to shed light on the many different dynamics at play in any given situation.

Table 10.1

Criteria for evaluating right and wrong actions.

Approach	Proponent	Keywords	Key Question
Utilitarianism	Jeremy Bentham (1748-1832)	Benefits and Harms	What are the anticipated benefits and harms to all of the affected parties?
Social Contract Theory	Jean-Jacques Rousseau (1712-1778)	Rights	Does this action violate the inherent rights of any of the involved parties?
Kantian Ethics	Immanuel Kant (1724-1804)	Duties and Obligations	Are any of the involved parties being treated as a means to an end?
Virtue Theory	Aristotle (384-322 BC)	Character	Does this action violate any moral virtues?

This section will provide a brief overview of each philosophy and then apply them to the following fictional ethical dilemma to help illustrate their unique contributions.

Bob's Ethical Dilemma

Bob has been enjoying his cyber classes at the University. After learning so much about hacking he decided to start a penetration testing business so that he could legally exercise his newfound skills. In a recent client engagement for a non-profit, he encountered an ethical dilemma, and he needs some help resolving it.

Before the pentesting began, Bob had done everything by the book. He had the client sign a permission memo and a limited liability agreement, he carefully defined the scope of the tests and the rules of engagement, and he signed a non-disclosure agreement.

During the tests, after he gained root access to a server, he pivoted to the local area network router, set up a packet sniffer, and let it run overnight. When he collected the PCAP file the next day for analysis, he was surprised to see that he had captured unencrypted VOIP traffic from the night before. At this point, he realized his testing scope was not as comprehensive as he thought it was. VOIP traffic was not expressly forbidden, but then again, the client may not have realized that their agreeing to the collection of all network traffic might include plaintext phone calls. Bob's curiosity got the best of him, however, and as his heart raced as it often did during his hacking escapades, he listened to the late night call.

It turned out to be a personal conversation between the president of the organization, who was the one who hired Bob, and the president's wife. They were discussing a serious and private medical condition. Immediately afterwards, Bob's heart sank. He realized that he had crossed a line and violated the trust of his client—even if it was in scope to listen to VOIP traffic, listening to a call that took place after normal business hours would be difficult to justify.

But now he was not sure what to do. Like much of what happens in cyberspace, Bob reasoned that there was no way anybody would ever discover what he had done. Even on the off chance that someone might realize he had access to VOIP traffic during the pentest, Bob could convincingly deny that he ever listened to any of the calls. Plus, he was pretty sure that he was under no legal obligation to report the incident. Besides, if he confessed what he had done, not only would it create awkwardness between him and the client, but it could result in him being fired from the project, and it could damage his reputation as an ethical hacker. But on the other hand, something seemed unethical about not coming clean about his lapse of judgement...

10.1.1.1 Affected Parties

"The principles of justice are chosen behind a veil of ignorance." - John Rawls

The first step in an ethical analysis is to pause and consider how the action will affect others and not just oneself. Ethical paradigms are concerned with the impact of the action on all of the *affected parties*. The affected parties are all the people who will be impacted by the decision. Some of the affected parties may not be obvious—it takes careful thought to consider who all may be impacted. Identifying the affected parties is a crucial step in formulating a robust ethical analysis.

In Bob's ethical dilemma, the main affected parties are Bob and the president of the organization. It could be argued that others are also impacted, including the president's wife and his employees. We can assume that the president represents the interests of all the parties on that side of the equation. Similarly, Bob's friends and family could also be impacted, but focusing on him will capture those wider impacts as well.

After identifying the affected parties, the *veil of ignorance principle* is a helpful thought experiment for eliminating personal bias. The veil of ignorance is *reasoning about the appropriate action to take while pretending to not know to which of the affected parties one belongs*. When children are fighting over a candybar, one way to resolve the conflict is by having one child divide the candy bar and allowing the other to choose the piece he wants. This helps ensure fairness because the child doing the dividing does not know which portion of the candybar he will end up with—he is behind a "veil of ignorance." The veil of ignorance principle helps one view the situation from the perspectives of all the affected parties, instead of just fixating on one's own perspective.

Therefore, as Bob reasons about what he should do, he needs to take seriously how the president will be impacted. If Bob were in the president's shoes, would he want to know what happened? To help see the other side clearly, Bob might want to consult with a wise and trusted mentor. This will help him achieve distance from the situation and gain objectivity so that he can think about it more fairly.

10.1.1.2 Utilitarianism

> *"An action then may be said to conform to the principle of utility…when its tendency to increase the happiness of the community is greater than any tendency it has to lessen it." - Jeremy Bentham*

Utilitarianism emphasizes the consequences of actions. In this type of analysis, the goal is to quantify the sum total benefit versus the sum total harm. If the benefits outweigh the harms, then the action is morally acceptable. It is also called the *greatest happiness principle* because it seeks to maximize benefits. In utilitarianism, happiness is understood to be everyone's ultimate goal. The analysis proceeds by trying to quantify the impact the decision will have on each of the affected parties. This is not an exact science because it is usually not possible to know exactly what impact a decision will have, but thinking it through sheds light on the consequences of different approaches. This is helpful to keep in mind when trying to reason about right and wrong.

Taking Bob's ethical dilemma as an example, what are the anticipated benefits and harms to all of the affected parties? By telling the president that he listened to the call, Bob would likely cause harm to himself (reprimand, lowered reputation, possible loss of wages) and harm to the president (humiliation, embarrassment, disappointment). Meanwhile, the benefits of confessing seem marginal for both parties. Ignorance may be considered bliss in this circumstance from the president's perspective, and both would avoid an awkward conversation. Therefore, the harms of telling the president arguably outweigh the benefits. In this framing of the utilitarian analysis, Bob should not tell the president that he listened to the call because it would do more harm than good.

10.1.1.3 Social Contract Theory

> *"The social pact, far from destroying natural equality, substitutes, on the contrary, a moral and lawful equality for whatever physical inequality that nature may have imposed on mankind; so that however unequal in strength and intelligence, men become equal by covenant and by right." - Jean-Jacques Rousseau*

Social contract theory emphasizes the preservation of basic human rights. For a social contract analysis, it is important to identify the commonly held rights of the society. Ethical decisions can be based on the standards that prevail in the society, and a decision can be judged to be unethical if it violates any mutually understood rights. Rights in the United States are explored at length in the next section.

For Bob's ethical dilemma, does the decision to tell or not tell violate the rights of any of the affected parties? Because the president is responsible for the organization, and he hired Bob to perform certain duties, most people would agree that he has a basic right to the information that Bob gleaned during the pentest. This includes, minimally, the right to know that VOIP calls are not encrypted—a highly relevant penetration test finding. Therefore, Bob should disclose this vulnerability. This would not necessarily mean that Bob would need to confess that he listened to the president's phone call, but there would be a temptation to hide this finding to avoid bringing up the question. The president also has a right to privacy. He clearly assumed that his phone call with his wife was a private conversation, and his privacy was violated by Bob (even if unintentionally). It is not clear that Bob has any relevant rights at stake in his decision to tell, so none of his rights would be violated either way. Therefore, from a social contract perspective, the president has a right to know and Bob should tell him what happened.

10.1.1.4 Kantian Ethics

> *"Act in such a way that you treat humanity, whether in your own person or in the person of any other, never merely as a means to an end, but always at the same time as an end." - Immanuel Kant*

Kantian ethics emphasizes the motivations behind actions. Kant wrote that people have a fundamental duty to never use others in accomplishing their objectives. This is the opposite of the well-known Machiavellian maxim, "The ends justify the means." To Kant, even if the ends are noble, there is a moral obligation to treat all people with dignity at all times, regardless of any extenuating circumstance. People are good at coming up with exceptions to justify actions that bend or break moral codes, but Kantian ethics says that exceptions should not be taken into account. People should act in the way they wish everybody else would act in the same circumstance.

In Bob's ethical dilemma, are any of the affected parties being treated as a means to an end? By not telling the president, Bob is preserving his own reputation and professional image at some potential cost to the president. In this way, Bob is using the president as a means to the end of preserving his own reputation. While Bob might be tempted to believe that it was not such a big deal and what the president does not know cannot hurt him, these mitigating factors, even if true, do not alter his obligation to tell the president what he did. Therefore, it is Bob's duty to tell him that he listened to the call.

10.1.1.5 Virtue Theory

"Knowing about virtue is not enough, but we must also try to possess and exercise virtue." - Aristotle

Virtue theory emphasizes living a life of moral excellence. Aristotle believed that virtue was the path to true happiness. Most societies throughout history have honored a similar set of virtues such as honesty, justice, courage, and patience. These character qualities are esteemed because they lead to societal flourishing, yet they are difficult to live by because they require discipline and self-sacrifice. Choosing character costs something. A life of virtue requires convictions and a belief in a greater good that is worth serving, despite short-term costs.

For Bob's ethical dilemma, are there any moral virtues at stake? Honesty is a virtue. Honesty is not only limited to what is said, but also to what is left unsaid. Telling partial truths or providing intentionally misleading information, even if technically accurate, is a form of dishonesty. Should Bob disclose his finding about VOIP calls being unencrypted? To not do so would be dishonest, but telling the president in a carefully crafted way could also be considered dishonest. Plus, Bob's conscience is bothering him, indicating deep down he knows what he did was not right. Therefore, Bob should choose character and confess to the president that he listened to the call. In the short term this means putting himself at the mercy of the president and whatever consequences may follow, but in the long term it is trusting that the path of virtue will lead to greater future happiness and well-being.

10.1.1.6 Summary

"Whoever can be trusted with very little can also be trusted with much, and whoever is dishonest with very little will also be dishonest with much." - Luke 16:10

Even though it was not unanimous, the weight of our analysis concludes that Bob should confess to the president that he listened to the personal phone call. A key insight is that even though personal negative consequences are first and foremost in Bob's mind, the fear of personal negative consequences are largely immaterial when it comes to evaluating the morality of a situation. By identifying all of the affected parties, applying the veil of ignorance, and viewing the situation through multiple ethical paradigms, Bob's bias towards self-protection is removed from the analysis.

There are many ways to view the ethical dimensions of any situation. Performing an ethical analysis such as this is just as much art as science—honest, wise, and well-meaning people may highlight different details and come to different ethical conclusions. We walked through one series of analyses for a single ethical dilemma that people may or may not find convincing. Importantly, our analysis can be debated on its merits, but there is still a fundamental weakness in this type of ethical analysis.

Because the tally was three to one we concluded that it would be unethical for Bob not to tell the president that he listened to the call. However, a strict utilitarian might still disagree with this stance, even though he is outnumbered. This shows that considering multiple viewpoints is not an algorithm that leads to foolproof conclusions. It cannot resolve all disputes. In the end, a person's *worldview* is the trump card that determines the final verdict. A worldview is *a person's beliefs about ultimate reality.* A person's worldview answers life's biggest questions such as, "Why are we here?" and, "What matters most in life?" The answers to these questions determine values, and values are what gives weight to different ethical factors.

A worldview is also where personal religious beliefs fit into the moral calculus. Religious convictions help individuals determine right from wrong and choose to do what is right. Each of the four ethical paradigms seeks to determine what is best for Bob, the president, and society through rational argument, but they, like religion, are ultimately based on unprovable presuppositions about the transcendent nature of reality. Each must be accepted by faith. Their goal is human and societal flourishing, but they do not all agree on what that means nor on how to achieve it. In the end, even if he is unable to persuade others, each individual must cast his lot according to his worldview and live with the consequences of his actions.

There is also a danger in an academic treatment of ethics in a textbook such as this. The reader may be tempted to compartmentalize ethics and view it as an emergency toolkit that can be kept on the shelf until it is needed. But even in the field of cybersecurity where ethical grey areas are relatively common, most days on the job cybersecurity professionals will not be faced with thorny ethical dilemmas. What is more likely is that they will be regularly faced with smaller ethical decisions where they know the right thing to do. It is in these seemingly low stakes and maybe even presumed "victimless" situations where they will be tempted to compromise.

Ethics is not something that can be compartmentalized; it is not an emergency toolkit. Ethical convictions are lived out every day. Cybersecurity professionals must possess the character to do the right thing every time. If they are faithful in the little things, they will be faithful in the big things when they do arise. The next section elaborates on why this is important for society.

10.1.2 Ethics and Social Responsibility

"Computing professionals' actions change the world. To act responsibly, they should reflect upon the wider impacts of their work, consistently supporting the public good. The ACM Code of Ethics and Professional Conduct ('the Code') expresses the conscience of the profession."
- *"Preamble,"* The ACM Code of Ethics and Professional Conduct

Social responsibility is the proposition that professionals and organizations have an obligation to promote the welfare of society. The guidelines and motivation for social responsibility rests on ethical frameworks. Company slogans like "don't be evil" or "do the right thing" point to a commitment to social responsibility and help remind company employees that ethical concerns are more important than profits. Because of the increasingly critical role that software plays in our lives, many people feel that software engineers have a social responsibility to develop easy to use and high quality, reliable software. Do cybersecurity professionals bear any social responsibility for their actions?

The benefits of technology increase as it is networked together; therefore, all technology is moving into cyberspace. A secure cyberspace is becoming increasingly important to the proper functioning of the economy, the well-being of the citizenry, and to the safeguarding of our freedoms. Many people feel vulnerable and confused by technology and cyberspace even as they depend upon it on a daily basis. Cybersecurity seeks to protect the rights of individuals and organizations in cyberspace. Cybersecurity professionals and ethical hackers are in a trusted position of privilege and power. They have a social responsibility to use their knowledge, skills, and abilities to promote the welfare of society.

Professions in which people must place their trust and that carry a significant level of responsibility for the welfare of the public, often have codes of ethics that provide moral guidance to the professionals working in those fields. These codes are usually crafted and maintained by the professional societies of those fields. Codes of ethics are not laws, but, for some professions, like law and medicine, violations of the code could result in the forfeiture of one's license to practice. For professions not requiring a license to practice, a violation could result in being removed from the membership of professional societies of those fields. If professional misconduct does result in harm that leads to prosecution, attorneys might use a code of ethics to bolster their case, arguing that the actions of the accused were understood to be wrong by professionals in that field.

Having a code of ethics imbues a profession with dignity and elevates the status of those who work in that profession in the eyes of the society. A code of ethics provides moral guidance to professionals—it helps to clarify grey areas and fill in gaps where the law is silent. Having a prominent and highly visible code of ethics removes plausible deniability for those who practice unethical behaviors—they cannot reasonably say that they did not know certain actions were considered wrong by their peers.

For cybersecurity, no widely embraced, fully fleshed out code of ethics exists (yet).[1] However, more broad codes of ethics for computing professionals have stood the test of time and are applicable to cybersecurity professionals. The Association for Computing Machinery (ACM) is arguably the most reputable professional society for computing professionals. They first produced a code of ethics in 1966 and have updated it three times since then. The most recent revision from 2018, the *ACM Code of Ethics and Professional Conduct* (the *Code*), was written by a task force of professionals and academicians and included three major drafts, each undergoing extensive peer reviews.

The Code starts with a preamble and is organized into four sections: general ethical principles, professional responsibilities, professional leadership principles, and compliance with the Code. Table 10.2 lists the seven general ethical principles and nine professional responsibilities in Sections 1 and 2. In the Code, each item is followed by a detailed explanation.

Table 10.2

Sections 1 and 2 of the ACM Code of Ethics and Professional Conduct.

Section 1: General Ethical Principles	Section 2: Professional Responsibilities
1.1 Contribute to society and to human well-being, acknowledging that all people are stakeholders in computing.	2.1 Strive to achieve high quality in both the processes and products of professional work.
1.2 Avoid harm.	2.2 Maintain high standards of professional competence, conduct, and ethical practice.
1.3 Be honest and trustworthy.	2.3 Know and respect existing rules pertaining to professional work.
1.4 Be fair and take action not to discriminate.	2.4 Accept and provide appropriate professional review.
1.5 Respect the work required to produce new ideas, inventions, creative works, and computing artifacts.	2.5 Give comprehensive and thorough evaluations of computer systems and their impacts, including analysis of possible risks.
1.6 Respect privacy.	2.6 Perform work only in areas of competence.
1.7 Honor confidentiality.	2.7 Foster public awareness and understanding of computing, related technologies, and their consequences.
	2.8 Access computing and communication resources only when authorized or when compelled by the public good.
	2.9 Design and implement systems that are robustly and usably secure.

Many of the general ethical principles arise from the ethical paradigms outlined in the previous section. For example, *1.2 Avoid harm, 1.3 Be honest and trustworthy, 1.4 Be fair and take action not to discriminate,* and *1.6 Respect privacy,* pay homage to

---

[1] Information Systems Security Association (ISSA) and International Information System Security Certification Consortium (ISC2) are well known cybersecurity-related professional societies that each have a broad code of ethics.

utilitarianism, virtue theory, Kantian ethics, and social contract theory, respectively. The professional responsibilities are more specific. While all have some applicability to cybersecurity professionals, two are especially relevant: *2.8 Access computing and communication resources only when authorized or when compelled by the public good*, and *2.9 Design and implement systems that are robustly and usably secure.* For example, Section 2.9 provides the following guidance: "As threats can arise and change after a system is deployed, computing professionals should integrate mitigation techniques and policies, such as monitoring, patching, and vulnerability reporting."

To see how a code of ethics can be practically applied, we can consider which parts of the Code might relate to Bob's ethical dilemma. If Bob had been familiar with the Code, could it have helped him? Section 1.6 states, "Technology enables the collection, monitoring, and exchange of personal information quickly, inexpensively, and often without the knowledge of the people affected." This is exactly what Bob discovered. Bob violated the president's privacy without his knowledge. *1.3 Be honest and trustworthy*, is also applicable to Bob's situation. Reading a code of ethics is a good reminder of the high standards of behavior that a profession like cybersecurity requires, and could spur someone in Bob's position to strive to stay above reproach. It is possible that Bob may not have listened to the call in the first place had he been more in tune with the gravity of the situation and the position of trust he held.

It is important for cybersecurity professionals to recognize their social responsibility to do good and to abide by professional ethics. Because of the trust that must be placed in them and the vital nature of their work to society, they have a social responsibility to behave morally at all times. The next section examines the rights of individuals that cybersecurity practitioners have both a legal and ethical obligation to protect.

10.2 Rights

> *"The responsibility of respecting privacy applies to computing professionals in a particularly profound way...a computing professional should...understand the rights and responsibilities associated with the collection and use of personal information."*
> - Section 1.6, The ACM Code of Ethics and Professional Conduct

Individual rights are based on ethical paradigms and are protected by the government. Social contract theory promotes the government's involvement in codifying and enforcing rights to make for a just and free society. Part of being a citizen means entering into this social contract and respecting the rights of others. Rights are protected by laws. Laws give the courts the ability to punish individuals, organizations, and the government when they violate the rights of others. These laws will be explored more in depth in the next section.

Cybersecurity professionals and ethical hackers are involved either directly or indirectly in protecting the rights of individuals and organizations in cyberspace. Cyberspace rights are important because society increasingly depends on cyberspace, and violations of a person's rights in cyberspace have significant real-world impact. As we saw in the previous section, cybersecurity workers have a social responsibility to do good, and this includes respecting the rights of all persons at all times.

10.2.1 United States Constitution

> *"We hold these truths to be self-evident, that all men are created equal, that they are endowed by their Creator with certain unalienable Rights, that among these are Life, Liberty and the pursuit of Happiness.—That to secure these rights, Governments are instituted among Men, deriving their just powers from the consent of the governed."*
> *- The United States Declaration of Independence*

Cybersecurity is about protecting the rights of individuals and organizations in cyberspace. In order to do this effectively, ethical hackers must understand the basis for rights, what freedoms are protected by these rights, and in what circumstances they apply. The United States Constitution is the document that enumerates and guarantees the rights of United States citizens.

Written in 1787, the United States Constitution begins with a famous preamble:

> *"We the People of the United States, in Order to form a more perfect Union, establish Justice, insure domestic Tranquility, provide for the common defense, promote the general Welfare, and secure the Blessings of Liberty to ourselves and our Posterity, do ordain and establish this Constitution for the United States of America."*

The Constitution defines the purpose and structure of the United States government and the fundamental rights of United States citizens. It consists of seven articles. The first three outline the three branches of government: Article I:The Legislative Branch—responsible for creating the laws; Article II: The Presidency—responsible for enforcing the laws; and Article III: The Judiciary—responsible for interpreting the laws. The branches are carefully defined with special interdependencies to prevent the abuse of power. This is known as *checks and balances*. The Constitution codifies the ideals found in the Declaration of Independence, written on July 4, 1776, when the colonists declared themselves free from the "tyranny" of the King of Great Britain.

The United States government was established to protect the rights of its citizens, and to do so it enforces a type of social contract. The first ten amendments to the Constitution are known as the *Bill of Rights* (see Table 10.3). Many of these rights are in direct response to the abuses the colonists suffered under the King of Great Britain. As stated in the Declaration of Independence, "The history of the present King of Great Britain is a history of repeated injuries and usurpations, all having in direct object the establishment of an absolute Tyranny over these States."

Table 10.3

United States Bill of Rights summary.

Amendment	Summary
First	Provides freedoms of religion, speech, press, assembly
Second	The right to bear arms
Third	The right to not house soldiers in one's home
Fourth	The right to privacy
Fifth	The right to due process and to not be a witness against oneself
Sixth	The right to a speedy trial
Seventh	The right to a trial by jury
Eighth	Bans cruel and unusual punishment
Ninth	Protects natural rights not listed
Tenth	Reserves the rights for states to grant rights

The colonists were suspicious of governments and their tendency to abuse power and restrict freedoms. Therefore, the rights guaranteed by the Constitution exist to protect citizens from the government. For example, the First Amendment states, "Congress shall make no law respecting an establishment of religion, or prohibiting the free exercise thereof; or abridging the freedom of speech, or of the press; or the right of the people peaceably to assemble, and to petition the Government for a redress of grievances." These are impressive guarantees of freedom. Every prior form of government actively monitored these types of activities and cracked down on them the moment they were judged to be a threat to the government.

Strictly speaking, Constitutional rights can only be violated by *state actors* and only apply to citizens. State actors are *people acting under the authority of the government*. However, these rights over time became the public conscience of the nation, and violations of them in most contexts is considered unethical if not illegal. Private organizations, as non-state actors, are not technically bound to uphold the rights outlined in the Bill of Rights, but they often do. Also, laws have been passed by Congress to criminalize certain violations of these rights even by non-state actors.

The *right to due process* and the *right to privacy* are especially important for cybersecurity professionals because these rights are particularly vulnerable in cyberspace and require vigilance to protect.

10.2.2 The Right to Due Process

"No person shall...be deprived of life, liberty, or property, without due process of law."
- Fifth Amendment, United States Bill of Rights

The Fifth Amendments provide for the right to due process. The Fifth Amendment's due process clause is very similar to the Fourteenth Amendment's, but the Fifth is specific to the federal government whereas the Fourteenth applies to all state governments as well. *Due process requires that a person accused of a crime be provided a fair trial.* This means in part that they are guaranteed the right to know the crime they are being accused of and the evidence that is being used against them. Additional rights guaranteeing a fair trial are outlined in other Amendments.

The Fifth Amendment also protects an individual's right to not *incriminate* himself. Incriminate means *to provide evidence of a crime*. In the cyberspace age, many argue that this means a person cannot be compelled to hand over authentication tokens such as passwords and encryption keys to law enforcement agents. However, interpretations in this area are still maturing.

In the modern era, cyberspace significantly intersects most peoples' lives. Therefore, digital evidence can play a major part in criminal proceedings. Improperly disclosing, ignoring, mishandling, and destroying digital evidence could impact a person's right to due process. Cybersecurity personnel, especially cyber forensics experts, must treat cyber criminal wrongdoing seriously and handle cyber criminal investigations with the utmost care. Mishandling digital evidence is a serious violation of a person's rights. They must cooperate fully and diligently in any criminal investigations and make sure they are upholding the rights of the accused as codified in the Fifth and Fourteenth Amendments.

10.2.3 The Right to Privacy

"The right of the people to be secure in their persons, houses, papers, and effects, against unreasonable searches and seizures, shall not be violated, and no warrants shall issue, but upon probable cause, supported by oath or affirmation, and particularly describing the place to be searched, and the persons or things to be seized."

- Fourth Amendment, United States Bill of Rights

The right to privacy is not explicitly named in the Constitution but it can be inferred from the Fourth Amendment. The Fourth Amendment limits the government's ability to exercise power over individuals. The Fourth Amendment protects the right to privacy because it prohibits the government, including law enforcement, from general surveillance and from unreasonable intrusions. The following short story will help illustrate how individual privacy rights intersect with cybersecurity...

Charlie's Privacy at Work

Alice, a valued student worker for the University's IT department, had just received an email from her boss asking her to come to his office as soon as possible. As she approached his desk, she was alarmed by the grim look on his face. When she asked what was going on he told her that he had a secret special assignment for her that needed to be completed as quickly, thoroughly, and discreetly as possible. He informed her that the police had called that morning, and that they were conducting an investigation into an investment analyst who worked for the University named Charlie. He asked Alice to gather all of the information that she could on Charlie going back six months so that it could be turned over to the police.

Alice went right to work. First she grabbed the easy stuff, like all the emails he sent and received from his work email account, and the files stored in his directory on the network file server. Then she went a little deeper. She had remote admin login access to his work computer, so she logged in and copied down all of the files in his user directory and sub-directories, including a folder named "Personal"—her boss had said "all" data. Then she went to the VOIP server and ran a metadata report on all of the calls to and from Charlie's office phone—no recordings of phone calls were stored, or she would have grabbed those, too. After this, she logged into his department's copy machine, which did store copies of every document scanned, faxed, printed, and photocopied, and collected all the documents logged to his account over the past six months.

Then she started brainstorming a little more and realized that she had access to even more of Charlie's data. A user ID was needed to log on to the university's wireless network, even with personal devices like smartphones and laptops. So Alice accessed the firewall logs and produced a report of every IP address that Charlie had visited from the campus network, regardless of what device he had been using.

This is when she struck upon her best idea yet. Web browsing from university owned computers was proxied so that the network's intrusion detection system could perform deep packet inspection on all web traffic, including HTTPS traffic. Even better, all of the web data was cached for a week in case it was needed for further analysis. So Alice dug around in the cache and was able to retrieve the unencrypted contents of all of Charlie's web browsing from the past week, including all the URLs he visited, his social media activity, and the plaintext emails he sent and received from his personal email account. She considered that this might be a violation of Charlie's privacy, but then again, he was at work and using his work computer on the university's network—what could he expect?

Satisfied that she had done all that she had been asked to do, and likely a bit more, Alice stopped there and turned over her treasure trove of data to her boss, who dutifully handed it off to the police...

Because the Fourth Amendment was written in 1791, it does not explicitly address the privacy of digital information such as information on computers and smartphones, in emails, and posted on social media websites. However, the courts have found that information, even if not written on a physical object like a piece of paper, is protected, and this includes digital information. The judiciary has developed guidelines to determine when and whether the Fourth Amendment applies, the most important of which is the test of a *reasonable expectation of privacy*. The determination of whether a reasonable expectation of privacy exists is the crucial factor in deciding whether a citizen's rights have been violated by a search and seizure. A two-pronged test is typically applied:

> Did the individual exhibit an expectation of privacy?

> Would society recognize that his expectation of privacy was reasonable?

Some scenarios where an expectation of privacy is debatable include work email, cell phone calls, cell phone call metadata, cell phone location data, personal email written from work, personal files stored on a work computer, web browsing history, and posts to social media. In offices and at universities, digital and physical banners and labels are sometimes used to try to dispel any notion of an expectation of privacy. They typically state something like, "All uses of this system and all data may be intercepted, monitored, recorded, copied, audited, inspected, and disclosed to authorized sites and law enforcement personnel."

The *third-party doctrine* helps to determine if a person had a reasonable expectation of privacy. It states that *when a person shares information with another party, then he surrenders his expectation of privacy concerning that information*. The third party could turn it over to the government without violating any Fourth Amendment rights. The third-party doctrine applies in the digital age in a significant way because almost all information shared electronically is technically shared with a third party service provider. This is so problematic that the *Electronic Communications Privacy Act* (*ECPA*) specifically addresses the rights and responsibilities of electronic communication service providers regarding the handling of digital information (see next section for more details).

In the story above, all of Charlie's information collected by Alice was shared with the university in one form or another (i.e., the university was a third party). But that does not automatically mean that the courts would find that Charlie did not have a reasonable expectation of privacy. If we assume that there were no digital banners, labels, or notices posted around the university explicitly stating that no expectation of privacy existed and that no contract or agreement signed by Charlie included any clauses referencing the privacy of his information, then Charlie may have had a reasonable expectation of privacy.

How would a judge rule regarding Charlie's reasonable expectation of privacy regarding the data that Alice collected? It is generally not possible to know how a judge would rule—the details of every case are unique and every judge is different. The case for Charlie having a reasonable expectation of privacy generally increases as you go down the list (see Table 10.4), but perhaps surprisingly, it is possible that none of this data would be admissible as evidence. Charlie's lawyer would likely make a strong case for Charlie having a reasonable expectation of privacy, especially since no conspicuous banners, public notices, or labels existed. Courts generally rule on the side of protecting rights—the courts take the rights of citizens very seriously. In this situation we are assuming that law enforcement did not have a *warrant*; if a warrant had been issued, all of this data would be fair game. A warrant is *an order signed by a judge giving law enforcement the authority to search and seize evidence.* The Fourth Amendment does not say that a person's privacy can never be violated. It states that there must be probable cause and that a warrant must be issued before such a search can take place. When a warrant is issued, the right to privacy over the times and locations outlined in the warrant is forfeited.

Table 10.4

Charlie's reasonable expectation of privacy at work ratings.

Category of Data	Reasonable Expectation of Privacy
Work email	Low
Files on the network file server	Low
Files on his work computer, including the "Personal" folder	Medium
Office phone metadata	Low
Copy machine documents (scans, faxes, printed, copied)	Medium
Browsing metadata from work devices	Medium
Browsing metadata from personal devices	High
Personal social media activity at work	High
Personal email activity at work	High

Charlie's office phone metadata is the most likely to be admissible since courts have ruled phone call metadata does not require a warrant—however, recordings of the calls (if they had existed) definitely would not be admissible. Social media posts are shared in a very public forum, but with only select people—however, the way they were collected by Alice would likely make them inadmissible in this case.

So what about Alice's role in this? Did she do anything illegal? Alice as a university employee cannot violate Charlie's privacy rights, and she did not break any laws because she had authorized access to the computer systems. If Alice was officially deputized as an agent of the state, her actions would have violated Charlie's Constitutional rights. The bottom line is this was a bungled investigation by law enforcement—they should have requested only specific and limited information until they obtained a warrant necessary for a thorough search like the one conducted by Alice.

10.2.4 Summary

"If you give me six lines written by the hand of the most honest of men, I will find something in them which will hang him." - Cardinal Richelieu

The United States Constitution is the basis for our system of government and the foundation for our freedoms and rights. The Fifth and Fourteenth Amendments guarantee our right to due process, and the Fourth Amendment guarantees our right to privacy, and these rights must be protected in cyberspace as well as physical space.

For example, in the story about Charlie's privacy rights at work, how would you feel about Alice's search if you were in Charlie's shoes? If you did nothing wrong, then you would have nothing to hide, so why should you care? This is a popular argument for downplaying online privacy rights, but it is flawed. As we live more and more of our lives out in cyberspace, online privacy is becoming synonymous with personal privacy, especially with the popularity of smartphones and smart devices that come into our homes and personal spaces. It is not just law breakers that have something to hide; we all seek privacy and prefer to keep certain parts of our lives out of the public eye. For example, besides the physical privacy we obviously value, we value the ability to have private conversations with people we trust. Our friends know us well enough to properly contextualize our jokes and the random thoughts and feelings we express as we process ideas. As Cardinal Richelieu's quote makes clear, our words can be dangerously misconstrued when taken out of context—this goes for anybody. The right to privacy and due process are vital to a flourishing society, and increasingly, that includes our cyberspace interactions.

Cybersecurity professionals have a legal obligation and a moral responsibility to respect the rights of all citizens impacted by their work. In the next section, we will look at the law as it applies to cybercrime and cyber warfare

10.3 Laws

"Legal obligations and restrictions should be considered at the outset of any cybersecurity strategy, just as a company would consider reputational harm and budgetary issues. Failure to comply with the law could lead to significant financial harms, negative publicity, and, in some cases, criminal charges." - Cybersecurity Law by Jeff Kosseff

In this section, we will briefly examine some of the most important hacking-related laws. One pragmatic reason for cybersecurity professionals to know the law is so that they can avoid criminal prosecution and civil lawsuits. Ignorance of the law, especially for professionals, is not an acceptable defense in a court of law. Part of due diligence is learning the law and understanding how it applies.

10.3.1 The United States Code

Cybersecurity activities, especially ethical hacking, must be done in accordance with all applicable laws, otherwise criminal prosecution and civil litigation could result. The United States Code is an organized collection of every law created by the Legislative branch of the United States. It is organized into fifty-three titles covering all aspects of government from observed holidays (Title 36) to voting and elections (Title 52) to taxes (Title 26). *Title 18: Crimes and Criminal Procedures*, contains most of the laws for which a person can be charged with a federal crime and tried in a court of law (more on state laws later).

It could be argued that there is nothing new under the sun, and that cyberspace is just a different setting for the types of criminal behaviors that have existed since the dawn of time (e.g., stealing, fraud, destruction of property, etc.). If that is the case, then laws that predate cyberspace could in theory apply to cybercrimes as well. However, lawyers are experts at arguing about the strict definitions of terms, and pre-cyberspace laws are not precise enough to hold up in a court of law. Laws need to be specific. Therefore, even if a person does something that seems like it should be illegal, without reference to a specific law "on the books" no criminal prosecution can occur. Crimes involving computer technology have necessitated the creation of new laws that better fit the crime. The three main federal cybercrime laws are listed in Table 10.5. The following subsections review each law.

Table 10.5

The main federal cybercrime laws.

Law (Year)	Summary	Penalties for Individuals
The Computer Fraud and Abuse Act (1986)	The umbrella anti-hacking law	From one to ten years based on section, up to $5,000 in fines
The Electronic Communications Privacy Act (1986)	Protects the right to privacy for digital communications	Up to five years in prison and $250,000 in fines
The Economic Espionage Acts (1996)	Protects the intellectual property of companies	Up to fifteen years in prison and $500,000 in fines

10.3.1.1 Computer Fraud and Abuse Act

The *Computer Fraud and Abuse Act* (CFAA) was passed in 1986. It is the umbrella "anti-hacking" law. It is broad and is used to prosecute a huge variety of cases. It includes seven different sections, but the most often cited is Section 2 which reads in part, "Whoever intentionally accesses a computer without authorization or exceeds authorized access, and thereby obtains...information from any protected computer."

The definition of hacker we have been using in this text is *a person who attempts to gain unauthorized access to a computer system or data or deny access to authorized users.* It is no coincidence that the first part of this definition matches the language in the CFAA. This law and the many state laws that are similar to it, are why ethical hackers need to be wary. Well intentioned white hat hacking activities such as probing for vulnerabilities and penetration testing to improve security can violate the CFAA. This law is the main reason for the emphasis on procuring proper permission prior to performing pen testing is paramount! The first ever conviction of the CFAA was a case of white hat hacking going awry. Punishment under the CFAA can be minor (i.e., a misdemeanor conviction) such as a fine, or major (i.e., a felony conviction) such as five years in prison. Civil action can also be taken. Civil lawsuits are where one private party sues another party for causing harm—in this case, in connection with violating the CFAA.

Lawyers battle it out over the interpretation of terms, so each word and phrase in the law is important. As cases are brought to trial and lawyers try to persuade judges how the laws should be interpreted, judge's rulings set precedents. These precedents are then relied upon in future cases. However, the details of no two cases are identical, so it is not possible to know exactly how a judge will rule in any particular case. The CFAA has several key terms that are open to interpretation. Table 10.6 lists some of the most disputed terms from Section 2 of the CFAA and probable interpretations according to court precedents.

Table 10.6

Key terms and interpretations for Section 2 of the CFAA.

Term	Interpretation
intentionally	"stumbling into" does not count; the individual must show intent to gain unauthorized access or to exceed authorized access
exceeds authorized access	can include misusing information for which the individual did have proper access
obtains information	includes observing information, not just "stealing" it
protected computer	defined elsewhere in the CFAA, but it includes virtually any kind of computing device, and almost certainly any Internet-connected device

The following story will help shed light on how these terms might apply in a real-life case.

Alice's Brush with Cybercrime

Alice was on the University's registrar website looking over her grades on her transcript when something caught her eye in the address bar of her web browser. The URL included a familiar looking nine digit number. When she double-checked it against her student ID card, Alice confirmed that the number in the URL was indeed her student ID number.

From her web apps class she knew about URL query strings, and how arguments could be passed to web servers in the URL so that the appropriate dynamic web page could be retrieved for the user. In this case, she speculated that the query being sent to the back-end database was parameterized by the student ID number passed in the URL—that is how the web server knew to return her transcript and not some other student's.

But this led immediately to another thought: what if she changed the student ID number in the URL to another student's ID number? Would the website return the other student's transcript? If so, this would allow her to access information that she knew she was not authorized to see, but she could not resist indulging her curiosity. Her roommate was already in bed for the night, and like always, had left her student ID card out on her desk. Alice reached over, grabbed the ID, and replaced the nine digit number in the URL with the student ID number printed on the back of her roommate's card. With heart racing in anticipation, Alice hit the enter key, and voila, it worked! She was staring at her roommate's transcript! Not caring terribly much about the grades her roommate had received, Alice scanned the page quickly and closed her web browser, pleased that she had pulled off such a neat trick.

During breakfast the next morning at the dining hall, Alice could not wait to tell her classmate Bob, a pentester, about her discovery. When she told him, however, the concerned expression on his face quickly dampened her mood. Bob informed Alice that if school officials found out what she had done, she could not only face university disciplinary measures, but she could also be charged with a federal crime known as the Computer Fraud and Abuse Act. Alice hoped that Bob was overreacting, but she realized she needed to learn more about this law, and how it might apply to her actions.

In the story, Alice used her computer to read her roommate's transcript on the registrar's website. Did Alice violate CFAA? Alice demonstrated intentionality by grabbing her roommate's ID card and manually copying the numbers into the URL. Alice was not supposed to have access to her roommate's transcript, therefore, viewing her roommate's transcript exceeded her authorized access. Viewing the transcript is considered "obtaining information" even though Alice did not make a copy or take a picture of it. The university web server is an Internet-connected computer so it meets the "protected computer" criteria.

So on the surface, maybe Bob was right to be concerned. However, in her defense, Alice's lawyer would likely home in on the definition of "exceeds authorized access." The lawyer could argue that Alice had actually been given access to her roommate's transcript by the university IT department. The website should have had measures in place to safeguard the transcripts by making sure the viewed transcript matched the logged in user's ID;

because it didn't, they tacitly gave everyone access. Bottomline, charging Alice with the CFAA would likely not stick because of the university's poor cybersecurity posture. The *Family Educational and Privacy Rights Act* (*FERPA*) protects student educational records, including transcripts, from unauthorized disclosure. That may mean in this case that the university could have a major cyber risk management issue on their hands for failing to adequately safeguard that protected information. However, FERPA would have no bearing on Alice since she was not the steward of the protected data. Alice did not break a federal law, but she may have debatably crossed an ethical line (see Section 10.1). At the least, she should have asked her roommate first before trying her trick.

The first conviction under CFAA was in 1990 in the *United States v. Morris*. As we saw in Chapter 3, Robert Tappan Morris unleashed the first worm (the Morris Worm) on the Internet when he had just graduated from Harvard and was on his way to graduate school at Cornell. The Morris Worm marked the end of cyberspace innocence and foreshadowed a future of cyberspace insecurity. It crashed thousands of computers and cost organizations tens of thousands of man-hours to recover. Morris' case is especially interesting because his father was a high-ranking NSA official, and Morris was a brilliant young man who did not fit the stereotypical image of a felon. He showed a lack of judgment and did not think through the potential negative consequences of his actions. He had assumed that his malware would cause no actual damage, and had no intention of using it for any kind of profit. Rather, for him, his worm was an experiment to test the cybersecurity of the pre-Internet computer network. Unfortunately, Morris misjudged his worm's "infection rate" parameter, and even though he intended his worm to be innocuous, it ended up crashing the systems it infected. So it could be said that his worm had a bug in it, and that was the real cause of Morris' problems! Morris was rightly convicted of the CFAA and faced jail time, but the judge realized that probation, community service, and a fine better fit the full-context of the situation. Morris went on to become a professor at MIT.

Because the CFAA is broad, it is charged in a wide variety of computer-related crimes, and sometimes this means the charges do not stick. In the *United States v. Drew*, a famous cyberbullying case where an adult woman created a fake social media profile and used it to humiliate a teenage girl that sadly ended up committing suicide, the United States government charged Drew with violating the CFAA. However, the law did not really fit the crime and Drew was acquitted. The judge found that she did not "gain unauthorized access" and did not "exceed authorized access," therefore, she was innocent of the charges. This upset some people who felt that Drew had obviously committed a crime and should have been punished. Since the time of her case, several states have passed specific laws criminalizing cyberbullying. If those laws were on the books at that time in the state where Drew lived, she likely would have been found guilty. But since laws are not retroactive, they had no bearing on her.

10.3.1.2 Electronic Communications Privacy Act

The Electronic Communications Privacy Act (ECPA), also passed in 1986, complements the Fourth Amendment's right to privacy. It is the most comprehensive United States law relating to cyber surveillance. It limits the government's ability to obtain email, monitor networks, and obtain Internet traffic logs. The ECPA protects many types of Internet communications on the basis that people using the Internet have a reasonable expectation of privacy (see Section 10.2.3 above). The three sections of the ECPA are outlined in Table 10.7.

Table 10.7

The three sections of the ECPA.

Section	Summary
Stored Communications Act	Restricts the government's access to stored communications such as email and cloud content
Wiretap Act	Restricts the government's ability to monitor electronic communications while they are in transit (revised the 1968 statute that focused on tapping phone calls)
Pen Register Act	Restricts the government's access to telephone and Internet traffic metadata

The Stored Communications Act (SCA) is especially important because it protects consumer's privacy when dealing with third parties, and in the modern era, most computer users constantly rely on third parties. Cloud computing relies on third parties, including email, social media, and data backup services. Stored communications include all files stored on a third party's servers such as documents, emails, pictures, videos, and other personal records.

The SCA contains three general categories. The first category complements the CFAA. Criminal charges against hackers can include this section of the SCA when electronic communications like email are involved. Criminal penalties include fines and up to ten years in prison, and civil lawsuits are also permissible. The second category is the voluntary disclosure of stored communications by service providers. It is illegal for service providers to knowingly divulge the contents of electronic communication. Exceptions are made for exposing criminal activities and for helping to handle certain types of emergencies. The third category affects law enforcement's attempts to compel service providers to disclose stored communications. This does not supersede an individual's Fourth Amendment rights. It does, however, include a controversial clause making data older than 180 days less protected than newer data. The 180 day clause is not as meaningful by today's standards as it was in 1986. Back then data storage was expensive and older data was regularly purged. Today, data storage is inexpensive and the norm is to keep data around for a long time. Many people feel that there is no good reason to differentiate data based on its age, and if there is, 180 days is too short of a time period.

The Wiretap Act is similar to the SCA except that it protects electronic data in transit. That is, it protects live communication. *A wiretap is a device that can be attached to the telecommunications infrastructure to secretly listen to phone calls.* The term has been extended to mean the ability to secretly listen to any electronic communication. As we learned in Chapter 2, Internet data traverses over several hops in the core of the network as it flows between endpoints. Each hop, including at Internet service providers, provides a surveillance opportunity for the government and others. The Wiretap Act makes this illegal without a warrant, similar to how listening in on phone calls from within the telephone network was made illegal decades earlier. But even with a warrant, "tapping" electronic communications is not as valuable as it used to be because of the ubiquity of end-to-end encryption today. Encryption makes the data being "listened to" unintelligible. Section 10.4 discusses how encryption creates a perceived tension between individual privacy and security. The ECPA and the Foreign Intelligence Surveillance Act (FISA) (see Chapter 3) must be held in tension to protect national security while at the same time respecting individual rights.

The Pen Register Act is similar to the Wiretap Act. *A pen register is a device that records the metadata of electronic communications.* Therefore, the Pen Register Act protects the metadata of Internet communications from warrantless searches, including IP addresses, protocols, and timestamps. Metadata can be used to determine the identity of people and the websites they visited and can add substantial value to investigations.

10.3.1.3 Economic Espionage Act

The Economic Espionage Act prohibits the theft of a company's *trade secrets*, either by a foreign government or for economic benefit, through digital means. It was passed in 1996 when companies were starting to integrate the Internet into their daily business operations. A new federal law was needed to deter the theft of data (i.e., electronic trade secrets). Trade secrets are *confidential corporate data that the company has taken measures to protect and that has economic value.* A famous example of a trade secret is the secret formula for Coca-Cola. The Coca-Cola Company takes major precautions to protect the formula, including keeping it locked in a vault and limiting access to only a few employees.

The Economic Espionage Act contains two sections. Section 1831 prohibits stealing trade secrets and providing them to a foreign government (i.e., *foreign espionage*). Punishments for individuals include up to fifteen years in prison and a fine of up to five million dollars, and for organizations fines up to ten million dollars or three times the value of the trade secrets, whichever is greater. In the *United States v. Chung*, a former Boeing engineer was convicted of providing trade secrets to China regarding the designs of airplanes and rockets, and he was sentenced to fifteen years in prison. Section 1832 prohibits *cor-*

porate espionage or the theft of trade secrets to benefit one company at the expense of another. Punishments for individuals include fines and up to ten years in prison, and for organizations include fines up to five million dollars or three times the value of the trade secrets, whichever is greater.

10.3.2 State Laws

In addition to federal laws, every state in the union has anti-hacking laws. In many cases, state laws are more specific than federal laws because it is typically easier for states to pass legislation than it is for the federal government. State laws are important for cybersecurity professionals because jurisdiction in cyberspace is defined liberally; for example, a computer hacker operating in Ohio breaking into a computer located in California can run afoul of the laws not only in those two states, but in every state between where network traffic flows! It is possible to be charged with violating multiple different states laws in addition to federal laws. Ethical hackers in particular need to be well-versed with all applicable state laws.

For example, in the state of Ohio, the Ohio Revised Code Section 2913.04 defines crimes involving the "unauthorized use of computer, cable, or telecommunications property." In part, it states, "No person, in any manner and by any means, including, but not limited to, computer hacking, shall knowingly gain access to, attempt to gain access to, or cause access to be gained to any computer, computer system, computer network..." This provision was added to the law in 2011. Violations of this law could result in up to three years in prison.

Because states and the federal government are separate authorities, it is possible to be charged with both state and federal crimes for the same act.

10.3.3 International Laws

> *"In the face of the [cyber] threat, we must develop capabilities to defend ourselves in a manner that preserves international legal order. At the same time, it is the responsibility of the international community to ensure that peace, security, and stability are maintained, and that [cyber] capabilities are only used in accordance with international law."* - Tallinn Manual 2.0 *by the International Groups of Experts*

There are international laws that govern nation state conflict and warfare. In this context, the term *state* refers to a sovereign nation. The most important of these laws date to the early twentieth century and to the period immediately following World War II (WWII). They have been agreed on by most of the developed countries in the modern world. The context for these laws are obviously pre-Internet, but the principles apply in cyberspace because it is possible for nations to inflict harm on one another in and through cyberspace.

As we learned in Section 3.1.1.2, the United States Constitution provides the legal authorities and boundaries for conducting cyber operations. *Title 50* covers signals intelligence

(SIGINT) and *Title 10* covers military activity. These types of activity that cross the United States border are also governed by international laws. When a state conducts a cyber operation that affects another state, two main principles of law apply: *jus ad bellum* and *jus in bello* (*jus* is pronounced "yuse"). Both are Latin terms, and they mean "right to battle" and "right in war," respectively. *Jus ad bellum outlines the conditions for when it is justified for a state to enter into war with another state.* This concept is also known as *just war theory*. *Jus in bello outlines acceptable conduct in war* and is also known as *International Humanitarian Law (IHL).*

10.3.3.1 International Cyber Conflict

International cyber conflict has been ongoing for decades. There are many known examples of nation state-on-nation state cyber operations, but for everyone that is publicized, there are likely dozens that are not. This is due to the secret nature of cyber operations. Countries often do not acknowledge these activities whether they are the perpetrator or the victim. To help set the context for international law, this section will summarize two of the most well-known examples of cyber conflict, and even for them there remain many unanswered questions.

Arguably the world's first nation state cyber attack happened 2007 in Estonia in the aftermath of a Soviet-era statue known as the *Bronze Soldier* being moved to the outskirts of the capital city of *Tallinn*. Russians living in Estonia viewed the statue as a monument to the time when the Soviets liberated Estonia, but native Estonians viewed the statue as a monument to the time the Soviets invaded and brought their oppressive regime back to Estonia. (Estonia had gained their independence from Russia in 1918 and the Soviets returned in 1940.) The statue was a powerful symbol like statues are meant to be, and its symbolism was politically explosive. Some Russian news outlets misreported that in addition to the statue being moved to a less visible location, it had also been defaced along with the gravesites of soldiers. Passions were inflamed and riots erupted in the streets.

Meanwhile in cyberspace, Tallinn began to be subject to a relentless barrage of distributed denial of service (DDoS) attacks. Estonia, contrary to what many might believe, was one of the most technologically reliant nations in the world at that time. The DDoS attacks made online banking and commerce nearly impossible and shut the city government down. The ordeal lasted over twenty days.

Some high-level government officials in Russia allegedly helped to incite the riot by making inflammatory comments, but the Russian government claimed no responsibility for the incident. It is likely that patriotic hacking did occur (e.g., private Russian citizens engaged in military-style cyber operations). This gave the Russian government plausible deniability because they could acknowledge that attacks were coming from Russia but deny that they were sanctioned. It is difficult to determine to what extent this behavior was condoned or possibly funded by the Russian government. This is a crucial difference between physical space and cyberspace conflict—a relatively small group of loosely connected civilians with no chain of command could plausibly pull off an attack that could cripple a nation. This would be unimaginable in conventional warfare because of the high costs and complex logistics involved.

This event is remembered in history as the debut of cyber warfare, although it was not formally recognized as an act of war (i.e., there was no military retaliation against Russia).

Another important event in the history of nation state cyber conflict occurred around the same time. In 2010 a highly sophisticated cyber weapon was discovered and dubbed *Stuxnet* (the name was derived from two of the filenames found in its code). It is one of the most famous and complex pieces of malware ever discovered, leveraging multiple zero-days. Based on its extreme level of sophistication and its highly targeted nature (it would disable itself on machines that did not match specific rules), it was clearly the work of nation state actors that were concerned about collateral damage and legal issues surrounding cyber conflict.

The cyber weapon was targeted at industrial control systems used at an Iranian nuclear enrichment plant in Natanz. It was allegedly introduced into the heavily secured and air-gapped facility via a USB stick. Once inside the network, the malware remained undetected while autonomously pivoting to the industrial control systems controlling the operation of the nuclear centrifuges. The malware was a rootkit. It caused the centrifuge software to report that they were functioning normally when in fact the malware made them spin outside of their acceptable operating conditions, causing them to break down. The Iranians did not know why their centrifuges were breaking at a rate far surpassing normal wear and tear, and they suspected sabotage, but they could not determine the source.

It is widely believed that Stuxnet was a joint operation between the United States and Israel to sabotage Iran's nuclear weapons program, but neither nation has formally admitted they were involved. It was likely successful in setting the Iranians back multiple years. Some believe it also ushered in a new era in cyber conflict creating a precedent in cyberspace for nation states to conduct covert sabotage operations against other nation states. Despite the destructive nature of the attack, it did not trigger a military response from Iran perhaps because of its clandestine nature.

10.3.3.2 Jus ad bellum

Jus ad bellum addresses the question of when war is just. The *United Nations Charter* is an international treaty signed in 1945 right after WWII. In its preamble, it states that its purpose is to "save succeeding generations from the scourge of war." Article 2(4) states that, "All Members [of the United Nations] shall refrain in their international relations from the threat or use of force against the territorial integrity or political independence of any State." The key term in this sentence is *use of force* because when this threshold is reached, it can trigger a justified retaliatory response.

Cyber operations are sophisticated cyber attacks that involve significant planning and resources. Nation states engage in cyber operations to accomplish strategic ends. Cyber operations waged against the United States have never risen to the level of a use of force. However, the way these terms apply to cyberspace is not well-understood. *The Tallinn Manual 2.0* (the 2017 update of the 2013 original), is a noteworthy attempt to apply principles like use of force to cyber operations. The title pays homage to the attack on Estonia in 2007 when Tallinn became the cyber conflict capital of the world. *The Tallinn Manual* outlines a series of "rules" and accompanying analyses that provide insight into how cyber operations intersect with *jus ad bellum* and *jus in bello*. It is not tied to any one nation and is not legally binding on any nation. It has several authors that *The Tallinn Manual* refers to as the *International Groups of Experts*. *The Tallinn Manual* defines the use of force in Rule 69: "A cyber operation constitutes a use of force when its scale and effects are comparable to non-cyber operations rising to the level of a use of force." A cyber operation that has kinetic effects (i.e., like an explosion) would qualify as a use of force under this definition.

The authors consider the Stuxnet operation a use of force because it caused physical destruction. However, there are many types of cyber operations that are more difficult to categorize. For example, the installation of malware in a state's critical infrastructure that remains inactive unless triggered is a threatening action that causes no immediate impact. According to *The Tallinn Manual*, it would not qualify as a use of force because no damage has occurred, but this could be debated by others who may see it as a type of preemptive strike.

When it comes to state-on-state conflict, not only is it important to be able to clearly demarcate acceptable from unacceptable behavior, but it is also vital to be able to attribute activities to their origins. With lines drawn and attributions made, damaging activities can be deterred or at least appropriately retaliated against. But cyberspace suffers from *the attribution problem*. The inherent dynamics of cyberspace (e.g., distanceless, digital, dynamic) all help to create plausible deniability or even anonymity. It is difficult to pin any activity on a state in cyberspace with 100% confidence. Even if there are clues that indict a particular nation, it is possible that another state is actually the guilty party and carefully planted evidence to pin the blame on another nation.

10.3.3.3 Jus in bello

Jus in bello addressed proper conduct in war. The *Hague Conventions* (1899, 1907) and the *Geneva Conventions* (1949) are international agreements on the humane conduct of war. These conventions defined war crimes such as the inhumane treatment of prisoners of war, attacks on civilians, violating the *principle of proportionality*, and others for which states can be punished by an international court of law.

The Tallinn Manual defines a cyber attack as, "a cyber operation, whether offensive or defensive, that is reasonably expected to cause injury or death to persons or damage or destruction to objects." (Rule 92) Using this definition, a cyber attack against critical infrastructure that kills civilians would violate IHL. The Russian attack on Tallinn would not be considered a cyber attack even though it affected civilians, because it did not cause death or physical destruction.

The principle of proportionality is an important concept in state relations. It says that if a state suffers harm from another state, a response out of proportion to the harm inflicted is not justified. For example, the United States attributed the 2014 doxxing attack on Sony Pictures to North Korea (see Chapter 4). A response from the United States against North Korea would have been justified as long as it was in proportion to the harm North Korea caused. For example, ordering air strikes against North Korea that could result in death and destruction would not have been justified under the principle of proportionality. That would exceed the damages done by North Korea and escalate the conflict. It is not known for sure how the United States did respond, but at the time President Obama stated, "We will respond proportionately and in a space, time and manner that we choose."

10.3.3.4 Summary

"Cybersecurity is national security." - NSA saying

The concepts of *jus ad bellum* and *jus in bello* create important boundaries for international relations. They help to prevent nation state conflicts from escalating and seek to limit human suffering. They also impose justice that transcends the laws of any one nation.

In cyberspace nation state cyber operations almost never rise to the level of Stuxnet or the attack on Tallinn. Rather, the history of cyber operations has been one of nearly continuous cyber competition and maneuvering between nations. The operations invariably fall short of the use of force thresholds that could trigger a response according to international law. Over time, however, such operations can have a cumulative strategic effect.

Cybersecurity is vital to United States national security. In 2018, the United States adopted a *persistent engagement* approach to cyber operations characterized by *defending forward*. Defending forward means *gaining the initiative in cyberspace by persistently engaging adversaries on their own networks*. The goal is to create friction in adversary networks in order to prevent them from achieving their objectives. Prior to 2018, the United States counted on our ability to deter our adversaries from attacking us in cyberspace. We were conservative, attempting to set the precedent of respecting national sovereignty in cyberspace. This approach resulted in essentially waiting for a cyber incident to occur and then cleaning up afterwards. By then the damage was already done, and our ability to meaningfully respond was muted because the attacks, while consequential, did not qualify as uses of force or acts of war. *Deterrence*, or the threat of retaliation, cannot work if clear lines cannot be drawn and if plausible deniability can be maintained by the transgressor. Persistent engagement recognizes cyberspace is fundamentally different from physical space, that deterrence in cyberspace is ineffective, and that territorial sovereignty need not be respected in the same way it is in physical space. Persistent engagement has helped establish international norms of behavior in cyberspace that respect international law and also reflect the uniqueness of the cyber domain.

10.4 Civil Disobedience

> *"Rules that are judged unethical should be challenged. A rule may be unethical when it has an inadequate moral basis or causes recognizable harm. A computing professional should consider challenging the rule through existing channels before violating the rule. A computing professional who decides to violate a rule because it is unethical, or for any other reason, must consider potential consequences and accept responsibility for that action."*
> *- Section 2.3, The ACM Code of Ethics and Professional Conduct*

As we have seen in this chapter, ethics, rights, and laws are related. Cybersecurity professionals need to understand how they interact and what to do if they ever come into conflict. Figure 10.1, shows how laws, including those that protect rights, and ethics, including those that are based on rights, relate to each other. The top-left square (legal-ethical) and bottom-right square (illegal-unethical) are obvious relationships. Examples of activities that are both legal and ethical include treating people fairly and telling the truth, and examples of activities that are both illegal and unethical are theft, assault, and fraud.

Figure 10.1

The relationship between ethics and laws.

	Ethical	Unethical
Legal	Most legal activities	Ethical grey area candidate
Illegal	Civil disobedience candidate	Most illegal activities

The other two squares give rise to more interesting scenarios. The top-right square is the combination of legal-unethical. Many activities in this category are too trivial to demand the attention of law enforcement. Examples are lying and cheating. These are unethical but not illegal behaviors, although lying under oath (perjury) and cheating on one's taxes are illegal. Also in this category of legal-unethical are ethical gray areas. For example, lying to protect another person's feelings (a so-called "white lie") is not illegal, but is potentially unethical. There are also activities, especially in spheres like cyberspace where the law has not kept up with technology, that are unethical and should be illegal, but laws have yet to be passed to make them illegal. An example of this is cyberbullying, or more formally, harassment through cyberspace, which is unethical but was not illegal in most states until the 2000s.

The last category is the combination of ethical-illegal (the bottom-left square). This category should be rare in the United States because laws are passed only after a careful, deliberate, and democratic process. How could a law be passed to make an ethical behavior illegal? This would make the law unjust. Segregation laws are an example of unethical laws because they enforced divisions and discriminated against people based on skin color. Defying these restrictions was ethical but illegal.

Ethical but illegal activities like defying segregation laws are candidates for civil disobedience. Civil disobedience is refusing to obey a law, or actively disobeying a law, as an act of protest based on the belief that the law is unjust. In order for an illegal action to qualify as civil disobedience, it must be non-violent and morally justifiable, and the actor must submit to the authorities for punishment. Rosa Parks is an example of a person that committed civil disobedience when she was arrested in Montgomery, Alabama, on December 1, 1955, for refusing to obey bus segregation laws. The Montgomery City Code Chapter 6 Section 11 read, "...it shall be unlawful for any passenger to refuse or fail to take a seat among those assigned to the race to which he belongs..." See Figure 10.2 for the police report that was written after Parks knowingly violated this ordinance and was arrested.

Does hacktivism (see Chapter 3) qualify as civil disobedience? Some hacktivists help to keep the identity of political dissidents anonymous—this may be considered a form of civil disobedience. Other hacktivists expose corrupt leaders by linking them to illegal activities. This is a gray area because it ignores the rights of the accused and borders on *vigilante justice*. Vigilante justice is when individuals without the proper authority attempt to enforce the law. Vigilante justice is illegal and unethical, and hacktivists have themselves been arrested for trying to take the law into their own hands by hacking to expose evidence of crimes.

The *crypto wars* of the 1980s and 1990s provide a good illustration of civil disobedience in the domain of cybersecurity. During this era, strong computer encryption was considered a military munition and protected from export under the United States *International Traffic in Arms Regulations (ITAR)*. Therefore, exporting computer cryptography was illegal. While it may sound strange to group computer encryption with military-grade weapons like missiles and machine guns, up to this point in history, strong encryption was considered a military advantage that needed to be protected from falling into enemy hands. But once cryptographic algorithms were implemented in software and could be run on any computer, the idea of trying to keep cryptography contained was considered ludicrous by many. Effectively, the ITAR made posting encryption algorithms on a computer network bulletin board illegal because it could then be accessed by people in different countries—a form of exporting.

Figure 10.2

Rosa Parks arrest police report.

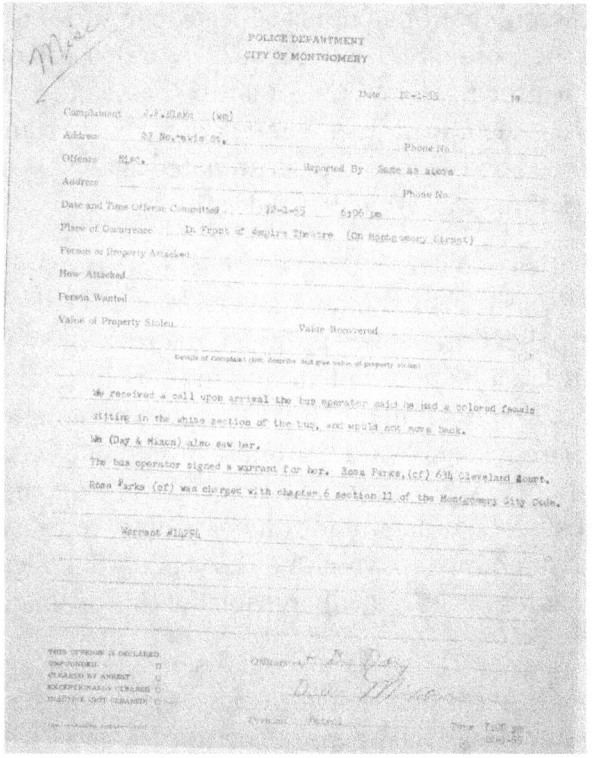

Meanwhile, many believed that people had a right to protect their computer network communications from eavesdropping, and argued that preventing people from using cryptography was unethical. *Phil Zimmermann* was a proponent of this view, and he created free software to encrypt and authenticate digital communications. He called his software *Pretty Good Privacy* (*PGP*) and made it available online in 1991, violating ITAR. He faced a serious criminal investigation but was never charged with a crime. Zimmerman became a cult hero for his actions. Some of his supporters made shirts with the RSA encryption algorithm printed on them as a form of protest, pointing out the absurdity that a shirt then somehow qualified as an export-controlled munition (see Figure 10.3).

The export restrictions around computer cryptography were eliminated in the late 1990s, but the crypto wars have never completely been resolved. The United States government has an uncomfortable relationship with computer cryptography, at times arguing that it is a threat to national security and hampers criminal investigations. Many technologists, on the other hand, point out its benefits to individuals and organizations, the futility of trying to regulate it, and the dangers of incorporating *cryptographic backdoors* into cryptosystems. The idea of a cryptographic backdoor is that cryptosystems would include some mechanism that would allow messages that were encrypted by them to be decrypted by law enforcement under the proper legal authority. Many cyber experts argue that trying to create a backdoor that can only be used by people with the proper legal authority is a fool's errand and dangerous. Either cryptography is strong and protects us all equally (i.e., the good and bad guys alike), or cryptography is weak and we all forfeit the assurance of genuine cryptographic privacy.

Figure 10.3

Export-controlled munitions shirt from the crypto wars.

Civil disobedience is morally justifiable if it is done in the right way. Most forms of hacktivism are forms of vigilante justice (i.e., taking the law into one's own hands) and not civil disobedience. Cybersecurity professionals must understand the legal and ethical implications of their actions.

10.5 Conclusion

To protect the reputation and advance the profession of cybersecurity, it is vital that ethical hackers and cybersecurity professionals behave ethically at all times, respect the rights of all citizens, and obey all applicable laws and legal authorities. Applying multiple ethical paradigms and the veil of ignorance principle can help ethical hackers proceed with moral clarity when confronted with ethical gray areas. The United States Constitution is the basis for our system of government and the foundation for our freedoms and rights, such as the rights to due process and privacy. Federal and state statutes, the United States Constitution and international laws define the legal boundaries of cyberspace activities, including cyber warfare.

 **Chapter 11**

11. Conclusion: The Impact of Cybersecurity

"'Begin at the beginning,' the King said gravely, 'and go on till you come to the end: then stop.'" - Alice's Adventures in Wonderland *by Lewis Carroll*

In the summer of 2016 a hacker group calling themselves The Shadow Brokers appeared out of nowhere on the Internet peddling what they claimed were super-secret hacking tools stolen from the NSA. They released a few of the tools to prove that they had the goods and promised to sell the rest to the highest bidder. Eventually, they just released all the tools on the Internet, giving everybody free access to them. Amidst this treasure trove of cyber weapons were several powerful zero-days, including the EternalBlue exploit that was later used in the NotPetya and WannaCry attacks of 2017—two of the most damaging cyber attacks in history till that time. The tools were definitely for real and probably belonged to the Equation Group, the APT many believe is actually the NSA's Tailored Access Operations (TAO). Soon after this The Shadow Brokers went dark and were never heard from again.

Who were The Shadow Brokers and how did they obtain these hacking tools? It is not known for sure who they are, but they are almost certainly a nation state APT. The prime suspects are China, North Korea, and Russia, three of the United States' major cyber adversaries. The Chinese have undertaken massive data collection operations against the United States, including the theft of millions of highly sensitive personnel records from the United States Office of Personnel Management data breach in 2015. They have also targeted private industry IP and have profited enormously through their cyber espionage activities, saving themselves billions in research and development costs. The North Koreans have also been a significant threat in cyberspace, with their activities tending to center around financing their government through cyber crime. They have stolen cryptocurrency from crypto exchanges, siphoned funds from the international banking system, and installed ransomware on critical infrastructure systems. But many believe The Shadow Brokers are the Russians and that the theft may even have been enabled by

Kaspersky Labs, a Russian company. One theory is that a TAO contractor with TS/SCI clearance managed to sneak the tools out of a SCIF and put them on an Internet-connected computer that was running a Kaspersky antivirus product. The antivirus program, using its trusted root access, was allegedly secretly programmed to look for signatures of cyber weapons and exfiltrate them back to corporate headquarters. As evidence for this claim, some have cited the fact that in 2017 the United States government banned the use of Kaspersky software in government computers, sending a warning to everyone that they could not be trusted. Of course, these accusations are not proven and are vehemently denied by Kaspersky. Why were the tools released for free on the Internet? Besides to embarrass the victim, the tools were likely released to neutralize their capabilities. As soon as a zero-day is exposed, it becomes an n-day and loses most of its potency and value. The United States has done the same with Russian malware, in effect issuing a security bulletin so the vulnerabilities can be patched, defusing the cyber weapons.

This story about The Shadow Brokers was shared as this text closes to make a couple of concluding points. First, it illustrates how cyber news stories can make for good reading! Due to core cyberspace dynamics like the attribution problem and the three Ds (see the Introduction), mystery and intrigue abound, and experts often differ in their theories—meanwhile the actors almost always remain silent or deny involvement. Second, it was intentionally laden with cyber jargon to give the reader confidence that you are an insider and can follow the cyber news. The story would not have registered the same had you read it at the beginning instead of the end of this textbook. I hope you will study the terms in the Glossary and continue learning to add more words to your cyber vocabulary, some of which would not yet have been coined at the time of this writing.

We have come to the end of the book so it is time to stop, but we are nowhere near the end of the cybersecurity story. My hope is that this textbook has motivated and equipped you to follow current events in cybersecurity, educate others, and continue on this journey. It is important because cyber has a significant impact on our lives. It impacts our way of life and our freedoms on a national level through nation state cyber operations. This era has been compared to the Cold War in terms of how a new technology can fundamentally and forever alter how nations interact with one another. The balance of power is at stake and cyber is a game changer, enabling significant cumulative strategic gains all while staying below military response thresholds. Cybersecurity also impacts us on a personal level. The organizations we interface with are under constant attack from hackers of all stripes, and we are reminded of this whenever we receive yet another data breach notification. Our privacy in cyberspace is far from secure—we are all vulnerable. Cybersecurity will only become more important as we become ever more technologically dependent. I hope that you will accept the invitation to continue deepening your cyber understanding, and that many of you will join the white hats in the fight!

Acknowledgements

"I, from the influence of thy looks, receive access in every virtue; in thy sight more wise, more watchful, stronger…" Adam to Eve in Paradise Lost *by John Milton*

First, I want to thank the Ohio Cyber Range Institute (OCRI) for believing in me and investing in this project. Their support of me and many other faculty members throughout the state has helped to advance cybersecurity in Ohio and has had an even farther reaching exponential impact through the students we teach. A special thank you to the visionary behind the OCRI, Dr. Richard Harknett. His name will go down in the history books for his contributions to cyber persistence theory and the study of nation state cyber competition. In the cyberspace age, cybersecurity is national security, and Dr. Harknett's influence has without question strengthened the cybersecurity of our nation—it is difficult to overstate the impact of such a contribution to the greatest and most influential country in the world. He is a treasure to the state of Ohio, and because of his extraordinary humility, is also a friend and mentor to me.

Thank you to Tricia Clark and the Cedarville University Library for publishing this textbook. Thank you to the administration at Cedarville, Dr. White, Dr. Mach, and Dean Chasnov, for enabling me to pursue cyber education beyond our campus making projects like this possible.

Thank you to my computer science and cyber operations faculty colleagues. Our common vision for the Word of God and the testimony of Jesus Christ is our secret weapon that helps us punch far above our weight class. Our exceptional collegiality has enabled us to do so much more together than we ever could have done alone. A special thank you to my mentor, Dr. Dave Gallagher, who was the first to recognize the opportunity for cybersecurity education at Cedarville. Through his wise counsel I enrolled at the Air Force Institute of Technology for my PhD to study cybersecurity, launching me on a blessing-filled adventure that now includes the writing of this textbook.

Thank you to all my Center for the Advancement of Cybersecurity Student Fellows who work side-by-side with me in advancing cybersecurity for the benefit of our nation. They have helped to shape cybersecurity in the academy. A special thank you to my three-time Chief Student Fellow, David Reid, my right-hand man who has left a legacy at Cedarville that will pay dividends for decades, and to Kaicheng Ye, our resident elite hacker who helped build the lab exercises and VMs that accompany this textbook.

294 INVITATION TO CYBERSECURITY

I want to thank all of my CY-1000: Introduction to Cybersecurity students. They were the first readers of this material as it took shape over the years and have helped me organize it in a coherent manner. They were in the forefront of my mind with every word and sentence I typed. There is so much about cybersecurity that I want them to know that I am afraid I tried to pack too much into this one volume! Hopefully they will forgive me. Students, never forget: WLBLFHVVZKVIHLMHPROOVWRMSRHDLIPSVDROOHG-ZMWRMGSVKIVHVMXVLUPRMTH

Last, I want to thank my wife and our four children. Maggie, Kenny, Teddy, and Rebekah, you are my beloved ones with whom I am well pleased. Emily, because of you and your influence, I am ever stronger, wiser, and more virtuous. I rejoice in you, the wife of my youth.

Glossary

"'What's the use of their having names,' the Gnat said, 'if they won't answer to them?' 'No use to them,' said Alice; 'but it's useful to the people who name them, I suppose. If not, why do things have names at all?'" - Through the Looking-Glass by Lewis Carroll

To name something is to know something. This textbook highlights numerous names, or vocabulary words, that are important terms and concepts in the field of cybersecurity. Understanding these terms and using them in their proper context is what makes you a cybersecurity insider. These definitions are the author's own unless otherwise indicated. The number in parenthesis after the definition is the chapter where the term is defined.

ob: binary number prefix (2)

ox: hexadecimal number prefix (2)

2600 Hz: the tone that triggered operator mode when whistled into a phone (3)

2600: The Hacker Quarterly: a hacker periodical whose name comes from 2600 Hz (3)

32-bit architecture: a computer with a word size of 32 bits (2)

64-bit architecture: a computer with a word size of 64 bits (2)

A Secret History of Hacking: a Discovery Channel documentary that highlights the folk heroes of hacking (3)

AAA: the three components of access control: authentication, authorization, and accounting (8)

abstraction: a high-level summary that retains the essential elements (2)

accepting risk: a deliberate decision to live with a risk (5)

access control: monitoring and controlling access to computer systems and data (8)

access control list (ACL): the set of user permissions associated with an object (8)

accounting: recording who did what when (8)

actions on objectives: the last phase of a cyber attack where the hacker accomplishes his goals on the target (4)

Active Directory (AD): a Microsoft application that provides IAM and other access control services (8)

additive: using substitution and transposition together in some proportion in a cryptographic scheme (7)

Adleman, Leonard: a pioneer of cryptography who along with Adi Shamir and Ron Rivest discovered the RSA cryptosystem (7)

administrator access: the highest level of user access available on a Windows OS (8)

advanced persistent threats (APTs): a team of elite hackers (3)

adversarial thinking: the ability to embody the technological capabilities, the unconventional perspectives, and the strategic reasoning of hackers (6)

adversarial thinking (principle): a principle of cybersecurity that states one must never forget about the existence of intelligent human hackers (9)

affected parties: all the people who will be impacted by an ethical decision (10)

Aleph One: the hacker nick used by Elias Levy, the hacker wrote the famous paper, "Smashing the Stack for Fun and Profit" (3)

alert fatigue: the problem of ignoring alerts because of too many false positives (8)

algorithm: an abstract, step-by-step recipe for solving a well-defined problem (2)

Allen, Paul: a pioneer of computing who along with Bill Gates co-founded Microsoft (2)

alphabetic shift cipher: a monoalphabetic substitution cipher that uses a shift of the alphabet for the ciphertext alphabet (7)

alteration: modifying or creating data without authorization (4)

analytical game theory: another name for traditional game theory (i.e., not behavioral game theory) (6)

Android: a mobile OS developed by Google (2)

annualized loss expectancy (ALE): the projected losses to a cyber asset due to a cyber risk over the course of a year (5)

annualized rate of occurrence (ARO): the expected annual frequency of the incident occurring (5)

annualized safeguard costs (ASC): the cost of a safeguard over the course of a year (5)

anomaly detection: monitoring the behavior of software looking for unusual or suspicious behavior (9)

Anonymous: a notorious international and decentralized hacktivist group (3)

antivirus software: a program that scans files to identify malware (9)

application firewall: scanning for sensitive information in outbound packets and blocking them from leaving the network (8)

application logs: logs that record events associated with programs installed on the OS (8)

argument: input into a procedure call (2)

Armitage: a GUI wrapper for Metasploit (3)

ARPANET: a United States' Department of Defense-sponsored project to create a computer communications network that evolved into the Internet (2)

artificial intelligence (AI): cutting edge computing technology that performs at human-like levels or better (2)

AI for cyber: using AI to enhance cybersecurity (2)

ASCII (American Standard Code for Information Interchange): the standard binary encoding of the English character set (2)

ASCII art: artistic images created from ASCII characters (3)

assembly language: a human-readable form of machine code (2)

asset value (AV): the value of the cyber asset (5)

Atbash cipher: a monoalphabetic substitution cipher devised by the Hebrews that uses the letters at the opposite end of the alphabet for the ciphertext alphabet (7)

attack and defend CTF: a CTF where competitors earn points based on the amount of time systems and services are online (3)

attribution problem:: the difficulty of determining who is responsible for a cyber attack (3)

authentication: verifying an identity (8)

authentication cookie: a text string assigned to a web browser by a web server for the purposes of authenticating a user (8)

authentication token: an artifact that uniquely identifies a user (8)

authorization: permitting or denying access to a resource (8)

availability: ensuring authorized users have access to their data and computer systems (4)

avoiding risk: eliminating the risk as a possibility (5)

Babbage, Charles: the first person to envision a general-purpose computer he called the Analytical Engine (2)

backdoor: an unauthorized access point (4)

bandwidth: the amount of data that a link can carry (2)

base two-base ten conversion rule: ten binary places are approximately equal to three decimal places (2)

behavioral game theory: an alternative to traditional game theory based on observing what people actually do when faced with strategic choices (6)

Bell-LaPadula model: a rule for managing access to information primarily concerned with the confidentiality of information based on the high water mark principle (8)

Berners-Lee, Tim: the inventor of the World Wide Web (2)

best response analysis: a technique for solving games by starting with the options of the other player and determining how to respond in each case (6)

Biba model: a rule for managing access to information primarily concerned with the integrity of information based on the low water mark principle (8)

bigram: a pair of letters (7)

Bill of Rights: the first ten amendments to the United States Constitution that protect the rights of citizens from the government (10)

binary: a base two number system that uses the symbols 1 and 0 (2)

biometric authentication: a method for authenticating a person based on physical characteristics (8)

birthday paradox: only twenty-three people are required to make it probable that any two people will have the same birthday (7)

bit: binary digit (1 or 0) (2)

Bitcoin: a cryptocurrency that is built on a blockchain (7)

Bitlocker: Microsoft Windows built-in encryption utility (9)

Black Hat: a major hacking conference held every summer in Las Vegas focused on industry and vendors (3)

black hat hackers: are individuals who engage in illegal hacking (3)

black market: a marketplace for stolen and illegal goods (5)

blacklist: a list of explicitly denied resources (8)

bloatware: a pejorative term for applications that come pre-installed on devices (9)

block: a block size number of consecutive bits that are grouped and processed together in a block cipher cryptosystem (7)

block chaining: a technique used in block cipher cryptosystems that incorporates the previous encrypted block into the next block to obscure the relationship between plaintext blocks (7)

block cipher: a cryptographic system that encrypts and decrypts a block of bits at a time (7)

block size: the number of bits grouped together in a block in a block cipher cryptosystem (7)

blockchain: a technology used to create a trusted public record in a low-trust environment (7)

blue team: the defenders in a CTF (3)

Boole, George: the founder of binary logic (2)

Boolean logic: a system that accepts true and false inputs and produces a true or false output (2)

booting: loading an operating system into memory when computers are powered on (2)

bot: a pwned computer that can be remotely controlled (3)

botnet: a collection of bots that can remotely controlled as a group by a hacker (3)

botnet: a collection of "slave" computers that respond to the commands from a "master" computer (4)

broken: the state of a cryptographic technique for which attackers can reveal ciphertext messages through cryptanalysis in a shorter amount of time than it would take to perform a brute-force key search attack (7)

Brooks, Frederic: a pioneer of software engineering (2)

browser-based application: a program that runs in the cloud that is accessed via a web browser (2)

brute-force key search attack: an attack on cryptography that tries every key in the keyspace until it finds the one that unlocks the ciphertext (7)

buffer overflow attack: an exploit where the attacker is able to send code to the target computer and force it to execute it (4)

bug: a programming mistake (2)

bug bounty: a payment made by an organization for finding a vulnerability in one of their products (3)

bulletin board system (BBS): online discussion forums that predate the Internet (3)

business continuity planning (BCP): ensuring that a business can continue to operate in the wake of a disruption (5)

business email compromise: an attack where an employee is tricked into making a fraudulent funds transfer (9)

business impact analysis (BIA): a method for determining how a cybersecurity incident will impact the organization (5)

Butler, Max (AKA Max Vision): a gray hat hacker who was the subject of the book Kingpin by Kevin Poulsen (3)

byte: a group of eight bits (2)

C-suite: the topmost leaders of an organization (5)

Caesar cipher: a monoalphabetic substitution cipher devised by Julius Caesar that uses a three letter shift of the alphabet for the ciphertext alphabet (7)

calling card: a signature deliberately left behind as evidence after a successful cyber attack so the hacker can prove his involvement (3)

capture-the-flag (CTF): a hacking contest (3)

carders: cybercriminals who sell stolen credit card information (3)

catphishing: a social engineering attack where a bad actor creates trust through online interactions and then manipulates victims financially, emotionally, or otherwise (9)

caveat emptor: the buyer accepts the risk for a purchase (9)

central processing unit (CPU): the unit of a computer that performs all the calculations (2)

Cerf, Vint: a pioneer of the Internet who along with Bob Kahn created the TCP/IP protocols (2)

certificate authority (CA): a third party that is trusted to verify real-life identities (7)

challenge coin: a custom coin presented by a leader as a commendation for a job well done (3)

Chaos Computer Club: a pioneering European hacktivist group (3)

checks and balances: the designed interdependencies between the executive, legislative, and judicial branches of the United States government to prevent the abuse of power (10)

Chief Information Security Officer (CISO): the officer in charge of cybersecurity (5)

Children's Online Privacy Protection Act (COPPA): a United States data privacy law that protects the data of children under thirteen (5)

chosen-plaintext attack: an attack on ciphertext where the adversary can generate messages using the same key and cryptosystem that produced the ciphertext (7)

Chrome: Google's web browser (2)

CIA triad: the three goals of cyber defense: confidentiality, integrity, and availability (4)

cipher: a cryptographic scheme that operates on the level of letters (7)

ciphertext: the scrambled message (7)

ciphertext block: a block of encrypted bits (7)

ciphertext-only attack: an attack on ciphertext where the adversary starts only with the ciphertext (7)

circuit: a path through which electricity flows (2)

civil disobedience: refusing to obey a law, or actively disobeying a law, as an act of protest based on the belief that the law is unjust (10)

classifications: in the United States Department of Defense MLS system the permission levels assigned to objects (8)

clearances: in the United States Department of Defense MLS system the permission levels assigned to subjects (8)

client program: a program that initiates connections to a server (2)

client program: a program that initiates network connections to servers (2)

closed-source software: software for which the source code is not published (9)

cloud computing: the practice of using third-party servers over the Internet for business purposes (5)

code cracking: an attack against ciphertext to reveal the encrypted message (7)

codebook: a cryptographic scheme that operates on the level of words (7)

collision resistance: a feature of hash functions that make hash collisions highly improbable (7)

column player: the player in a normal form game listed on the top (the columns) (6)

command and control (C2): the ability to remotely issue commands to the compromised device (4)

command-line interface (CLI): a purely text-based interface that enables users to type in commands and read typed output (2)

Common Vulnerabilities and Exposures (CVE): the main classifier and catalog of cybersecurity vulnerabilities (9)

Common Weakness Enumeration (CWE): a publicly available catalog of software and hardware-related flaws that can lead to vulnerabilities (5)

compartment (DOD classification): a category of sensitive information (9)

compartmentalization: a principle of cybersecurity that states access to resources should be segmented (9)

compiler: a program that transforms source code into machine code (2)

Computer Fraud and Abuse Act (CFAA): a United States umbrella anti-hacking law (10)

conditional statements: statements that define alternate execution paths (2)

Condor: the hacker nicked used by Kevin Mitnick, a phone phreaker and famous social engineer (3)

Conficker Worm: a worm that exploited the MS08-067 vulnerability to create a large-scale botnet (4)

Confidential (C): a United States Department of Defense classification for information that could reasonably be expected to cause identifiable damage if it was disclosed (8)

confidentiality: preventing the unauthorized reading of data [credit: Mark Stamp] (4)

connected graph: a graph where there exists at least one path between any pair of nodes (2)

connection handshake: part of a network protocol that establishes the parameters for a connection (8)

control: a measure taken to reduce risk (5)

cookie stealing attack: an attack where a user's web browser authentication cookies are copied, allowing an attacker to impersonate the user (8)

cooperating: a game theory strategy of being loyal to the other player (6)

core: a processing unit controlled by the CPU capable of executing instructions (2)

corporate espionage: spying conducted by a business competitor (3)

corporate sabotage: impairing a business competitor's ability to operate (3)

corrective controls: measures taken to recover after a cyber incident (5)

cost center: a part of a business that is not revenue producing (5)

covert channel: a hidden communication path (4)

crash: the abrupt termination of a running program (2)

credential stealing: fraudulently obtaining a valid set of user credentials to computer systems and data (4)

credential stuffing: an attack where a hacker attempts to login to many different websites using a known valid username and password pair (4)

criminal hacking: hacking motivated by making money (3)

critical: a log level that indicates an event that needs to be reviewed (8)

cross-site request forgery (CSRF): a website attack that tricks the target's web browser into making a website request crafted by the attacker (4)

cross-site scripting (XSS): a website attack that tricks the JavaScript engine into executing code provided by the attacker (4)

cryptanalysis: the art and science of decrypting ciphertext when not in possession of the secret key (7)

crypto wars: a period in the 1980s and 1990s when the United States government tried to restrict the use of strong cryptography to protect digital communications (10)

cryptographic backdoor: a mechanism that would allow messages that were encrypted with a cryptosystem to be decrypted by law enforcement under the proper legal authority (10)

cryptographic hashing: a form of cryptography that creates digital fingerprints for a data object (7)

cryptography: the art and science of scrambling and unscrambling information using a secret to keep it private (7)

cryptosystem: all the components needed to implement a specific type of computer cryptography (7)

Cult of the Dead Cow (cDc): a pioneering United States hacktivist group (3)

cyber 9/11: the potential for a large-scale cyber attack motivated by terrorism (3)

cyber assets: computer systems and data of value (5)

cyber forensics: the process of collecting, analyzing, and preserving cyberspace evidence (3)

cyber insurance: a way to transfer cyber risk to an insurance company (5)

cyber operations: intelligence and military operations that take place in and through cyberspace (3)

cyber Pearl Harbor: the potential for a large-scale cyber attack carried out by an enemy nation state (3)

cyber range: a safe online space for practicing cybersecurity (2)

cyber risk: the potential for a cyber threat actor to exploit a vulnerability that allows him to disclose, alter, or deny access to a cyber asset (5)

cyber risk management: a detailed process of identifying cyber assets, enumerating how threats and vulnerabilities pose risks to assets, analyzing the severity of the risks, and then choosing how to handle the risks (5)

cyber threat: an action taken with malicious intent that discloses, alters, or denies access to a cyber asset (5)

cyber threat actor: a person that poses a cyber threat (5)

cyber threat modeling: a systematic approach to identifying cyber vulnerabilities (5)

cyber warriors: individuals that hack with the authorization of the government (3)

cyberbullying: harrassment through cyberspace (10)

cybersecurity: the practice of protecting and respecting the rights of every individual and organization in cyberspace (1)

Cybersecurity and Infrastructure Agency (CISA): a United States government agency devoted to the cyber defense of our nation that provides resources to organizations (5)

cybersecurity audit: an accounting of how an organization's cybersecurity complies with a standard (5)

Cybersecurity Framework (CSF): a NIST standard that focuses on the functions of cyber risk management (5)

cybersecurity governance: the oversight of the security risks of an organization (5)

cyberspace: an electronic world composed of computer devices that transmit, receive, and process data (digital information) (2)

cyberspace perimeter: the domain names and IP addresses that belong to the target and all the software and hardware accessible via those domain names and IP addresses (4)

DAD: the three goals of a cyber attack: disclosure, alteration, and denial (4)

dancing men cipher: a monoalphabetic substitution cipher that uses stick figures for the ciphertext alphabet that was devised by Arthur Conan Doyle (7)

Dark Dante: the hacker nick used by Kevin Poulsen, a phone phreaker turned hacker journalist (3)

dark web: a collection of websites accessible via specialized web browsers designed to protect the anonymity of the website hosts and clients (5)

data breach: a type of cyber attack where hackers obtain unauthorized access to an organization's data (3)

data encoding: representing information using bit strings (2)

data exfiltration (exfil): copying data from a victim to the attacker (4)

data leak: the unauthorized disclosure of private information (3)

data loss prevention: scan for sensitive and proprietary information in outbound packets and block them from leaving the network (8)

data retention: the practice of storing data (9)

Debian: a popular Linux distribution known for being stable and efficient (2)

decimal: a base ten number system that uses the symbols 0-9 (2)

decryption: the process that undoes the ciphertext scrambling to recover the plaintext (7)

deductible: the amount owed by the insured party for a covered incident (5)

DEF CON: a major hacking conference held every summer in Las Vegas focused on hackers and competitions (3)

default gateway: the first hop for all of a computer's Internet traffic (2)

defecting: a game theory strategy of betraying the other player (6)

Defender Antivirus: the Windows OS built-in antivirus software (9)

Defender Firewall: the Windows OS built-in firewall (8)

defending forward: gaining the initiative in cyberspace by persistently engaging adversaries on their own networks (10)

defense in depth: a principle of cybersecurity that states security should be implemented in layers (9)

delegation: the property of running with the permissions of the initiating user (8)

denial: denying authorized users access to their data and computer systems (4)

denial of service (DoS) attack: an attack that denies authorized users access to a resource (4)

Department of Defense (DOD): the division of the United States government in charge of military and intelligence operations (3)

depth wins: a principle of cybersecurity that states the success of cyber attacks and cyber defense often comes down to who knows more (9)

desktop application: a program that runs locally on a computer (2)

detective controls: measures taken to detect incidents (5)

deterministic: the property of producing the same output for the same input every time (7)

deterrence: using the threat of retaliation to deter (dissuade) adversaries from attacking (10)

deterrent controls: measures taken to discourage cyber threat actors from acting (5)

device driver: a low-level program that manages communication between a peripheral device and the CPU (2)

dictionary attack: a password hashing attack that draws base words from a wordlist (e.g., a dictionary) and applies string mangling (9)

Diffie, Whitfield: a pioneering computer cryptographer who along with Martin Hellman discovered public key cryptography (7)

dig: a Linux command line utility for using DNS to resolve domain names to IP addresses (2)

digital: a feature of cyberspace where all data is composed of 1s and 0s (1)

Digital Millennium Copyright Act (DMCA): a United States law that provides intellectual property protections for companies (3)

directory (folder): a container for files (8)

disassembler: a tool that helps convert machine code into a higher-level language so that it can be dissected (3)

disaster recovery plan (DRP): a formal document that details an organization's incident response process (5)

disclosure: obtaining unauthorized access to data (4)

discretionary access control (DAC): an access control model where users can assign permissions to other users (8)

distanceless: a feature of cyberspace where all devices are within instant reach of every other device (1)

distributed denial of service (DDoS) attack: a DoS attack where a deluge of network traffic is directed against a target from multiple sources (4)

distributions (distros): Linux-based operating systems that are packaged with software, tools, and user interfaces (2)

DMZ: a segmented portion of a computer network that contains Internet facing servers (9)

DNS filtering: a service that provides firewall-like functionality by preventing access to black-listed websites (9)

domain controller (DC): the server used by Active Directory where user credentials are stored (8)

domain name: the part of the URL that ties it to an IP address (2)

Domain Name System (DNS): a public distributed database that maps domain names to IP addresses (2)

dominated strategy: a strategy that will never be chosen (6)

double ROT-13 cipher: a facetious cipher used to call attention to the fact that the communicating parties should have used cryptography but did not (7)

double transposition cipher: a cipher that uses a two-dimensional table to transpose letters (7)

doxxing attack: an attack where a hacker publishes private data to embarrass or otherwise harm the victim (4)

Draper, John (AKA Captain Crunch): a hacking fold hero who pioneered phone phreaking (3)

drive-by-download: an attack that exploits vulnerabilities in web browsers and is triggered by just visiting a malicious website (4)

dropped packet: a packet that never reaches its destination (2)

due diligence: a threshold based on what a "prudent man" would do to safeguard an organization (5)

dumpster diving: a low-tech recon tactic that involves looking through discarded trash for insider information about the target (4)

dwell time: the amount of time that an unauthorized actor remains undetected on a system or network (9)

dynamic: a feature of cyberspace where devices and data are continually in flux (1)

e-gold: an early online currency that was shut down by law enforcement because of its use in illegal transactions (3)

Economic Espionage Act: a United States law that protects the intellectual property of companies (10)

Edge: Microsoft's web browser (2)

Electronic Communications Privacy Act (ECPA): a United States law that protects the right to privacy for digital communications (3)

elite hackers: people who possess an enormous amount of technical understanding and expertise (3)

Elk Cloner virus: one of the first computer viruses and an example of nuisance hacking that displayed a silly poem on the victim's screen (3)

email spoofing attack: an attack where a hacker changes the From: field in an email to make it appear like the email came from someone else (4)

encapsulation: the process of grouping functionality into a single simple unit (2)

encryption: the process that scrambles plaintext into ciphertext (7)

end-user license agreement (EULA): a type of contract that software providers make their end users accept before providing the users access to their services (9)

endpoint: the computers, smartphones, and other devices on the network (8)

enrollment phase: the part of the authentication process when the user's access credentials are registered and stored in an authentication database (8)

equal error rate: the level of specificity where the fraud rate and the error rate are set equal to one another for a biometric authentication method (8)

escalate privileges: increasing access privileges on a device (4)

EternalBlue: a Windows vulnerability that was disclosed within a cache of cyber weapons purportedly belonging to the NSA (9)

Ethernet: the main Layer 2 protocol used to connect computers to the Internet (2)

ethical hacking: hacking while behaving ethically at all times, respecting the rights of all citizens, and obeying all applicable laws and legal authorities (3)

Event Viewer: the Windows OS built-in log viewer and manager (8)

evil maid attack: an attack where a hacker gains physical access to an unattended computer and compromises it (4)

evil twin attack: an attack that tricks users into connecting to a wireless network that is administered by a hacker (9)

exclusions: a loss explicitly not covered by the policy (5)

Exclusive OR (XOR): a Boolean function that outputs 1 when exactly one of the input bits is 1 and 0 otherwise (7)

executable: a compiled program that can be run by the operating system (2)

execute: the permission to run a program or script (8)

experimental game theory: another name for behavioral game theory (6)

exploit: an action that takes advantage of a vulnerability to compromise security (3)

exposure: the potential losses that could result from an incident (5)

exposure factor (EF): the percentage of the asset's value that will be compromised if the risk is realized (5)

externality: a cost borne by external parties that exceeds the cost borne by the party responsible for causing or preventing it (5)

extortion: threatening to carry out a harmful action unless a payment is made (3)

false negative alert error: an error that occurs when there is malicious activity but no alerts are generated (8)

false negative error: an authentication error that occurs when the right person fails authentication (8)

false positive alert error: an alert that turns out to be normal activity (8)

false positive error: an authentication error that occurs when the wrong person passes authentication (8)

Family Educational and Privacy Rights Act (FERPA): a United States data privacy law that protects student educational records (5)

fetch-decode-execute cycle: the process performed by the CPU of fetching a block of memory, decoding it, and then executing it (2)

file system: the organization of the files in an OS (8)

Firefox: Mozilla's web browser (2)

firewall: a software application or hardware appliance that allows or denies network traffic based on a set of rules (5)

five domains of warfare: land, air, sea, outer space, and cyberspace (3)

five Ds: disrupting, degrading, denying, destroying, and deceiving our adversary's capabilities in and through cyberspace (3)

five layer model: the design of the Internet to coordinate communication (2)

floating point numbers: the way computers represent base ten numbers with decimal points (2)

floppy disks: an old form of removable media used for storage (2)

follow the leader: choosing the same cybersecurity solutions that a peer organization in the same sector uses (5)

for-statement: along with the while-statement, the typical looping statement keyword (2)

foreign espionage: spying conducted by a foreign nation (10)

Foreign Intelligence Surveillance Act (FISA): a United States law that allows the government to collect foreign intelligence signals on domestic soil under specific circumstances (3)

forward search attack: an attack against a hash to reveal the string it represents by generating putative strings, hashing them, and comparing the outputs to the hash (7)

fraud rate: the probability of false positives for a biometric authentication method (8)

free and open-source software (FOSS): software that is free to use and whose source code is publicly available (3)

full disk encryption: a technology that stores data in encrypted form and decrypts and encrypts it transparently as needed (9)

fully qualified: the full directory path is provided (8)

gaining unauthorized access: the phase of a cyber attack where the hacker exploits a vulnerability to gain access to the target (4)

game: a strategic contest (6)

game theory: the study of interdependent decision making between multiple players where each player strives to maximize his own utility (6)

garbage data: data in memory that holds outdated or random values (2)

Gates, Bill: a pioneer of computing who along with Paul Allen co-founded Microsoft (2)

generative AI: computer technology capable of generating coherent media such as text, audio, images, and video (2)

genesis block: the first block in a blockchain (7)

Geneva Conventions: international peace treaties conducted in 1949 after World War II that define the humane conduct of war (10)

gigabyte (GB): 2^30 bytes (approximately a billion) (2)

gigahertz (GHz): a billion cycles per second (2)

Google hacking: a high-tech recon tactic that involves advanced uses of Google to search for vulnerabilities and sensitive information about the target (4)

Gramm-Leach-Bliley Act (GLBA): a United States data privacy law that protects customer records of financial institutions (5)

graphical user interface (GUI): an interface with graphics such buttons, icons, and text fields that allow users to point and click as well as to type (2)

gray hat hackers: individuals who blur the lines between legal and illegal hacking and engage in both kinds (3)

greatest happiness principle: another name for utilitarianism (10)

gross negligence: the willful disregard and failure to comply with best practices (5)

group: a collection of users (8)

guest OS: a virtual machine OS running on a computer (2)

hack: an action that is allowed by the system but that undermines the intent of the system (4)

hacker: a person who attempts to gain unauthorized access to a computer system or deny access to authorized users (3)

Hackers **(1995 film)**: Prodigy hacker Dade Murphy (Jonny Lee Miller), AKA Crash Override, leads a group of teenage hackers including Kate Libby (Angelina Jolie), AKA Acid Burn, in a battle against law enforcement and other hackers (3)

hacking back: the practice of trying to hack somebody who has hacked you (3)

hacktivism (hacktivist hacking): hacking motivated by a political or ideological cause (3)

Hague Conventions: international peace treaties conducted in The Hague, Netherlands, in 1899 and 1907 that define the humane conduct of war (10)

Hamming distance: the difference between two equal length bit strings (7)

handle: another name for nick (3)

hard disk drive (HDD): storage that uses spinning metal platters (2)

hardened: made secure (9)

hardware: the physical computational components of a computer (2)

hash: a short, fixed-length binary string that uniquely represents a data object (7)

hash chain: a chain of hash values created by hashing a string, then hashing the hash of the string, and so on (7)

hash collision: different data objects having the same hash (7)

hash dump: a file that contains password hashes (4)

hash function: a cryptographic algorithm that takes as input a data object and outputs a hash (7)

header: a prefix added to a packet that specifies delivery-related information (2)

Health Insurance Portability and Accountability Act (HIPAA): a United States data privacy law that protects patient medical information (5)

Heartbleed: a famous vulnerability in the OpenSSL software library that was discovered in 2014 (3)

Hellman, Martin: a pioneering computer cryptographer who along with Whitfield Diffie discovered public key cryptography (7)

Hello, World! program: the traditional first program written in a new language that outputs "Hello, World!" (2)

hex editor: a program that can be used to view and edit the raw bytes of a file (2)

hexadecimal: a base sixteen number system that uses the symbols 0-9 and a-f (2)

hide-and-seek game: a game theory game that illustrates focal point biases and level-k reasoning (6)

high-water mark principle: a principle of access control that states the highest level of information that a subject is exposed to sets the bar (8)

HMAC (hashed message authentication code): the hash of a message that has been combined with a shared secret (7)

hop: shorthand for a communication link (2)

host OS: the base OS running on a computer (2)

HTML (Hypertext Markup Language): the syntax for web pages (2)

HTTP (Hypertext Transfer Protocol): the application layer network protocol web browsers and web servers use to communicate (2)

HTTPS (Hypertext Transfer Protocol Secure): a secure form of HTTP that uses encryption and authentication (2)

hypertext: formatted text that enables linking to URLs (2)

Iceman: the hacker nick used by Max Butler, AKA Max Vision, the hacker whose story is told in Kingpin: How One Hacker Took Over the Billion-Dollar Cybercrime Underground by Kevin Poulsen (3)

identity and access management (IAM): software that is centralized and accessed over a network to provide authentication across an organization (8)

identity theft: fraudulent actions taken in someone else's name to obtain a financial benefit (5)

if-statement: the typical conditional statement keyword (2)

in-depth: reusing one-time pad keystream characters (7)

incident response: measures taken after a cyber incident to contain the damage, determine what happened, and recover (5)

incriminate: to provide evidence of a crime (10)

indemnify: to compensate for a loss (5)

indicators of compromise (IOCs): detectable evidence that indicate a device has been compromised (4)

influencing: the final step in a social engineering attack (4)

information superhighway: a nickname for the Internet (2)

information-theoretic secure: a cryptographic scheme that reveals no information about the plaintext and is therefore impervious to cryptanalysis and brute-force key search attacks (7)

initialization vector (IV): the first block used in cipher block chaining (7)

injection attack: an attack where hackers provide code as user input and trick the target computer into executing their input (4)

input/output (I/O) devices: devices for interacting with a computer (2)

insider threat: a person that works for the organization they attack (3)

insult rate: the probability of false negatives for a biometric authentication method (8)

insurance policy: a contract stating what is covered by the insurance agency and under what circumstances (5)

integrity: preventing, or at least detecting, the unauthorized writing of data [credit: Mark Stamp] (4)

intellectual property (IP): proprietary information that provides a competitive advantage to an organization (3)

interdependent choices: the outcome for each player depends in part on the choices made by the other players (6)

interdiction: a supply chain attack where hardware is intercepted in route to the target and compromised before it is delivered (4)

International Humanitarian Law (IHL): another name for jus in bello (10)

International Traffic in Arms Regulations (ITAR): a regulation that prohibits exporting technologies that provide a military advantage to the United States (10)

Internet (the Net): the network of networks that became the global computer network (2)

Internet Engineering Task Force (IETF): the group that manages the Internet's protocols (2)

Internet of Things (IOT): devices not normally thought of as computers that connect to the Internet (2)

Internet Service Provider (ISP): a company that provides Internet access to homes and organizations (2)

interpreted language: a programming language that is executed by an interpreter (2)

interpreter: an execution engine program for interpreted languages (2)

intractable problems: computable problems that would take a computer a prohibitively long time to solve (2)

intrusion detection and prevention system (IDPS): a program that performs anomaly detection to identify malicious behavior (9)

iOS: a mobile OS developed by Apple (2)

IP (Internet Protocol): the main Layer 3 protocol used in the Internet (2)

IP address: the Layer 3 address for computers on the Internet (2)

Jeopardy style CTF: a CTF where competitors earn points by finding flags and solving challenges (3)

Jobs, Steve: a phone phreaker as a young person who went on to co-find Apple with Steve Wozniak (3)

John the Ripper: a command-line tool for Linux that is used to crack password hashes (4)

jus ad bellum: a principle of international law that outlines the conditions for when it is justified for a state to enter into war with another state (10)

jus in bello: a principle of international law that outlines acceptable conduct in war (10)

just war theory: another name for jus ad bellum (10)

Kahn, Bob: a pioneer of the Internet who along with Vint Cerf created the TCP/IP protocols (2)

Kali Linux: an open-source OS designed for hacking that is frequently installed on a VM (2)

Kantian ethics: an ethical paradigm that emphasizes the motivations behind actions and focuses on duties and obligations (10)

Kerckoffs's principle: a principle of cryptography that states it needs to be assumed that the adversary has access to the underlying cryptosystem (7)

kernel: the core low-level component of an OS (2)

key: the secret that is input into the encryption and decryption processes (7)

key binding problem: the difficulty associated with associating public keys with real-life identities when using public key cryptography (7)

key distribution problem: the difficulty associated with securely distributing keys to the communicating parties when using symmetric key cryptography (7)

key stretching: a technique for slowing down password cracking attempts by using the last link in a hash chain as the authenticating hash (7)

keys to the kingdom dilemma: it is convenient for users to have fewer keys, but if those keys open more doors, then they become bigger targets and do more damage if compromised (8)

keyspace: the number of possible keys (7)

keystream: a sequence of characters used to encrypt and decrypt messages on a per character basis (7)

keystroke logging (keylogging): a technique for stealing passwords by recording the keys typed into a machine (4)

keyword: a reserved word in the programming language with a predefined meaning (2)

kill chain: the chain of events leading up to a successful attack (4)

kilobyte (KB): 2^{10} bytes (approximately a thousand) (2)

Known Exploited Vulnerabilities (KEV): a catalog of CVEs that have been exploited by threat actors (9)

Knuth, Donald: a pioneer of algorithmic analysis (2)

Lopht: a pioneering group of cybersecurity researchers (3)

LAND (local area network denial) attack: an early computer network attack that crashed victim computers by sending them a packet with a source IP address that matched the destination system's IP address (5)

least privilege: a principle of cybersecurity that states permissions should be granted only up to the level needed and only for as long as necessary (9)

leet speak (1337): a system of character substitutions in writing that is popular in hacking culture (3)

level: the attribute in a log that records the importance of the event (8)

level-k reasoning: the process of iteratively thinking about what the other player might do and how to best respond (6)

Lightweight Directory Access Protocol (LDAP): a protocol used to securely send authentication tokens over a network (8)

limit of liability: the maximum amount the insurance company will pay in case of a loss (5)

links: a direct connection between nodes (2)

Linux: a free and open-source operating system developed by Linus Torvalds (2)

log analysis: a review of system and network logs to identify malicious activity (5)

logging: recording cyberspace events (8)

logic bomb attack: an attack where a hacker plants malware and sets it to "detonate" at a later date (4)

looping statements: statements that define a repeated path of execution (2)

lossless: a transformation process where none of the original information is lost and is recoverable, like in cryptography (7)

lossy: a transformation process where original information is lost and is unrecoverable, like in hashing (7)

Lovelace, Ada: the world's first computer programmer who worked with Charles Babbage on the Analytical Engine (2)

low-water mark principle: a principle of access control that states the lowest level of information that a subject is exposed to sets the bar (8)

LSB RGB steganography: an image steganography technique that encodes secret message bits in the least significant bits of the pixel colors (7)

MAC (Media Access Control) address: the Layer 2 address for NICs (2)

machine code: instructions that a CPU can execute (2)

macOS: Apple's operating system (2)

MafiaBoy: the hacker nick used by Michael Calce, as a fifteen year old he made national headlines for taking down yahoo.com (3)

malware: malicious software used in cyber attacks (4)

malware signature: a specific sequence of 1s and 0s in known malware (9)

man-in-the-middle attack: an attack where an adversary undetectably intercepts communications between a communicating parties (7)

mandatory access control (MAC): an access control model where permissions are managed exclusively by administrators (8)

MasterSplyntr: the hacker nick used by an FBI agent to infiltrate a cybercriminal network (3)

maximum allowable downtime (MAD): the maximum amount of time an asset can be unavailable before the organization is severely impacted (5)

MD5 (Message Digest 5): a 128 bit hash function created in 1991 that has not been recommended for use since 1996 but is still used in CTFs (7)

megabyte (MB): 2^{20} bytes (approximately a million) (2)

memory: volatile data storage where the CPU access instructions and data (2)

metadata: the attributes of an item of data (9)

Metasploit: a hacking tool that enables users to pair exploits with payloads and configure them for specific target machines (3)

Metcalfe, Robert: a pioneer of the Internet who created the Ethernet protocol (2)

Microsoft Security Bulletin: a notice of vulnerabilities discovered in Microsoft software (9)

Mimikatz: a hacking tool that extracts plaintext passwords from a computer's memory (3)

Mirai botnet: a botnet composed mostly of IoT devices that was responsible for some large-scale and high-profile DDoS attacks in 2016 (4)

mitigating risk: reducing the risk (5)

Mitnick, Kevin: a hacking folk hero famous for his social engineering exploits (3)

mobile operating systems: an OS that runs on smartphones (2)

monoalphabetic substitution ciphers: a cipher that uses one plaintext-to-ciphertext alphabet mapping (7)

Morris Worm: the first worm on the Internet and took many computers offline (3)

Morris, Robert Tappan (AKA RTM): an ethical hacker who crossed legal boundaries and was convicted of the Computer Fraud and Abuse Act for unleashing the Morris worm on the Internet (3)

MS08-067: a notorious security bulletin from Microsoft that disclosed a major RCE vulnerability in Windows (4)

Mudge: the hacker nick used by Peiter Zatko, a member of Lopht that testified before Congress (3)

multi-factor authentication (MFA): authentication based on tokens from two or more different categories (8)

multi-level security (MLS): an access control scheme that assigns permissions based on information sensitivity levels (8)

n-day: an exploit that targets an n-day vulnerability (4)

n-day vulnerability: a known vulnerability (4)

name collision: a situation where two objects have the same name (8)

namespace: a domain in which no name collisions are permitted (8)

Nash equilibrium: a stable point in a game where neither player can unilaterally change his choice and end up with a more preferred outcome (6)

Nash, John: a game theoretician who proved all finite games have an equilibrium point (6)

National Institute of Standards and Technology (NIST): a United States government agency that provides cybersecurity risk management guidance to the federal government and organizations (5)

National Security Agency (NSA): an organization within the United States Department of Defense authorized to conduct signal intelligence on foreign adversaries (3)

National Vulnerabilities Database (NVD): a catalog that provides additional guidance for CVEs including criticality scores and remediation (9)

need-to-know: a rule that manages access to information based on its relevance to job duties (8)

negative permission: an explicit denial of an action (8)

network interface card (NIC): the standard I/O device for sending and receiving data over a computer network (2)

network latency: the time it takes network traffic to travel from the source to final destination (2)

network scanning: sending packets to probe for devices and processes running on the network (4)

ng: next generation, often used in software names (3)

nibble: a group of four bits (2)

nick: an online identity used by a hacker (3)

Nmap: a popular Linux-based command-line tool used to perform network scanning (4)

no-click attack: an attack that exploits vulnerabilities in smartphones and is triggered by just receiving a text message (4)

nodes: computer devices (2)

non-computer problems: well-defined problems that are not solvable by computers (2)

non-disclosure agreement (NDA): a legal document that binds pentesters to confidentiality (3)

non-repudiation: the property of not being able to deny having produced a message (7)

nonce: a randomly generated string of bits also known as a "number used once" (7)

normal form: a representation of a game that uses rows and columns to capture the essential elements of the game (6)

NotPetya: a Russian wiperware virus that targeted Ukraine but caused a massive amount of collateral damage when it occurred in 2017 (4)

nslookup: a Windows command line utility for using DNS to resolve domain names to IP addresses (2)

nuisance hacking: hacking motivated by curiosity, fun, and bragging rights (3)

object: resources in a computer system (8)

one-time pad: an information-theoretic secure polyalphabetic substitution cipher that uses a random key of the same length as the message (7)

one-time passcode: a randomly generated number that is only used once and has a short expiration that is used as an authentication token (8)

one-way property: a feature of hash functions that make it impossible to start with a hash and generate the original data object (7)

open source intelligence (OSINT): identifying and collecting information that is available to the public (4)

open-source software: software for which the source code is published (9)

OpenSSL: a free command line utility that performs a large variety of cryptographic operations (9)

OpenVAS: a vulnerability scanner that identifies weaknesses in computer systems and networks (3)

operating system (OS): the program that runs continuously and underlies all of the other computer programs that run on the computer (2)

packet: data sent over the Internet (2)

packet capture (PCAP): a file that records network packets (8)

packet filter firewall: a type of firewall that focuses only on metadata in the TCP and IP headers of individual packets (8)

packet sniffer: an application that logs the network traffic processed by a computer's NIC (2)

packet switching: the process of directing packets one link at a time towards their destination (2)

padding: extra bits added to round out a group of bits to the block size (7)

passing the hash: an attack that takes advantage of trust relationships of a network to pivot to other machines (4)

password: a secret string used as an authentication token based on the assumption that only the user knows it (8)

password cracking: an attack against password hashes to reveal user passwords (4)

password guessing: an attack where a hacker attempts to login as a user by guessing his password (4)

password hash: the hash of a user's password (4)

password manager: a software solution that stores user credentials in an encrypted file (a.k.a. vault) (9)

password spraying: an attack where a hacker attempts to login as any user by using the same few password guesses for many different usernames (4)

patch Tuesday: the second Tuesday of every month when Microsoft and other software companies release their security bulletins (9)

path: a set of links that leads from one node to another (2)

patriotic hackers: hacking motivated by national pride (3)

payload: the data portion of a packet (2)

pen register: a device that records the metadata of electronic communications (10)

Pen Register Act: a United States law that restricts the government's access to telephone and Internet traffic metadata (10)

penetration testing (pentesting): the active probing of the cybersecurity defenses of an organization for the purpose of improving security (3)

pentest report: the deliverable from a pentest describing findings and recommendations (3)

people, processes, technology, and facilities: the functional underpinnings of any organization (5)

peripheral devices: outside devices that are plugged into a computer (2)

permission memo: a legal document that explicitly grants pentesters permission to hack an organization (3)

permissions matrix: a two-dimensional table that captures the permissions that subjects can take on objects (8)

persistence: the ability to retain access to the compromised device (4)

persistent engagement: a strategic approach to cyber operations adopted by the United States in 2018 that is characterized by defending forward (10)

personal identification number (PIN): a short easy-to-remember password that is typically used as a second authentication token (8)

personally identifiable information (PII): data that can be used to identify a person and commit identity theft (3)

phishing emails: a type of social engineering over email (4)

phone phreakers: individuals who engaged in phone phreaking (3)

phone phreaking: an early form of nuisance hacking that exploited vulnerabilities in the landline, and later cellular, telephone system (3)

Phrack: a hacker periodical whose name comes from the terms phreak and hack (3)

pigpen cipher: a monoalphabetic substitution cipher that uses geometric shapes for the ciphertext alphabet (7)

Ping of Death attack: an early computer network attack that crashed victim computers by sending them a single malformed packet (4)

pivoting: using access to one device to gain access to another device (4)

pixel: a picture element used to illumine computer screens (2)

plaintext: the original unencrypted message (7)

plaintext block: a block of unencrypted bits (7)

planning for failures: a principle of cybersecurity that states organizations must assume that cyber incidents will occur (9)

player perfect rationality: the assumption in analytical game theory that players behave perfectly rationally when making choices (6)

players: the actors in the game (6)

policies: written guidance that define how actions are to be performed (5)

polyalphabetic substitution cipher: a cipher that uses multiple plaintext-to-ciphertext alphabet mappings (7)

port number: the Layer 4 address for computer processes (2)

post exploitation: the phase of a cyber attack where the hacker has gained access to the target and begins commandeering it (4)

post-mortem: a review of an incident after the fact to improve the process going forward (5)

Poulsen, Kevin: a hacking folk hero famous for phone phreaking who later because a journalist (3)

pre-mortem: a thought experiment where one imagines a failure has occurred and explores how and why it could have happened (9)

premium: the cost to purchase the insurance (5)

pretexting: a social engineering step that involves creating a believable background story for contacting the target (4)

Pretty Good Privacy (PGP): encryption software written by Phil Zimmermann in 1991 to protect digital communications (10)

preventative controls: measures taken to prevent a risk from being realized (5)

primary domain: the part of the URL that is registered and ties it to a real-life entity (2)

principle: a high-level guideline that informs daily priorities and decisions (9)

principle of proportionality: a principle that declares if a state suffers harm at the hands of another state, a response out of proportion to the harm inflicted is not justified (10)

private key: in public key cryptography the key that is kept private (7)

procedure: an encapsulation of a set of programming statements that perform a task (2)

process: a running program (2)

processing delay: the amount of time it takes a router to receive, process, and resend a packet (2)

processor interrupt: a signal sent to the CPU to prompt it to handle a new action (2)

product: using multiple cryptographic schemes in sequence (7)

prompt injection attack: an attack where malicious prompts are fed into large language models to manipulate their behavior (9)

proof of work: a technique used in blockchains for ensuring that computational effort is required to produce a valid block (7)

propagation delay: the amount of time it takes a signal to travel a distance (2)

proportional allocation strategy: allocating soldiers in the Colonel Blotto game in direct proportion to the values of the battlefields (6)

protocol: a specification for communicating over a network (2)

pseudo one-time pad: a one-time pad (or keystream) produced by a deterministic algorithm and is therefore not truly random (7)

pseudo-random number generator (PRNG): a deterministic algorithm that generates a stream of numbers that appear random (8)

public key: in public key cryptography the key that is made public (7)

public key certificate: a signed message tying a real-life identity to a public key (7)

public key cryptography: a cryptographic system that uses different keys for encryption and decryption (7)

public key infrastructure (PKI): a trust hierarchy that uses signed certificates to bind keys to real-life identities (7)

pwn: taking over a victim's computer through hacking (3)

qualitative risk assessment: an approach to analyzing risk severity based on coarse-grained categories (5)

quantitative risk assessment: an approach to analyzing risk severity based on fine-grained numerical values (5)

quantum computing: a paradigm of computation based on qubits (2)

quantum mechanics: the behavior of atoms and subatomic particles (2)

quantum superposition: the ability to contain a range of values simultaneously (2)

qubit: quantum bit (2)

raising awareness: small actions taken to regularly expose employees to cybersecurity threats and best practices (5)

RAM (random access memory): memory that can be access directly and in any order (2)

ransomware: malware that encrypts data on a victim's computer and demands a ransom in exchange for the decryption key (4)

read: the view permission (8)

read-only: a permission that allows viewing but not modifying access (8)

reasonable expectation of privacy: a guideline developed by the courts to determine whether the right to privacy applies (10)

recognition phase: the part of the authentication process when the user's identity is validated (8)

reconnaissance (recon): the first phase in a cyber attack where the hacker obtains information about the target (4)

red team: the attackers in a CTF (3)

registers: a type of extremely fast memory that act as a scratchpad for CPU calculations (2)

remote access trojan (RAT): malware that provides the ability to remotely command a compromised computer (4)

remote code execution (RCE) attack: an attack where hackers are able to execute their code on a victim's computer from over the network (4)

residual risk: the risk that remains after being mitigated or transferred (5)

responsible disclosure: the steps taken to report a discovered vulnerability to an organization (3)

retainer: a fee paid in advance to secure future services if and when they are needed (5)

return on investment (ROI): the net savings that result from an investment (5)

reverse shell:: a connection made from a compromised machine out to the hacker providing C2 access to the victim (4)

RFCs (requests for comments): the standards documents used by the IETF (2)

RGB (red, green, blue) color model: a 24-bit encoding scheme for colors that uses eight bits for each part of red, green, and blue (2)

right to due process: a person accused of a crime must be provided a fair trial (10)

right to privacy: a person cannot be subjected to unreasonable searches or seizures without a warrant (10)

Risk Management Framework (RMF): a NIST standard that defines the steps needed to implement cyber risk management (5)

Ritchie, Dennis: a pioneer of computing who along with Ken Thompson created the UNIX operating system (2)

Rivest, Ron: a pioneer of cryptography who along with Adi Shamir and Leonard Adleman discovered the RSA cryptosystem (7)

role-based access control (RBAC): an access control model where subjects are able to perform actions and access data based on their user role (8)

root (rooted): gaining root level, or administrative access, to a computer through hacking (3)

root access: the highest level of user access available on a Linux-based OS (8)

rootkit: malware that compromises the operating system providing an extreme level of access and persistence to the compromised device (4)

router: computing devices that process and route packets toward their destinations (2)

row player: the player in a normal form game listed on the left (the rows) (6)

Rowhammer attack: a hardware attack that exploits a vulnerability in memory chips that can be used to gain unauthorized modification access to data (5)

RSA: the most widely used public key cryptosystem (7)

rubber ducky attack: an attack that uses a special-purpose USB stick to open a command prompt and quickly execute a series of commands by "typing" at computer speeds (4)

rules of engagement (ROE): a legal document that defines what is allowed and disallowed during a pentest (3)

Safari: Apple's web browser (2)

salami attack: an attack that accumulates gains over time by taking several small slices (6)

salt: a short random string that is combined with a user's password before it is hashed (7)

sandbox: a compartmented safe space for testing and exploring (2)

Saudi Aramco attack: a wiperware attack against a Saudi Arabian oil company that destroyed tens of thousands of machines and hard drives in 2012 (4)

Schneier, Bruce: a modern cybersecurity thought leader (7)

Schneier's law: a principle of cryptography that states anybody can create a cryptosystem that he himself cannot break (7)

SCIF (sensitive compartmented information facility): a specially designed space to contain and isolate classified information (9)

scope of work (SOW): a legal document that outlines the scope of a pentest (3)

script: an interpreted program (2)

script kiddies: unskilled individuals who utilize user-friendly tools and scripts developed by others to hack into computer systems (3)

Secret (S): a United States Department of Defense classification for information that could reasonably be expected to cause serious damage if it was disclosed (8)

secure AI: securing AI technology (2)

security as a process: a principle of cybersecurity that states cybersecurity must permeate all aspects of an organization and be continually monitored and improved (9)

security game: a type of game theoretical game involving an attacker and a defender (6)

security information and event management (SIEM): the process of aggregating logs and analyzing them for suspicious activity (8)

security key: a USB stick that can perform cryptographically-secure authentication (8)

security logs: logs that record information related to user accounts and file accesses (8)

security questions: an authentication token based on answers to personal questions (8)

sequential statements: statements executed one after the other (2)

server: a specialized computer optimized to rapidly process requests for data (2)

server program: a program that listens for incoming network connections (2)

SHA-1 (Secure Hash Algorithm 1): a 160 bit hash function created in 1993 that has not been recommended for use since the early 2000s (7)

SHA-2 (Secure Hash Algorithm 2 - 256 bit output): a 256 bit hash function that has been the standard recommended hash function in the United States since the early 2000s (7)

Shamir, Adi: a pioneer of cryptography who along with Ron Rivest and Leonard Adleman discovered the RSA cryptosystem (7)

Shannon, Claude: a pioneer of computing and founder of information theory (2)

SHAttered attack: a 2017 attack by Google against the SHA-1 hashing algorithm that produced the first documented hash collision (7)

ShellShock: a vulnerability in Bash, a command-line interpreter used in many Linux systems (9)

Shimomura, Tsutomu: a white hat hacker famous for helping the FBI capture Kevin Mitnick (3)

Shodan Search Engine: a website that continually scans the Internet and catalogs accessible software and hardware devices (4)

shoulder surfing: a technique for stealing passwords by observing users entering them (4)

side-channel attack: a hardware attack that exploits incidental information leakage (5)

Signals intelligence (SIGINT): an intelligence gathering mission focused on electronic communications (3)

signature detection: scanning software looking for malware signatures (9)

signing: using one's private key to "encrypt" a message to provide authentication and non-repudiation (7)

SIM (subscriber identity module): a unique ID used by mobile carriers to identify customers when they change phones (8)

SIM swapping: an attack on authentication tokens sent via text messages that hijacks a user's phone number so that the attacker receives the one-time passcode (8)

Simple Mail Transfer Protocol (SMTP): a protocol that is used for transmitting email messages over the Internet (9)

simplicity: a principle of cybersecurity that states simplicity should always be pursued (9)

single loss expectancy (SLE): the projected losses to a cyber asset due to a cyber risk as a result of a single incident (5)

single sign-on (SSO): an authentication scheme that allows a user to sign-in once, be granted an authentication token, and then use that token to be automatically logged-in to other websites (8)

Skoudis, Ed: a white hat hacker who identified five phases of a cyber attack (4)

smart card: a plastic ID card with an embedded integrated circuit that can perform cryptographically-secure authentication (8)

snapshot: a capture of the complete state of a VM at a point in time (2)

***Sneakers* (1992 film)**: Martin Bishop (Robert Redford) leads a team of pentesters including Darren Roskow (Dan Aykroyd), AKA Mother, who work with the NSA to recover a Russian code-cracking device. (3)

social contract theory: an ethical paradigm that emphasizes the preservation of basic human rights (10)

social distance: the state of being removed from the feelings of others (1)

social engineering: the practice of deceiving people into divulging sensitive information or performing actions that undermine security (4)

social responsibility: the proposition that professionals and organizations have an obligation to promote the welfare of society (10)

socket programming: programming for communicating over a network (2)

software: the instructions that the computer hardware executes (2)

software library: a group of related procedures that can be imported into a program (2)

software patching: fixing vulnerabilities in software by installing an updated version (4)

software reverse engineering: the skill of dissecting an executable program to determine how it works and where it may have vulnerabilities (3)

SolarWinds hack: a supply chain attack that compromised many United States government agencies in 2019 through an IT trust relationship (4)

solid state drive (SSD): storage in cells that uses no moving parts (2)

something you are: a category of authentication token that is based on physical characteristics (8)

something you have: a category of authentication token that is based on being in possession of a unique device (8)

something you know: a category of authentication token that is based on secret information (8)

Sony Pictures hack: a doxxing attack conducted by North Korea in retribution for releasing a comedy film that made fun of them in 2014 (4)

source code: a text document that contains a high-level program (2)

spam: an unsolicited and unwanted communication (8)

spear phishing: a highly targeted phishing attack (9)

spoofed website: a fake version of a legitimate website (4)

SQL injection: a website attack against that tricks the database server into executing code and queries provided by the attacker (4)

SQL Slammer Worm: a worm that infected seventy-five thousand computers in less than ten minutes in 2003 (4)

SSID (service set identifier): the name of a wireless network that users see when they perform a wireless network scan (9)

state (politics): a sovereign nation (10)

state actors: people acting under the authority of the government (10)

stateful firewall: a type of firewall that maintains a memory of inbound and outbound packets for a window of time and uses that context to determine a packet's fate (8)

stateless: the property of not retaining information about previous interactions (8)

steganography (stego): the art and science of hiding information in plain sight (4)

storage: non-volatile disk space for storing programs and data (2)

Stored Communications Act (SCA): a United States law that restricts the government's access to stored communications such as email and cloud content (10)

stream cipher: a cryptographic system that uses a keystream and the XOR function to encrypt and decrypt on a per bit basis (7)

string mangling: modifying base words in formulaic ways by changing capitalization, using character substitutions, and adding prefixes and postfixes (9)

Stuxnet : a cyber attack against an Iranian nuclear enrichment facility in 2010 (4)

subject: actors in a computer system (8)

successive elimination of dominated strategies: removing dominated strategies iteratively as choices are eliminated (6)

sudo: a command in Linux OSs that allow a user to run a command as another user (9)

supply chain attack: an attack where hackers first compromise a supplier of their target in order to get access to their target (4)

surreptitious account access: logging into another person's account to spy on them (8)

symmetric key cryptography: a cryptographic system that uses the same key for encryption and decryption (7)

syntax: the rules for writing a program in a given language (2)

system calls (syscalls): OS-defined procedures that allow user programs to exercise some control over OS functionality (2)

system logs: logs that record actions taken by the OS and issues related to the operating system (8)

system users: non-human users that can take actions in a computer system (8)

Tallinn: the capital city of Estonia and location of the world's first nation state cyber attack in 2007 (10)

Target data breach: a supply chain attack against Target that caused a large-scale data breach in 2013 through a contractor trust relationship (4)

TCP (Transmission Control Protocol): the main Layer 4 protocol used in the Internet (2)

techniques, tactics, and procedures (TTPs): a signature of a certain sequence of steps and types of tools used by a hacker (3)

THC Hydra: a hacking tool used for conducting password guessing attacks (4)

the cloud: an expression for the computer servers we access online (2)

the Colonel Blotto game: a game theoretical game that captures the strategic dynamics of allocating scarce resources as efficiently as possible (6)

The Equation Group: an advanced persistent threat that many believe is USCYBERCOM or the NSA (3)

the hacker mindset: a mentality of creativity that leads to achieving objectives in unconventional ways (6)

the prisoner's dilemma: a famous game theory game where the players are drawn into to a bad mutual outcome (6)

The Social Engineering Toolkit (SET): a hacking tool that provides a step-by-step wizard for configuring and deploying social engineering attacks (3)

The Tallinn Manual 2.0: a manual compiled by an international team of experts that attempts to apply international law principles to the cyber domain (10)

the traveler's dilemma: a famous game theory game that illustrates the successive elimination of dominated strategies (6)

the triarchic theory: a theory of intelligence devised by Robert Sternberg that divides the intellect into three areas: the analytical, the creative, and the practical (6)

third-party governance: oversight imposed on an organization by an outside organization (5)

third-party doctrine: when a person shares information with another party, he surrenders his expectation of privacy concerning that information (10)

Thompson, Ken: a pioneer of computing who along with Dennis Ritchie created the UNIX operating system (2)

time-to-exploit: the time between vulnerability disclosure to exploitation (4)

Title 10: the section of the United States Code that covers signals intelligence (10)

Title 18: the section of the United States Code that contains most of the laws for which a person can be charged with a federal crime (10)

Title 50: the section of the United States Code that covers military activity (10)

Top Secret (TS): a United States Department of Defense classification for information that could reasonably be expected to cause exceptionally grave damage if it was disclosed (8)

top-level domain (TLD): the final part of the domain name that separates URLs into high-level categories (2)

Torvalds, Linus: the original and primary developer of the Linux OS (2)

traceroute: a command line utility that shows the hops to a destination on the Internet (implemented as tracert in Windows) (2)

tracking cookies: a text string assigned to a web browser by a web server for the purposes of uniquely identifying a user (9)

trade secret: confidential corporate data that the company has taken measures to protect and that has economic value (10)

transferring risk: passing the risk to another organization (5)

transistor: a device that can control the flow of electricity (2)

transposition: a cryptographic technique where letters or words are rearranged (7)

trigram: a triplet of letters (7)

trojan: trojan horse program (3)

trojan horse malware: malware that appears to be and functions like a normal program, but it comes bundled with malware that creates a backdoor into the victim's machine (4)

trusting trust: a principle of cybersecurity that states trust relationships should be explicitly identified and examined so that they can be managed appropriately (9)

TS/SCI (sensitive compartmented information): a United States Department of Defense clearance that provides TS subjects with access to more categories of sensitive information (9)

Turing Award: the "Nobel Prize" of computer science (4)

Turing, Alan: a pioneer of computer science (2)

two-factor authentication: MFA when exactly two tokens are used (8)

Ubuntu: a popular Linux distribution known for being user-friendly (2)

Unclassified (U): a United States Department of Defense classification for information that would pose no risk if it was disclosed (8)

Unified Combatant Command (UCC): a division of the United States Department of Defense with broad, continuing missions and coordinate military activity either by geography or by specialty domain area (3)

uniform resource locator (URL): a unique name for a resource on the world wide web (2)

United Nations Charter: an international treaty signed in 1945 in the immediate aftermath of World War II (10)

United States Computer Emergency Readiness Team (US-CERT): a United States department devoted to protecting the nation's Internet infrastructure by coordinating defense against and response to cyber attacks (5)

United States Cyber Command (USCYBERCOM): the unified combatant command that coordinates cyber-related activities (3)

United States v. Chung: a case of foreign espionage where a Boeing engineer was convicted under the Economic Espionage Act for providing trade secrets to China (10)

United States v. Drew: a case of cyberbullying where Lori Drew was charged with violating the CFAA but was acquitted because the law did not fit the crime (10)

United States v. Morris: a case of nuisance hacking where Robert Tappan Morris became the first person convicted under the CFAA for the Morris Worm in 1990 (10)

UNIX: one of the most influential operating systems ever created (2)

use of force: a threshold in international conflict that can trigger a justified retaliatory response (10)

utilitarianism: an ethical paradigm that emphasizes the consequences of actions and focuses on benefits and harms (10)

utility preferences: an ordering of the outcomes from least to most desirable (6)

variable: a placeholder for data that is provided when the program is run (2)

veil of ignorance principle: a thought experiment where a person reasons about the appropriate action to take while pretending to not know to which of the affected parties he belongs (10)

Venona project: a United States counterintelligence operation that collected and analyzed Soviet Union nuclear espionage correspondence (7)

Vernam, Gilbert: a pioneer of cryptography that operationalized the one-time pad cryptographic scheme (7)

Vigenère cipher: a polyalphabetic substitution cipher that uses a two-dimensional alphabetic table and a codeword to select the ciphertext alphabets (7)

vigilante justice: situations where individuals without the proper authority attempt to enforce the law (10)

virtual machine (VM): an OS that runs as a program on top of a computer's actual OS (2)

virtue ethics: an ethical paradigm that emphasizes living a life of moral excellence and focuses on character (10)

virus: a generic term for malware that infects victim machines and is capable of spreading (4)

virus scan: a software-based inspection of a computer to detect if it has been infected with malware (5)

VM pod: multiple VMs networked together and isolated from the Internet in a sandboxed computer network (2)

VOIP (voice over Internet Protocol): a technology that carries phone call traffic over computer networks (10)

volatile: loses its value when the power is off (as in volatile and non-volatile memory) (2)

von Neumann architecture: the standard model for the hardware components of a computer (2)

von Neumann, John: a pioneer of computer architecture (2)

VPN (virtual private network): a technology that creates an encrypted channel between an endpoint and a trusted server used for accessing resources and web browsing (9)

vulnerability: a weakness (3)

vulnerability assessment: a scan for known vulnerabilities on a computer system or network (5)

WannaCry: a worldwide ransomware attack that exploited the EternalBlue vulnerability in 2017 and was attributed to the North Korean government by the United States (9)

warez: (pronounced "wares" as in goods) pirated software (3)

WarGames **(1983 film)**: Seattle high schooler David Lightman (Matthew Broderick) impresses his classmate Jennifer Mack (Alley Sheely) by hacking into a computer he believes belongs to a video game company but is actually an AI-driven Department of Defense computer and almost triggers a nuclear war with the Soviet Union (3)

warrant: an order signed by a judge giving law enforcement the authority to search and seize evidence (10)

watering hole attack: a supply chain attack where a hacker compromises a web server that the target visits to attack the target (4)

watermark: a conspicuous marking in the background of a document (8)

weakest link: a principle of cybersecurity that states hackers will take the easiest path towards accomplishing their objectives (9)

web browser: the main program used to access Internet resources (2)

web exploitation: an attack on a website (4)

website cloning: creating a copy of a website (4)

website defacing: taking over a website and replacing it with a defaced version or a new webpage promoting the hacker (3)

while-statement: along with the for-statement, the typical looping statement keyword (2)

white hat hackers: individuals who hack legally and ethically (3)

whitelist: a list of explicitly approved resources (8)

Wikileaks: a website that publishes documents from data leaks (3)

Windows: Microsoft's operating system (2)

wiperware: malware that deletes data or causes physical damage to a computer system (4)

Wireshark: a free and open-source packet sniffer (2)

wiretap: a device that can be attached to the telecommunications infrastructure to secretly listen to phone calls (10)

Wiretap Act: a United States law that restricts the government's ability to monitor electronic communications while they are in transit (10)

word size: the number of bits that can be processed at a time by the CPU (2)

world wide web (the Web): the Internet resources hosted on web servers and accessible via web browsers (2)

worldview: a person's beliefs about ultimate reality (10)

worm: a type of computer virus that is programmed to propagate on its own (4)

Wozniak, Steve: a phone phreaker as a young person who went on to co-find Apple with Steve Jobs (3)

wrapper program: software that binds two different programs together and is useful for creating trojan horse malware (4)

write: the modify permission (8)

www: the most common subdomain and stands for world wide web (2)

XProtect: the macOS built-in antivirus software (9)

zero day: an exploit that targets a zero-day vulnerability (4)

zero trust: a security strategy that assumes internal systems may be compromised (9)

zero-day vulnerability: an unknown and unpatched vulnerability (4)

Zimmermann, Phil: a cult hero of computer cryptography who published encryption software on the Internet in defiance of the Arms Export Control Act (10)

Zuckerberg, Mark: a pioneer of social media and the founder of Facebook (now Meta) (9)

Bibliography

"It said what he would have said, if it had been possible for him to set his scattered thoughts in order. It was the product of a mind similar to his own, but enormously more powerful, more systematic, less fear-ridden. The best books, he perceived, are those that tell you what you know already." - 1984 by George Orwell

This textbook is a synthesis of many books that have shaped my understanding of cybersecurity. Listed below are several that have had varying levels of influence on this textbook.

Bishop, M. A. (2003). *Computer Security: Art and Science*. Addison-Wesley Professional.

Bowden, M. (2012). *Worm: The First Digital World War*. Grove Press.

Brams, S. J. (2003). *Biblical Games: Game Theory and the Hebrew Bible*. MIT Press.

Brooks, F. P. (1995). *The Mythical Man-Month: Essays on Software Engineering*. Addison-Wesley Professional.

Camerer, C. (2003). *Behavioral Game Theory: Experiments in Strategic Interaction*. Princeton University Press.

Chapple, M., Stewart, J. M., & Gibson, D. (2018). *(ISC)2 CISSP Certified Information Systems Security Professional Official Study Guide*. John Wiley & Sons.

Colman, A. M. (1995). *Game Theory and its Applications in the Social and Biological Sciences*. Psychology Press.

Conrad, E., Misenar, S., & Feldman, J. (2015). *CISSP Study Guide*. Syngress Publishing.

Du, W. (2022). *Computer & Internet Security: A Hands-on Approach* (3rd ed.). Self-published.

Erickson, J. (2008). *Hacking: The Art of Exploitation, 2nd Edition*. No Starch Press.

Fischerkeller, M. P., Goldman, E. O., & Harknett, R. J. (2022). *Cyber Persistence Theory: Redefining National Security in Cyberspace*. Oxford University Press, USA.

Goodman, M. (2015). *Future Crimes: Everything is Connected, Everyone is Vulnerable and What We Can Do about it*. Doubleday Books.

Hafner, K., & Lyon, M. (1998). *Where Wizards Stay Up Late: The Origins Of The Internet*. Simon and Schuster.

Hafner, K., & Markoff, J. (1995). *Cyberpunk: Outlaws and Hackers on the Computer Frontier, Revised*. Simon and Schuster.

Harel, D. (2004). *Computers Ltd: What They REALLY Can't Do*. Oxford University Press.

Hillis, W. D. (1998). *The Pattern on the Stone: The Simple Ideas That Make Computers Work*. Basic Books.

Hofstadter, D. R. (1999). *Godel, Escher, Bach: An Eternal Golden Braid*. Basic Books.

Hogland, G., & McGraw, G. (2004). *Software Security: Building Security In*. Addison-Wesley.

Isaacson, W. (2014). *The Innovators: How a Group of Hackers, Geniuses, and Geeks Created the Digital Revolution*. Simon and Schuster.

Kahn, D. (1996). *The Codebreakers: The Comprehensive History of Secret Communication from Ancient Times to the Internet*. Scribner.

Kaplan, F. M. (2016). *Dark Territory: The Secret History of Cyber War*. Simon and Schuster.

Kerr, O. S. (2012). *Computer Crime Law*. West Academic Publishing.

Kosseff, J. (2017). *Cybersecurity Law*. John Wiley & Sons.

Krebs, B. (2014). *Spam Nation: The Inside Story of Organized Cybercrime--from Global Epidemic to Your Front Door*. Sourcebooks.

Kurose, J. F., & Ross, K. W. (2013). *Computer Networking: A Top-down Approach* (6th ed.). Addison-Wesley.

Levy, S. (2001). *Crypto: How the Code Rebels Beat the Government, Saving Privacy in the Digital Age*. Viking.

Mitnick, K. (2012). *Ghost in the Wires: My Adventures as the World's Most Wanted Hacker*. Back Bay Books.

Mitnick, K. D., & Simon, W. L. (2003). *The Art of Deception: Controlling the Human Element of Security*. John Wiley & Sons.

Moore, D. (2022). *Offensive Cyber Operations: Understanding Intangible Warfare*. Oxford University Press, USA.

Payne, B. (2022). *Go H*ck Yourself: A Simple Introduction to Cyber Attacks and Defense.* No Starch Press.

Petzold, C. (2022). *Code: The Hidden Language of Computer Hardware and Software.* Microsoft Press.

Poulsen, K. (2011). *Kingpin: How One Hacker Took Over the Billion-dollar Cybercrime Underground.* Crown.

Quinn, M. J. (2017). *Ethics for the Information Age.* Pearson.

Raymond, E. S. (1999). *The Cathedral and the Bazaar: Musings on Linux and Open Source by an Accidental Revolutionary.* O'Reilly Media.

Richards, J. W. (2018). *The Human Advantage: The Future of American Work in an Age of Smart Machines.* Forum Books.

Rosenberg, S. (2008). *Dreaming in Code: Two Dozen Programmers, Three Years, 4,732 Bugs, and One Quest for Transcendent Software.* Three Rivers Press.

Sanger, D. E. (2018). *The Perfect Weapon: War, Sabotage, and Fear in the Cyber Age.* Crown.

Schmitt, M. N. (Ed.). (2017). *Tallinn Manual 2.0 on the International Law Applicable to Cyber Operations.* Cambridge University Press.

Schneider, G., & Gersting, J. (2015). *Invitation to Computer Science* (7th ed.). Cengage Learning.

Schneier, B. (1996). *Applied Cryptography: Protocols, Algorithms, and Source Code in C.* John Wiley & Sons.

Schneier, B. (2000). *Secrets and Lies: Digital Security in a Networked World.* John Wiley & Sons.

Schneier, B. (2015). *Data and Goliath: The Hidden Battles To Capture Your Data And Control Your World.* National Geographic Books.

Schneier, B. (2019). *Click Here to Kill Everybody: Security and Survival in a Hyper-connected World.* National Geographic Books.

Schneier, B. (2023). *A Hacker's Mind: How the Powerful Bend Society's Rules, and How to Bend them Back.* National Geographic Books.

Silberschatz, A., Gagne, G., & Galvin, P. B. (2011). *Operating System Concepts* (8th ed.). Wiley.

Singh, S. (2000). *The Code Book: The Science of Secrecy from Ancient Egypt to Quantum Cryptography*. Anchor Books.

Skoudis, E., & Liston, T. (2006). *Counter Hack Reloaded: A Step-by-step Guide to Computer Attacks and Effective Defenses*. Pearson.

Soni, J., & Goodman, R. (2017). *A Mind at Play: How Claude Shannon invented the Information Age*. Simon & Schuster.

Stallings, W., & Brown, L. (2015). *Computer Security: Principles and Practice*. Prentice Hall.

Stamp, M. (2011). *Information Security: Principles and Practice*. John Wiley & Sons.

Stamp, M., & Low, R. M. (2007). *Applied Cryptanalysis: Breaking Ciphers in the Real World*. John Wiley & Sons.

Stoll, C. (1990). *The Cuckoo's Egg: Tracking a Spy Through the Maze of Computer Espionage*. Pocket Books.

Trappe, W., & Washington, L. C. (2006). *Introduction to Cryptography: With Coding Theory*. Pearson.

Tzu, S. (1971). *The Art of War* (S. Griffith, Trans.). Oxford: Clarendon Press.

Walker, M. (2016). *CEH Certified Ethical Hacker All-in-One Exam Guide, Third edition*. McGraw-Hill Education.

Wang, W. (2006). *Steal this Computer Book 4.0: What They Won't Tell You About the Internet*. No Starch Press.

Figure Sources

Figure 2.3: *A picture of computer pioneers Ken Thompson and Dennis Ritchie (pic. bmp) opened with an image viewer (left) and a hex editor (right).*
https://commons.wikimedia.org/wiki/File:Ken_Thompson_and_Dennis_ Ritchie--1973.jpg

Figure 3.1: *Roy Rogers wearing a white hat in an old western movie poster.*
https://commons.wikimedia.org/wiki/File:The_Arizona_Kid.jpg

Figure 3.2: *USCYBERCOM seal.*
https://commons.wikimedia.org/wiki/File:Seal_of_the_United_States_Cyber_Com- mand.svg

Figure 3.4: *A cyber forensics lab.*
https://commons.wikimedia.org/wiki/File:Digital_forensics_lab.jpg

Figure 3.7: *A portion of the title page from an issue of Phrack.*
https://commons.wikimedia.org/wiki/File:PwnPhrack.jpg

Figure 3.8: *The Cap'n Crunch toy whistle that triggered operator mode.*
https://commons.wikimedia.org/wiki/File:Cap%E2%80%99n_Crunch,_Spielzeugp- feife_(2600_Hz).jpg

Figure 3.9: *Gray hat and later white hat hackers Kevin Mitnick (left) and Kevin Poulsen (right).*
https://commons.wikimedia.org/wiki/File:Lamo-Mitnick-Poulsen.png

Figure 3.10: *An electronic DEF CON conference badge.*
https://commons.wikimedia.org/wiki/File:DEFCON_22_ Badge_%2814891107355%29.jpg

Figure 4.1: *A hardware-based keylogger.*
https://commons.wikimedia.org/wiki/File:USB_Hardware_Keylogger.jpg

Figure 5.2: *A pair of shoes concealing explosives that were successfully snuck through airport security.*
https://commons.wikimedia.org/wiki/File:Richard_Reid%27s_Shoes.jpg

Figure 5.5: *A screenshot of a ransomware note.*
https://commons.wikimedia.org/wiki/File:Ransom_Note_of_Petwrap.jpg

Figure 6.2: *The triarchic theory applied to hackers.*
https://commons.wikimedia.org/wiki/File:Gros_-_First_Consul_Bonaparte_(Detail).png
https://commons.wikimedia.org/wiki/File:Self-portrait_-_Vincent_van_Gogh.jpg
https://commons.wikimedia.org/wiki/File:%22Cabeza_de_Einstein%22.jpg

Figure 6.7: *The Colonel Blotto game.*
https://commons.wikimedia.org/wiki/File:General_Charles_George_Gordon_statue,_Embankment,_London_(2).JPG
https://commons.wikimedia.org/wiki/File:Gouverneur_K._Warren_Gettysburg_statue.jpg

Figure 7.15: *Prisoner of war Jeremiah Denton saying all the right words while blinking the word TORTURE in Morse code.*
https://commons.wikimedia.org/wiki/File:JamesDentonHanoi1966NatArchiv.jpg

Figure 8.3: *A physical authentication token that displays a series of random numbers.*
https://commons.wikimedia.org/wiki/File:RSA_SecurID_SID800.jpg

Figure 9.6: *A diagram of the Titanic illustrating the compartmentalization principle.*
https://commons.wikimedia.org/wiki/File:Titanic_side_plan_annotated_English.png

Figure 10.2: *Rosa Parks police report.*
https://commons.wikimedia.org/wiki/File:Rosaparks_policereport.jpg

Figure 10.3: *Export-controlled munitions shirt from the Crypto Wars.*
https://commons.wikimedia.org/wiki/File:Munitions_T-shirt_(front).jpg

About the Author

"Who has ever given to God, that God should repay them? For from him and through him and for him are all things. To him be the glory forever! Amen." - Romans 11:35-36

Seth Hamman, PhD, is a tenured Professor of Cyber Operations and Computer Science at Cedarville University. He founded and directs the Center for the Advancement of Cybersecurity at Cedarville and designed Ohio's first Cyber Operations BS degree program. In addition to teaching his Introduction to Cybersecurity and Cyber Defense courses, he heads the Knowledge Units criterion for the National Security Agency's National Centers of Academic Excellence designations, co-leads the Cyber Operations Community of Practice, and is a member of the Cybersecurity Curriculum Taskforce. He holds a BA in Religion from Duke University, an MS in Computer Science from Yale University, and a PhD in Computer Science with a research focus in Cybersecurity from the Air Force Institute of Technology at Wright-Patterson Air Force Base. Dr. Hamman is passionate about shaping the growing discipline of cybersecurity education, developing the next generation of cyber leaders, and integrating his Christian faith into his teaching.

www.ingramcontent.com/pod-product-compliance
Lightning Source LLC
Chambersburg PA
CBHW082036290526
45791CB00015B/2173